Big Data Analytics and Machine Learning for Smart Agriculture

Big Data Analytics and Machine Learning for Smart Agriculture

Guest Editors

Maciej Zaborowicz
Jakub Frankowski

MDPI

Basel • Beijing • Wuhan • Barcelona • Belgrade • Novi Sad • Cluj • Manchester

Guest Editors
Maciej Zaborowicz
Department of Biosystems Engineering
Poznan University of Life Sciences
Poznań
Poland

Jakub Frankowski
Department of Bioeconomy
Institute of Natural Fibres and Medicinal Plants—National Research Institute
Poznań
Poland

Editorial Office
MDPI AG
Grosspeteranlage 5
4052 Basel, Switzerland

This is a reprint of the Special Issue, published open access by the journal *Agriculture* (ISSN 2077-0472), freely accessible at: https://www.mdpi.com/journal/agriculture/special_issues/E2T86A70A9.

For citation purposes, cite each article independently as indicated on the article page online and as indicated below:

Lastname, A.A.; Lastname, B.B. Article Title. *Journal Name* **Year**, *Volume Number*, Page Range.

ISBN 978-3-7258-4877-5 (Hbk)
ISBN 978-3-7258-4878-2 (PDF)
https://doi.org/10.3390/books978-3-7258-4878-2

Contents

About the Editors

Maciej Zaborowicz

Maciej Zaborowicz works as a scientist and academic teacher, serving as the Head of the Department of Biosystems Engineering at the University of Life Sciences in Poznań. He is a specialist in data analysis and neural modelling, and he applies these skills in his daily work, including in projects funded by ministerial, EU, and commercial sources. He is actively involved in research and development activities, including data processing and analysis, statistical analysis and visualization, image processing and analysis, as well as the application of artificial intelligence methods in the natural, exact, and medical sciences. He is proficient in C# programming and has advanced skills in Python, the Pandas library, and R. In his everyday work, he also uses the STATISTICA and H2O packages. He has strong expertise in team and project management and considers effective communication essential for successful collaboration. Creativity and innovative problem-solving are at the core of his professional approach.

Jakub Frankowski

Jakub Frankowski works as a scientist, serving as the Head of the Department of Bioeconomy at the Institute of Natural Fibres and Medicinal Plants—National Research Institute in Poznań. He has a PhD in engineering and technical sciences in the discipline of mechanical engineering. He is a specialist in bioeconomy, waste management, rural development, and biofuels production. He is also an applicant, executor, reviewer, and manager of grants funded by the EU and ministerial funds. He is the author of patents, a laureate of awards for implementing progress in agriculture, and the author of numerous scientific publications. His current research focuses on implementing the cultivation of sorghum in temperate climates, producing biofuels from agricultural waste biomass, and using artificial intelligence in agriculture. Dr. Frankowski combines scientific work with practice in the area of R&D. He completed postgraduate studies in energy management, project management, and business processes. Currently, he works as a scientist, serving as the Head of the Department of Bioeconomy at the Institute of Natural Fibres and Medicinal Plants—National Research Institute in Poznań.

agriculture

Editorial

Big Data Analytics and Machine Learning for Smart Agriculture

Maciej Zaborowicz [1,*] and Jakub Frankowski [2]

1 Department of Biosystems Engineering, Poznań University of Life Sciences, Wojska Polskiego 50, 60-627 Poznań, Poland
2 Department of Bioeconomy, Institute of Natural Fibers and Medicinal Plants–National Research Institute, Wojska Polskiego 71B, 60-630 Poznań, Poland; jakub.frankowski@iwnirz.pl
* Correspondence: maciej.zaborowicz@up.poznan.pl

Modern technologies are continuously entering every aspect of our lives. Today, we can no longer do without smartphones, intelligent applications, and devices that significantly contribute to the improvement and comfort of our lives. In addition to functions related to the acceleration of everyday activities and household processes, aspects related to artificial intelligence and machine learning support industry and the production of goods and services. One such branch of the economy is agriculture, the primary goal of which is to meet human nutritional needs. Taking actions against hunger and striving for greater yields and better-quality harvests are some of the most important goals of current agriculture. The era of industrialization and the implementation of mechanization, including the mechanization of agriculture, is followed by the digital era [1,2].

It is widely believed that digitization is revolutionizing the world. However, it should be noted that it will not replace machines and devices in agriculture, but it can modify and optimize plant and animal production process. Since the first edition of "Big Data Analytics and Machine Learning for Smart Agriculture" in 2023, many technologies have been implemented into agricultural technology, changing it from Agriculture 4.0 to 5.0 [3,4].

Over the past few years, big data analysis and machine learning have revolutionized the management of agricultural systems and farms [5–7]. The large amounts of data collected daily during the observation of vegetation processes, the harvesting of crops, and their processing into food have contributed to the new knowledge published in this Special Issue.

The published works reveal how collecting and storing data contributed to the creation of an algorithm for detecting weeds, which was created based on YOLOv8. He's team presented an improvement in the basic network by adding the so-called attention mechanisms and using dynamic convolution [8]. In another paper, the authors used deep learning, also based on YOLO, RT-DETR, and Mask-RCNN technologies, to detect and classify the ripeness of blueberries. Aguilera's team emphasized the importance of model optimization, and their results suggest that new algorithms and their correlation with empirical studies increase the effectiveness of the created systems, thus increasing the effectiveness of crop monitoring [9].

Artificial intelligence methods based on analyzed data can also be used to create a fertilization recommendation system, which optimizes the use of agricultural production means. Musanase et al. revealed that implementing such solutions in precision agriculture can not only increase yields, but also reduce fertilizer losses and contribute to the implementation of sustainable agricultural practices [10]. Properly collected and processed data allow for the development of empirical system models supported by AI algorithms for forecasting industrial hemp seed yields. After considering data on climatic conditions,

https://doi.org/10.3390/agriculture15070757

agrotechnics, and seed quality, Sieracka and her colleagues created predictive models to assess crop efficiency, determining the factors that have the greatest impact on seed efficiency and yield [11].

Xie's team converted empirical data into digital data, which can be used to develop algorithms that identify pathological changes in fruits and vegetables, such as peppers. Systems based on AI solutions can improve the precision and efficiency of diagnosing plant diseases, which is important in Agriculture 5.0 [12]. Similarly, Bai's and Amin's teams described the use of AI techniques and technological achievements such as drones and other devices that not only identify plant diseases, but also detect pests, both of which cause plant damage and reduce the quantity and quality of crops [13,14].

Nazir et al. indicated that models supporting the identification of diseases occurring on potato leaves and allowing for the classification of disease stages can generally affect the response time and the application of appropriate measures, thus reducing crop losses [15]. Leaf diseases are often the first prognosticator of a more complex problem related to the proper vegetation of plants. Tomatoes are a popular plant characterized by beneficial nutritional and health-promoting properties for humans. By collecting appropriate data and processing them into digital form, Ullah's team showed that it is possible to develop a classification model defining pathological disease changes in these plants [16].

Guava diseases (leaf blight) can also be detected based on the identification and classification of leaves. Depending on the plant and the type of problem and its complexity, various technologies can be used, such as convolutional networks or deep learning methods, e.g., those developed by Mumtaz et al. [17]. The basis for plant vegetation and crop quality is the environment in which the plant grows, especially the quality of the soil. In this respect, Shahare's team showed that data can also be used to develop appropriate models based on machine learning methods to assess and forecast the activity of soil enzymes, which are key to biological processes occurring in the soil, to help farmers optimize agricultural production [18].

Plant production, including field or greenhouse crops, is just one branch of modern agricultural production. The second largest aspect of agricultural production is the animal production branch. Artificial intelligence can be used to optimize meat production, e.g., beef from dairy cattle, as shown by Addis et al. Such systems can increase production efficiency regarding the use of animals for dairy or meat production and in providing unified food to consumers [19].

Forecasting and advisory systems are also based on artificial intelligence algorithms, which are based on large data sets, allowing for broad diagnosis and the prediction of failures in increasingly popular smart farms. Choe's team showed that such systems can be used to detect irregularities in data obtained from sensors and can predict potential failures of agricultural equipment [20].

Notably, machine learning models, neural models, or ordinary linear forecasting models cannot be created without previously collected data that have been appropriately processed and adapted for analysis. It is very difficult to compare research results concerning the same object. Data are collected using different devices and sensors. Additionally, there are varying frequencies of data collection, and devices have different operating conditions. Similarly, the varieties of measured plants and the fields or buildings in which livestock are kept are different.

This is a significant problem in establishing a methodology and indicating objective conclusions. We are surrounded by a multitude of data. We continuously collect and try to systematize data. It is challenging to not only collect and store data for a long time, but also to effectively process and analyze them to obtain valuable information. Modern IT tools allow us to systematize data, discover patterns, and generate new scientific knowledge, supporting innovations in digital agriculture. However, data are the basis, and they represent the most time-consuming and cost-intensive part of the research process. Nevertheless, obtaining a large amount of data will allow for the creation of new methods and technologies supporting agricultural production.

In this Special Issue, particular emphasis is placed on big data in agriculture and machine learning, i.e., on the collection, management, and analysis of large data sets, analysis and prediction, decision support systems, and automation based on AI IoT methods. It also focuses on the integration of sensor networks and intelligent monitoring systems, which allows for the transition to a new era, Digital Agriculture 5.0, in which automation, robotics, and artificial intelligence support modern precision agriculture and enable sustainable development through resource optimization, loss reduction, and improved efficiency.

Author Contributions: Conceptualization, M.Z. and J.F.; methodology, M.Z. and J.F.; software, M.Z. and J.F.; validation, M.Z. and J.F.; formal analysis, M.Z. and J.F.; investigation, M.Z. and J.F.; resources, M.Z. and J.F.; data curation, M.Z. and J.F.; writing—original draft preparation, M.Z. and J.F.; writing—review and editing, M.Z. and J.F.; visualization, M.Z. and J.F.; supervision, M.Z.; project administration and funding acquisition, M.Z. and J.F. All authors have read and agreed to the published version of the manuscript.

Funding: This research received no external funding.

Conflicts of Interest: The authors declare no conflicts of interest.

References

1. Shaikh, T.A.; Rasool, T.; Lone, F.R. Towards leveraging the role of machine learning and artificial intelligence in precision agriculture and smart farming. *Comput. Electron. Agric.* **2022**, *198*, 107119.
2. Frankowski, J.; Zaborowicz, M.; Sieracka, D.; Łochyńska, M.; Czeszak, W. Prediction of the hemp yield using artificial intelligence methods. *J. Nat. Fibers* **2022**, *19*, 13725–13735.
3. Osinga, S.A.; Paudel, D.; Mouzakitis, S.A.; Athanasiadis, I.N. Big data in agriculture: Between opportunity and solution. *Agric. Syst.* **2022**, *195*, 103298.
4. Pinna, D.; Pezzuolo, A.; Cogato, A.; Pornaro, C.; Macolino, S.; Marinello, F. Applications of satellite platforms and machine learning for mapping and monitoring grasslands and pastures: A systematic and comprehensive review. *Smart Agric. Technol.* **2024**, *9*, 100571.
5. Huang, Q.; Wang, X.; Gao, Q.; Carraro, A.; Pezzuolo, A.; Marinello, F. How to assess the digitization and digital effort: A framework for Digitization Footprint (Part 1). *Comput. Electron. Agric.* **2025**, *230*, 109875. [CrossRef]
6. Carraro, A.; Saurio, G.; Marinello, F. Towards rigorous dataset quality standards for deep learning tasks in precision agriculture: A case study exploration. *Smart Agric. Technol.* **2025**, *10*, 100721.
7. Parenti, O.; Guerrini, L.; Zanoni, B. Techniques and technologies for the breadmaking process with unrefined wheat flours. *Trends Food Sci. Technol.* **2020**, *99*, 152–166.
8. He, C.; Wan, F.; Ma, G.; Mou, X.; Zhang, K.; Wu, X.; Huang, X. Analysis of the Impact of Different Improvement Methods Based on YOLOV8 for Weed Detection. *Agriculture* **2024**, *14*, 674. [CrossRef]
9. Aguilera, C.A.; Figueroa-Flores, C.; Aguilera, C.; Navarrete, C. Comprehensive Analysis of Model Errors in Blueberry Detection and Maturity Classification: Identifying Limitations and Proposing Future Improvements in Agricultural Monitoring. *Agriculture* **2024**, *14*, 18. [CrossRef]
10. Musanase, C.; Vodacek, A.; Hanyurwimfura, D.; Uwitonze, A.; Kabandana, I. Data-Driven Analysis and Machine Learning-Based Crop and Fertilizer Recommendation System for Revolutionizing Farming Practices. *Agriculture* **2023**, *13*, 2141. [CrossRef]
11. Sieracka, D.; Zaborowicz, M.; Frankowski, J. Identification of Characteristic Parameters in Seed Yielding of Selected Varieties of Industrial Hemp (*Cannabis sativa* L.) Using Artificial Intelligence Methods. *Agriculture* **2023**, *13*, 1097. [CrossRef]
12. Xie, B.; Su, Q.; Tang, B.; Li, Y.; Yang, Z.; Wang, J.; Wang, C.; Lin, J.; Li, L. Combining Neural Architecture Search with Knowledge Graphs in Transformer: Advancing Chili Disease Detection. *Agriculture* **2023**, *13*, 2025. [CrossRef]
13. Bai, Y.; Hou, F.; Fan, X.; Lin, W.; Lu, J.; Zhou, J.; Fan, D.; Li, L. A Lightweight Pest Detection Model for Drones Based on Transformer and Super-Resolution Sampling Techniques. *Agriculture* **2023**, *13*, 1812. [CrossRef]
14. Amin, J.; Anjum, M.A.; Zahra, R.; Sharif, M.I.; Kadry, S.; Sevcik, L. Pest Localization Using YOLOv5 and Classification Based on Quantum Convolutional Network. *Agriculture* **2023**, *13*, 662. [CrossRef]

15. Nazir, T.; Iqbal, M.M.; Jabbar, S.; Hussain, A.; Albathan, M. EfficientPNet—An Optimized and Efficient Deep Learning Approach for Classifying Disease of Potato Plant Leaves. *Agriculture* **2023**, *13*, 841. [CrossRef]
16. Ullah, Z.; Alsubaie, N.; Jamjoom, M.; Alajmani, S.H.; Saleem, F. EffiMob-Net: A Deep Learning-Based Hybrid Model for Detection and Identification of Tomato Diseases Using Leaf Images. *Agriculture* **2023**, *13*, 737. [CrossRef]
17. Mumtaz, S.; Raza, M.; Okon, O.D.; Rehman, S.U.; Ragab, A.E.; Rauf, H.T. A Hybrid Framework for Detection and Analysis of Leaf Blight Using Guava Leaves Imaging. *Agriculture* **2023**, *13*, 667. [CrossRef]
18. Shahare, Y.; Singh, M.P.; Singh, P.; Diwakar, M.; Singh, V.; Kadry, S.; Sevcik, L. A Comprehensive Analysis of Machine Learning-Based Assessment and Prediction of Soil Enzyme Activity. *Agriculture* **2023**, *13*, 1323. [CrossRef]
19. Addis, A.H.; Blair, H.T.; Kenyon, P.R.; Morris, S.T.; Schreurs, N.M.; Garrick, D.J. Agent-Based Modelling to Improve Beef Production from Dairy Cattle: Young Beef Production. *Agriculture* **2023**, *13*, 898. [CrossRef]
20. Choe, H.O.; Lee, M.-H. Artificial Intelligence-Based Fault Diagnosis and Prediction for Smart Farm Information and Communication Technology Equipment. *Agriculture* **2023**, *13*, 2124. [CrossRef]

agriculture

MDPI

Article

Pest Localization Using YOLOv5 and Classification Based on Quantum Convolutional Network

Javeria Amin [1,*], Muhammad Almas Anjum [2], Rida Zahra [1], Muhammad Imran Sharif [3], Seifedine Kadry [4,5,6] and Lukas Sevcik [7]

1 Computer Science Department, University of Wah, Rawalpindi 47040, Pakistan
2 National University of Technology (NUTECH), Islamabad 44000, Pakistan
3 Department Computer Science, COMSATS University Islamabad, Wah Campus, Rawalpindi 47040, Pakistan
4 Department of Applied Data Science, Noroff University College, 4612 Kristiansand, Norway
5 Artificial Intelligence Research Center (AIRC), Ajman University, Ajman 346, United Arab Emirates
6 Department of Electrical and Computer Engineering, Lebanese American University, Byblos 13-5053, Lebanon
7 University of Zilina, 01026 Zilina, Slovakia
* Correspondence: javeria.amin@uow.edu.pk

Abstract: Pests are always the main source of field damage and severe crop output losses in agriculture. Currently, manually classifying and counting pests is time consuming, and enumeration of population accuracy might be affected by a variety of subjective measures. Additionally, due to pests' various scales and behaviors, the current pest localization algorithms based on CNN are unsuitable for effective pest management in agriculture. To overcome the existing challenges, in this study, a method is developed for the localization and classification of pests. For localization purposes, the YOLOv5 is trained using the optimal learning hyperparameters which more accurately localize the pest region in plant images with 0.93 F1 scores. After localization, pest images are classified into Paddy with pest/Paddy without pest using the proposed quantum machine learning model, which consists of fifteen layers with two-qubit nodes. The proposed network is trained from scratch with optimal parameters that provide 99.9% classification accuracy. The achieved results are compared to the existing recent methods, which are performed on the same datasets to prove the novelty of the developed model.

Keywords: localization; qubits; quantum; YOLOv5; pest

1. Introduction

Crop pests are among the key factors that lower the productivity and quality of the crop. Therefore, both academics and businesses are paying close attention to the efficient prevention and management of pest species. The most effective method used to control crop pests is the use of agrochemicals. However, farmers who lack the knowledge to identify pests recklessly apply vast quantities of pesticides, endangering not only their health but also the environment and soil [1]. The real-time monitoring of agricultural pests at an early stage, according to the integrated pest management (IPM) theory, could lessen the damage caused by careless use of pesticides. To improve the situation, forecasting, proper identification, and localization of crop pests are crucial first steps [2]. Pest classification initially relied heavily on agricultural professionals with technical knowledge, which was time consuming and caused lags in farmers' information [3]. The disparity between the growing need for precise and real-time insect identification and the dearth of professionals who can meet it is indeed growing. Therefore, fast and accurate automatic detection of pest systems must be proposed [4]. Three methods can be used to assess the number of agricultural pests in a field, per an examination of recent literature: (1) manual monitoring and measuring, which takes time and delays the release of agricultural statistics; (2) multi-dimensional data, such as field temperature and moisture levels, which can be utilized to calculate the level of pest prevalence in the absence of precise data; (3) employing trapping devices to capture photos of agricultural pests, then counting the pests using object tracking detection methods [5]. The third method was adopted as the primary location for crop pest monitoring studies. Manual monitoring and identification of pest diseases are time-consuming and error-prone assignments [6]. Computerized methods play a significant role in pest disease detection. Currently, more attention is paid to ML methods to conquer challenges relating to the detection of pest disease. In ML models SVM, ANN, AdaBoost, MLR, and decision trees are utilized based on the hand-crafted features used for the analysis of plant pests [7,8]. Deep learning methodologies rely on convolutional kernels that provide innovative results to overcome the problems of ML methods. Compared to ML models, DL ascertains automatically significant features from training plant data and allows the users to develop end-to-end systems, avoiding processing the input images separately [9]. Achieving good results in a variety of object detection and classification tasks using DL has been made possible by exceptional CNNs performance [10]. K-means clustering is employed to segment the pest disease after the custom 2D-CNN model is used to predict pest/normal images. Moreover, pre-trained models such as VGG-16,19, Xception, Mobile-Net, Dense-Net201, NAS-Net-Mobile, and InceptionResNet-v2 are fine-tuned for pest disease classification [11]. The empirical research observes that k-means performed better compared to watershed and thresholding methods. In the classification experiment, the custom CNN model provides 0.96 prediction accuracy, while transfer learning Mobile-Net and InceptionRes-Net-V2 provides 0.82 and 0.81 prediction accuracy [12]. To improve the detection accuracy, we investigate the segmentation based on DL. The size of the pest dataset needs to be increased and must include the grading of the pest infections [13]. The Faster RCNN model is applied for the recognition of five types of pests such as Cicadellidae, Flea Beetles, Aphids, Red spiders, and Flax Budworm. The results are computed in terms of precision of 0.50 on mobile-net, 0.86 on SSD, and 0.98 on the F-RCNN model. After the experimental analysis, we conclude that F-RCNN performs better compared to Mobile-Net and single-shot detectors [14]. The deep convolutional network is used for pest disease detection. The transfer learning ResNet-50 model is also applied and fine-tuned using the optimal hyper-parameters for pest disease classification. This model provides average accuracy of 0.95 [15].

Although sufficient work has been conducted based on the detection of pest disease, there is still room for improvement [16]. Occlusion, one of the main challenges, is caused by changes in the position of the blade and external lighting. The DL model training is difficult due to the occlusion problem, which leads to false detection. To overcome the existing challenge, two models are proposed for more accurate localization and classification of pest diseases. The core contributions of this research are as follows:

- The YOLOv5 is designed based on optimal learning parameters for the recognition of pests in RGB images.
- The novel quantum machine learning model is designed on the selected layers and trained on the selected hyperparameters that help with the accurate classification of Paddy with/without pest images.

The article is organized as follows: Section 2 describes related work, while proposed methods are explained in Section 3. Section 4 provides the results, and finally, Section 5 provides the conclusion.

2. Related Work

In this method, an optimum super-resolution model is applied to enhance the quality of images. The pest region is localized using D2Det's model. The proposed model provides detection scores of the 0.78 mAp [17]. The regional proposal network is used for rich feature extraction, which provides an mAP of 0.78 on the AgriPest21 dataset [18]. The multi-features fusion network is used for pest classification, in which dilated convolution is applied for features extraction and deep features are also derived using the deep features extraction network. Finally, the extracted features are fused for classification. This model provides 98.2% accuracy for the classification of 12 different types of pest diseases [19]. The attention model based on activation mapping is used for pest classification. In this model, tiny pest regions are detected using weighted mapping of activation, prediction scores, and labels with an average accuracy of 68.3 ± 0.3 [20]. In the saliency discriminative guided model, two types of branches are utilized: raw and fine grained. In the raw branch, coarse-grained features such as global and features fine-grained modules are used as a raw branch. A salient object detection model is used for the localization of pest diseases [21]. The pretrained Mobile network is used for pest disease classification. This model is trained on optimal parameters, which are selected after extensive experimentation. The results are computed on the IP102 dataset, which provides an accuracy of 0.916 [22]. The regional CNN model is used for pest disease detection with 100% accuracy for palm-infected trees [23]. The multi-detection pest model is used to localize the pest region in which the feature pyramid multi-scale model and adaptive novel regional proposal model are included. The results are computed using a multi-pests MP-2021 dataset that provides a 0.67 precision rate and 0.89 recall [24]. The super-resolution multi-scale network and soft IoU models are used for enhancement. The results are computed on the LLPD-26 pest classes dataset, which contains 18,585 pest images with an mAP of 0.67 [25]. The four-step network is proposed for pest disease detection in this network acquired frames of videos are de-noised using the Bayesian model. The guided context residual model is applied for segmentation that is fed to the CNN model to establish a model for the detection of pest diseases with a 0.99 mAP [26]. Pre-trained Resnet-50 uses the backbone of mask-RCNN for pest disease segmentation. Three-dimensional coordinates are obtained precisely for the target points that provide the average rate of precision: 0.94 [27]. The pyramid attention features extraction model and fine-grained models are used for pest disease detection. This method is evaluated on D0 and IP102 datasets with 0.74 and 0.99 accuracy, respectively [28]. The plant images are augmented by applying a data augmentation approach. Then, four pre-trained models such as Google-net, Alex-net, VGG-net, and Res-Net are trained on the selected hyper-parameters. The proposed models provide accuracy of 0.96 on the NBAIR dataset, 0.97 on 24 classes Xie1 and 0.95 on the Xie2 dataset [29]. The SVM model is used with different functions of the kernel for the classification of parasites in strawberry plants [30]. A simple and new method is proposed in which blocks of original and feature reuse residual blocks are combined. This is known as the feature re-use residual model [31]. The spatial channel network is fused with the CNN model. The regional proposal model is adopted to detect pest illnesses. In this model, a position-sensitive score map is applied, in which the FC layer is replaced by a regression layer. The results are computed on the MPD-2018, which provides an mAP of 0.75 [32]. The transfer learning DenseNet-169 is used for pest detection in tomato plants. The results are computed on 859 images of 10 classes of pest that affect tomatoes, providing 0.88 accuracy [33]. The GAN model is used for the generation of synthetic pest images. The GAN results are visualized through the t-SNE method. The classification of the pest insect is performed using a CNN model that provides a 0.95 F1-score [34]. The Deep neural network is used for pest disease classification, in which modified ResNet-50 provides 0.95 accuracy [15]. The image contrast is improved using contrast enhancement, then k-means clustering is used for segmentation. The geometrical features such as GLCM and GLRLM are fed to the SVM and KNN for classification of normal/pest images of leaves with 0.93 prediction accuracy [35]. The HOG, GIST, and color features are used with SVM, NB, MLP, bagging, KNN, XGBoost and RF classifiers on 10 fold cross-validation for pest disease classification [36]. The CNN model is applied with softmax for the classification of the pest images, which provides 0.91 accuracy [37].

3. Material and Methods

The proposed method steps are localization and classification. In the localization step, the YOLOv5 model is trained on the selected parameters to localize the pest region in plant images. After localization, pest images are classified into Paddy with pest/Paddy without pest using the proposed quantum neural network, as presented in Figure 1.

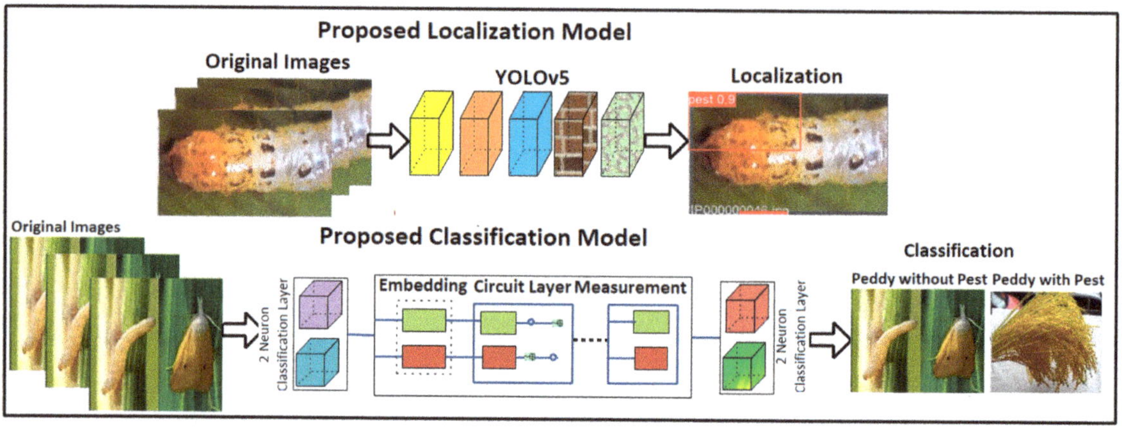

Figure 1. Proposed pest localization and classification models.

3.1. Localization of Pests Using the YOLOv5 Model

You look once (YOLO) models are widely used in different sectors for medical and agricultural purposes, as well as in parking meters, traffic signals, etc., In terms of recognition accuracy, YOLOv5 outperforms other detectors such as RCNN, YOLOv2, v3, etc. [38]. Therefore, the localization model is used to locate the actual region of pests in plant images more accurately.

The YOLOv5 detector contains three vital parts: the model backbone, the neck of the model, and the head model [39], in which the cross stage of the partial model is utilized for the extraction of rich features. The model neck is utilized for the generation of feature pyramids. Feature pyramids provide better generalizability when scaling the object. This helps with the identification of objects with distinct scales and sizes. Pyramid features help improve the performance of testing data. In the proposed model, FPN pyramids are used. The head of the model is utilized for the final part of the detection process. The anchor boxes are applied to the features that generate the final output vector with probabilities of class, scores of objectiveness, and rectangular boxes. The Leaky ReLU is applied in the hidden/middle layer and the sigmoid function is utilized in the detection of the final layer. The model is trained on the Sgdm optimizer function. YOLOv5 comprises the head and backbone.

The YOLOv5 head contains 16 convolutional, 1 focus, and 1 spatial pyramid pooling, while the YOLOv5 backbone comprises 23 layers: 16 convolutional, 2 up-sampling, 4 concatenation, and 1 detection layer. The proposed YOLOv5 architecture is shown in Figure 2. Table 1 lists the YOLOv5 model's training parameters.

The loss function of the YOLOv5 is mathematically explained as:

$$
\begin{aligned}
\gamma_{coordinate} \sum_{i=0}^{g^2} \sum_{j=0}^{box} \quad & 1_{ij}^{object}[(x_i - \hat{x}_i)^2 + (y_i - \hat{y}_i)^2] \\
& + \gamma_{coordinate} \sum_{i=0}^{g^2} \sum_{j=0}^{box} 1_{ij}^{object}[(weight_i - \widehat{weight}_i)^2 \\
& + (height_i - \widehat{height}_i)^2] + \gamma_{coordinate} \sum_{i=0}^{g^2} \sum_{j=0}^{box} 1_{ij}^{object}\left[(C_i - \hat{C}_i)^2\right] \\
& + \gamma_{noobj} \sum_{i=0}^{g^2} \sum_{j=0}^{box} 1_{ij}^{noobj}\left[(C_i - \hat{C}_i)^2\right] \\
& + \sum_{i=0}^{g^2} 1_i^{object} \sum_{c \in classes} [(p_i(c) - \hat{p}_i(c))^2]
\end{aligned}
$$

Here, 1_i^{object} represents the objects in i and 1_{ij}^{object} represent the jth bounding box, s denotes the grid and $x_i - y_i$ denote the center of the jth bounding box related to the i grid cell. $\hat{x}_i, \hat{y}_i$ represents the center of ground truth related to the i grid cell.

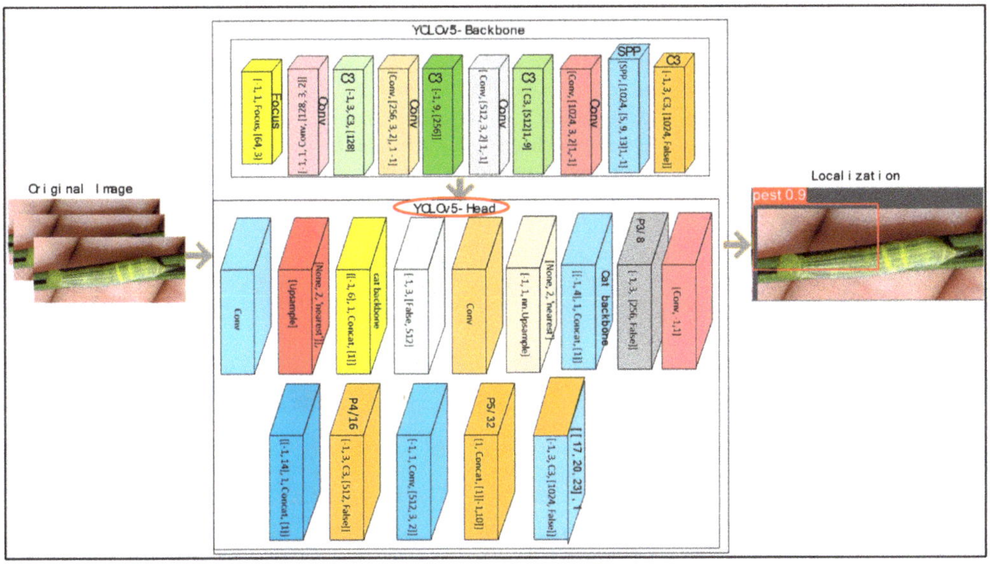

Figure 2. Proposed YOLOv5 model for localization.

Table 1. Parameters of YOLOv5 training.

Epochs	400
Batch size	8
Optimizer	Sgdm
Anchors	(10, 13, 16, 30, 33, 23) (30, 61, 62, 45, 59, 119) (116, 90, 156, 198, 373, 326)
Classes	2

Table 1 depicts training parameters in which 400 epochs, 8 batch-size, an Sgdm optimizer, (10, 13, 16, 30, 33, 23), (30, 61, 62, 45, 59, 119), (116, 90, 156, 198, 373, 326) and two classes are selected, providing better localization results.

3.2. Classification of Paddy with Pest/Paddy without Pest

The classical images are transformed through the computation of a matrix. The same images are presented as quantum states and encoded for the n qubits. The transformation of quantum images is achieved through the evolution of unitary ˆU under Hamiltonian. The qubits might store multiple values at the same time, which provides a huge advantage in terms of speed over the normal/classical algorithms. Therefore, in this article, the quantum model is proposed for the classification of the pest images. A qubit's state in quantum computing is represented by a unit vector in a complex two-dimensional vector space [40,41]. Consider a quantum system in two dimensions that are described by so-called computational states $|0>$ and $|1>$. The column vectors $\frac{0}{1}$, $\frac{1}{10}$ serve as a numerical representation of these reports. The basis of orthonormal for space of a qubit space is called a Hilbert. The primary distinction between quantum and classical bits is that bits of quantum are not limited to the binary states of $|0>$ and $|1>$. For instance, a qubit can exist in two states at once (a two-states superposition) [42].

$$|\text{þ}>= \acute{\alpha}|0> + \beta|1> \quad (1)$$

The model consists of three-layered hybrid architecture that contains two neurons of the classical fully connected (FC) layer. The two-qubit neurons are transformed into another two classical FC layers. A softmax activation function is applied for classification based on probability. The proposed model contains 15 layers and 2-qubit neurons, and it is trained with selected parameters such as the Adam optimizer, 16 batch size and 100 training epochs, and the data are divided into 5- and 10-fold cross-validation. The model layered architecture and training parameters are displayed in Tables 2 and 3.

Table 2. Training parameters of the proposed model.

Optimizer	Adam
Batch-size	16
Epochs	100
Split criteria	5- and 10-fold

Table 3. Proposed model layered architecture.

Type of Layer	Shape of Output
Dense	(None, 4)
Keras	(None, 4)
Dense	(None, 4)

Table 2 depicts training parameters that are selected after extensive experiments that provide better testing outcomes. Table 3 shows the layered design of the suggested model. The proposed architecture is presented in Figure 3.

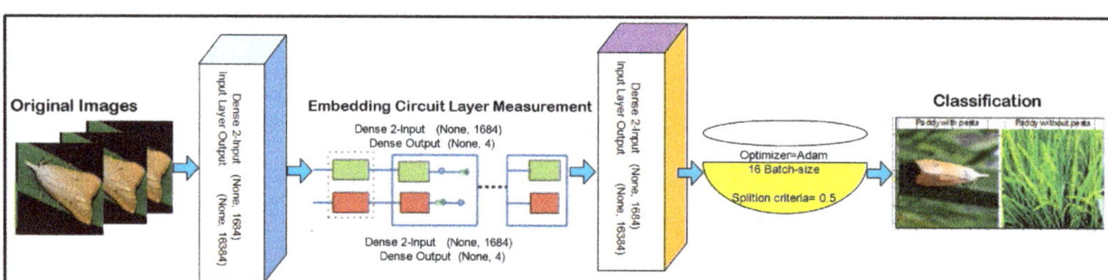

Figure 3. A proposed quantum model for classification.

4. Results and Discussion

Paddy pest data are downloaded from the Kaggle website "Paddy Pests Dataset | Kaggle". This dataset contains 135 files of Paddy with pests and 513 files of Paddy without pests. The IP102 dataset used for the recognition of pest insects contains 75,000 images split into 102 classes. Furthermore, 19,000 images are annotated to bounding boxes [43]. The proposed method is implemented on MATLAB toolbox 2022 on a Windows operating system with a 2070 RTX Nvidia Graphic Card. In this research, two experiments are performed: one for localization (Figure 4) and the second for the classification of pest images.

(**a**) (**b**)

Figure 4. (**a**,**b**) Shows the proposed method localization results.

4.1. Experiment#1: Localization of Pest Images Using the YOLOv5 Model

In this experiment, the localization method's performance is computed in terms of recall (R), precision (P), and mAP. The mathematical expression of performance metrics is as follows:

$$\mathrm{R} = \frac{\mathrm{Truepositive}(\partial)}{\partial + \mathrm{Falsenegative}}$$

$$\mathrm{P} = \frac{\partial}{\partial + \mathrm{Falsepositive}}$$

$$\mathrm{F1scores} = \frac{2 \times \mathrm{P} \times \mathrm{R}}{\mathrm{R} + \mathrm{P}}$$

The proposed localization model is trained/validated on the IP102 pest recognition dataset. The proposed model results are graphically presented in Figure 4. The statistical results are mentioned in Tables 4 and 5.

Table 4. Proposed method localization outcomes.

P	R	0.5-mAP	0.5:0.95-mAP
0.988	0.868	0.920	0.766
0.982	0.874	0.927	0.769
0.982	0.874	0.922	0.770
0.990	0.862	0.919	0.767
0.986	0.863	0.919	0.761
0.985	0.865	0.921	0.763
0.984	0.864	0.920	0.774
0.987	0.868	0.926	0.772
0.986	0.872	0.921	0.770
0.984	0.875	0.923	0.765
0.990	0.873	0.924	0.769
0.989	0.873	0.930	0.772
0.988	0.872	0.925	0.770
0.986	0.874	0.922	0.770
0.983	0.877	0.925	0.777
0.982	0.879	0.923	0.771
0.987	0.876	0.923	0.769
0.987	0.877	0.922	0.774
0.986	0.876	0.918	0.777
0.984	0.876	0.923	0.773
0.988	0.876	0.925	0.764
0.986	0.874	0.927	0.776
0.986	0.874	0.922	0.772
0.986	0.876	0.925	0.775
0.986	0.876	0.926	0.779
0.988	0.874	0.922	0.772
0.986	0.873	0.925	0.775
0.987	0.873	0.927	0.778
0.988	0.875	0.927	0.776
0.989	0.874	0.924	0.778
0.991	0.874	0.923	0.776
0.992	0.872	0.922	0.778
0.988	0.873	0.925	0.779
0.986	0.874	0.927	0.779
0.986	0.875	0.927	0.779
0.986	0.873	0.925	0.781
0.982	0.876	0.923	0.780
0.981	0.881	0.927	0.781
0.984	0.878	0.924	0.779
0.988	0.877	0.927	0.779
0.986	0.878	0.927	0.781
Mean: 0.987	**0.877**	**0.927**	**0.783**

Table 5. Localization results in comparison.

Ref	Year	Dataset	Results
[44]	2022	IP102	55.05 F1-score
[45]	2022		57.23 mAP
[46]	2022		67.82 F1-score
[47]	2022		77.04 mAP
[48]	2023		85.2 mAP
Proposed Model			0.93 F1-score 0.92 mAP

The proposed model training results with loss rate and validation outcomes are presented in Figure 5.

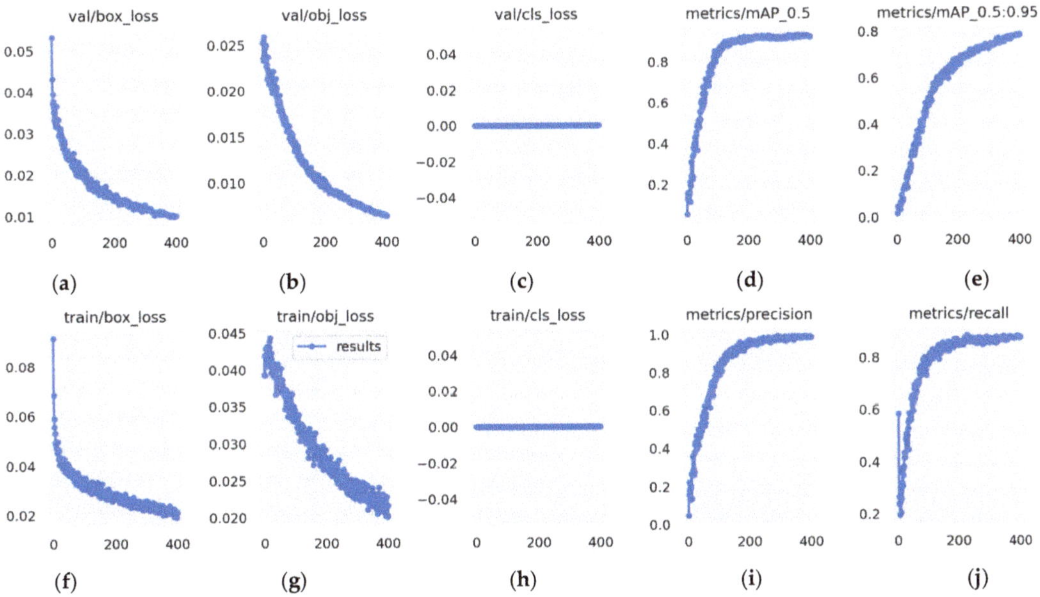

Figure 5. Training/validation results (**a**) loss of training box, (**b**) loss of training object, (**c**) loss of training cls, (**d**) metrics/precision, (**e**) metrics/recall, (**f**) loss of validation box, (**g**) loss of validation object, (**h**) loss of validation cls, (**i**) mAP (0.5), (**j**) mAP (0.95).

Figure 6 depicts the proposed model training/validation loss in terms of P, R, and mAP scores. In Figure 6, the proposed model validation results are visualized.

The validation scores of the proposed model are 0.86 precision, 0.92 recall, 0.93 F1-score and 0.92 mAP. The achieved outcomes prove that the proposed model more accurately localizes the pest regions. The tabular results are mentioned in Table 4.

Table 4 presents the results of the localization method, which achieves mean scores of 0.98764 precision, 0.87739 recall, 0.92728 mAP-0.5, and 0.78367 mAP-0.95. The proposed localization method's results are compared to the latest existing methods, as mentioned in Table 5.

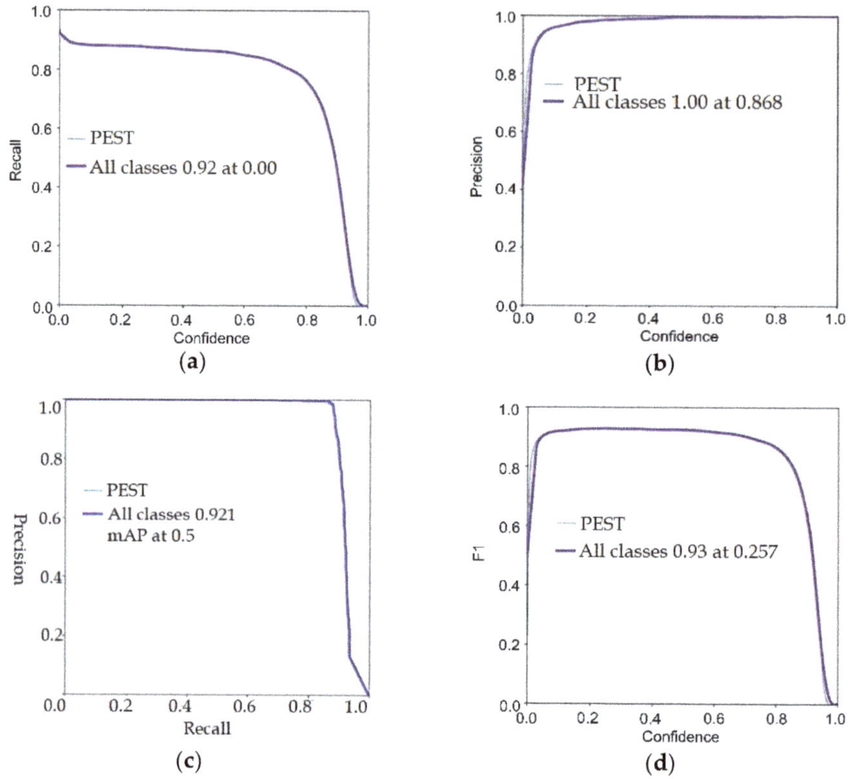

Figure 6. Proposed model validation results (**a**) precision, (**b**) mAP, (**c**) F1-score, (**d**) recall.

Table 5 presents the localization results that are compared to the existing methods [39–42]. An explainable neural model is used for localization, which provides 55.05 F1 scores [44]. CornerNet with DenseNet-100 is used for localization on the IP102 dataset, which provides an mAP of 57.23 [45]. The multi-scale attention model is utilized for localization on the IP102 dataset, which provides 67.28 F1 scores [46]. The multiple-scale attention method is utilized for the recognition of pest regions with 77.04 mAP [47].

As compared with the existing methods used in this research area, fine-tuned YOLOv5 provides better localization outcomes.

4.2. *Experiment#2: Classification of Pest Images Based on Proposed Quantum Neural Network*

The proposed model classifies the input images into two classes: pest/without pest. On 5/10-fold cross-validation, the performance of the suggested strategy is assessed in terms of various metrics, such as precision accuracy, F1 scores, and recall. Figure 7 shows a graphical representation of the suggested model training.

The classification results are computed using a confusion matrix, as shown in Figure 8.

Figure 8 shows the binary classification results of Paddy with pest and Paddy without pest. The achieved outcomes are mentioned in Table 6.

The proposed classification model achieves 99.31% accuracy, 0.99 precision, 1.00 recall, and a 0.99 F1 score with 5-fold cross-validation and 99.56% accuracy, 0.99 precision, 0.99 recall, and 0.99 F1 score with 10-fold cross-validation.

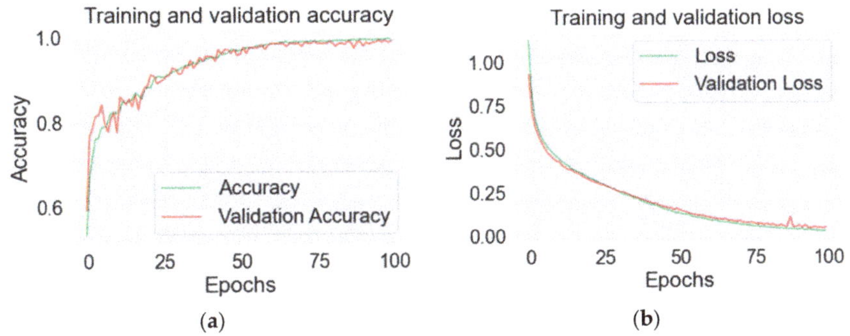

Figure 7. Proposed model results in (**a**) training and validation precision, and (**b**) loss of training and validation.

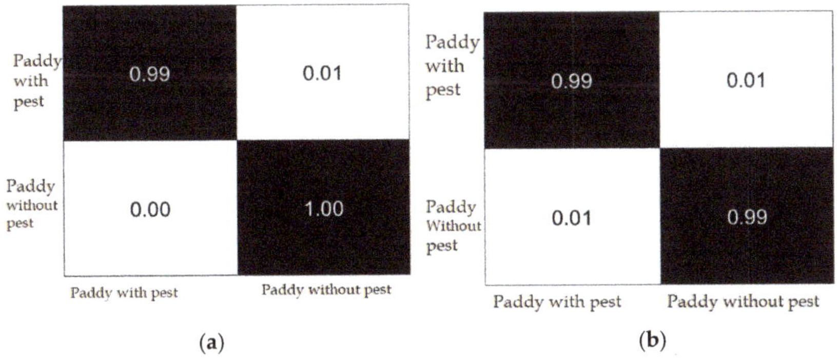

Figure 8. Confusion matrix (**a**) 5-fold; (**b**) 10-fold.

Table 6. Classification outcomes using the suggested model.

Cross-Validation	Classes	Accuracy	Precision	Recall	F1 Score
5-fold cross-validation	Paddy with pest	99.31%	0.99	1.00	0.99
	Paddy without pest	99.31%	1.00	0.99	0.99
10-fold cross-validation	Paddy with pest	99.56%	0.99	0.99	0.99
	Paddy without pest	99.56%	1.00	1.00	1.00

4.3. Statistical Analysis for Classification Model

The statistical analysis is performed in terms of mean/variation to compute the results of classification on 5- and 10-fold cross-validation. In this experiment, AUC values are computed on each fold, and the mean of AUC is measured using the deviation rate. The achieved results are graphically presented in Figure 9. The quantitative results are mentioned in Tables 7 and 8.

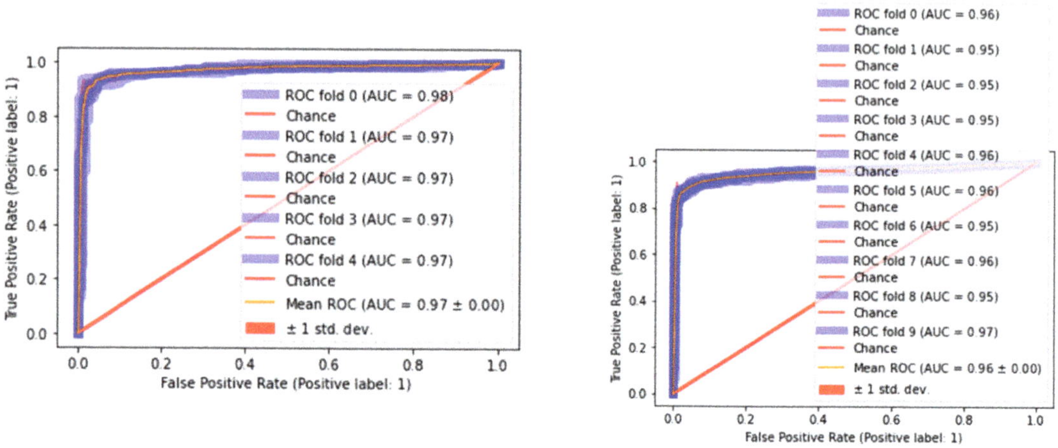

(**a**) (**b**)

Figure 9. Classification results (**a**) 5-fold and (**b**) 10-fold.

Table 7. Classification results in terms of mean/variance on 5-fold cross-validation.

0	1	2	3	4	Mean/Variation
0.98	0.97	0.97	0.97	0.97	0.97 ± 0.00

Table 8. Results of classification in terms of mean/variance on 10-fold cross-validation.

0-fold	1-fold	2-fold	3-fold	4-fold	5-fold	6-fold	7-fold	8-fold	9-fold	Mean/Variation
0.96	0.95	0.95	0.95	0.96	0.96	0.95	0.96	0.95	0.97	0.96 ± 0.00

Based on the five-fold cross-validation results, we achieve 0.98 ROC on 0-fold and 0.97 ROC on 1-, 2-, 3- and 4-fold, and 0.97 ± 0.00 mean/variation. The classification of pests is provided in Table 8.

The classification outcomes in Table 8 are 0.96 ROC for 0-fold, 0.95 ROC for 1-, 2-, 3-, 6- and 8-fold, 0.96 ROC for 4- and 7-fold, 0.97 ROC for 9-fold, and 0.96 ± 0.00 mean/variation. The proposed results comparison is provided in Table 9.

Table 9. Comparison of results.

Ref	Year	Dataset	Accuracy
[49]	2022	Kaggle	93.83%
[50]	2022		97.70%
[51]	2023		82.50%
[52]	2023		98.28%
[37]	2023		91.45%
Proposed Method			99.90%

Table 9 depicts the comparison of results with [37,49–52]. Transfer learning models such as MobileNet and VGG16 are used for classification, with 93.83% accuracy [49]. A Custom CNN model is developed for paddy classification with 97.70% accuracy [50]. The VGG19,16 and ResNet-50 are applied with a voting classifier for classification, which provides 82.50% accuracy [51]. The FCN model is applied to the conditional random field for refinement of the contour of the insect and localized boundaries. Finally, Densenet is applied with attention to the classification and provides accuracy of 98.28% [52]. The CNN model is designed for classification through softmax and SVM, which provides 91.45% accuracy [37]. However, compared to the existing method in this research, quantum machine learning is proposed which is trained from the scratch using the optimal parameters that provide 99.90% accuracy.

5. Conclusions

In agricultural domain, pests cause damage in fields, leading to significant losses in the crop yield. Therefore, it is necessary to obtain accurate and early detection of pests using plant images. Although researchers are working on accurate detection, there is still room for improvement because of several factors such as noise, illumination, occlusion, etc., That is why the end-to-end deep learning YOLOv5 model was designed based on a features pyramid (FPN) with optimal hyperparameters which provides a significant improvement in localization results, offering 0.987 Precision, 0.877 Recall, 0.927 mAP-0.5, 0.783 mAP-0.5:0.95 compared to existing methods.

The classification of the pest images is also a necessary part of this domain, which depends on optimal feature extraction and selection. Therefore, a fifteen-layered, two-qubit quantum model is designed and trained from scratch and based on selected learning parameters. The results achieved in terms of accuracy are 99.56%, 0.99 precision, 0.99 recall, and 0.99 F1 scores. The comparison between the proposed and the existing models authenticates that the proposed model performs better.

In this article, binary classification is performed; in the future, the work may be extended for multi-classification of pest images.

Author Contributions: Conceptualization, J.A. and M.A.A.; methodology, J.A.; software, R.Z.; validation, S.K. and M.A.A.; formal analysis, L.S. and M.I.S.; investigation, S.K.; resources, L.S.; data curation, R.Z. and M.I.S. All authors have read and agreed to the published version of the manuscript.

Funding: This work has been supported by the project of Operational Program Integrated Infrastructure: Independent re-search and development of technological kits based on wearable electronics products, as tools for raising hygienic standards in a society exposed to the virus causing the COVID-19 disease, ITMS2014+ code 313011ASK8.

Institutional Review Board Statement: Not applicable.

Informed Consent Statement: Not applicable.

Data Availability Statement: https://www.kaggle.com/datasets/zeeniye/paddy-pests-dataset, accessed on 15 August 2022.

Conflicts of Interest: All authors declare that there is no conflict of interest.

References

1. Lacey, L.; Grzywacz, D.; Shapiro-Ilan, D.; Frutos, R.; Brownbridge, M.; Goettel, M. Insect pathogens as biological control agents: Back to the future. *J. Invertebr. Pathol.* **2015**, *132*, 1–41. [CrossRef]
2. Vreysen, M.; Robinson, A.; Hendrichs, J.; Kenmore, P. Area-wide integrated pest management (AW-IPM): Principles, practice and prospects. In *Area-Wide Control of Insect Pests: From Research to Field Implementation*; Springer: Berlin/Heidelberg, Germany, 2007; pp. 3–33.
3. Tian, H.; Wang, T.; Liu, Y.; Qiao, X.; Li, Y. Computer vision technology in agricultural automation—A review. *Inf. Process. Agric.* **2020**, *7*, 1–19. [CrossRef]
4. Barbedo, J.G.A. Detecting and classifying pests in crops using proximal images and machine learning: A review. *AI* **2020**, *1*, 312–328. [CrossRef]
5. Smith, M. Scouting: The Tip of the IPM Spear. *EDIS* **2019**, *2019*, 24–27.
6. Orchi, H.; Sadik, M.; Khaldoun, M. On using artificial intelligence and the internet of things for crop disease detection: A contemporary survey. *Agriculture* **2022**, *12*, 9. [CrossRef]
7. Miranda, J.L.; Gerardo, B.D.; Tanguilig, B.T., III. Pest detection and extraction using image processing techniques. *Int. J. Comput. Commun. Eng.* **2014**, *3*, 189. [CrossRef]

8. Li, W.; Zheng, T.; Yang, Z.; Li, M.; Sun, C.; Yang, X. Classification and detection of insects from field images using deep learning for smart pest management: A systematic review. *Ecol. Inform.* **2021**, *66*, 101460. [CrossRef]
9. Shafkat, I. Intuitively Understanding Convolutions for Deep Learning. 2018. Available online: https://towardsdatascience.com/intuitively-understanding-convolutions-for-deep-learning-1f6f42faee1 (accessed on 6 June 2020).
10. Alzubaidi, L.; Zhang, J.; Humaidi, A.J.; Al-Dujaili, A.; Duan, Y.; Al-Shamma, O.; Santamaría, J.; Fadhel, M.A.; Al-Amidie, M.; Farhan, L. Review of deep learning: Concepts, CNN architectures, challenges, applications, future directions. *J. Big Data* **2021**, *8*, 53. [CrossRef]
11. Javed, M.H.; Noor, M.H.; Khan, B.Y.; Noor, N.; Arshad, T. K-means based automatic pests detection and classification for pesticides spraying. *Int. J. Adv. Comput. Sci. Appl.* **2017**, *8*, 236–240.
12. Thar, S.P.; Ramilan, T.; Farquharson, R.J.; Pang, A.; Chen, D. An empirical analysis of the use of agricultural mobile applications among smallholder farmers in Myanmar. *Electron. J. Inf. Syst. Dev. Ctries* **2021**, *87*, e12159. [CrossRef]
13. Singh, P.; Verma, A.; Alex, J.S.R. Disease and pest infection detection in coconut tree through deep learning techniques. *Comput. Electron. Agric.* **2021**, *182*, 105986. [CrossRef]
14. Karar, M.E.; Alsunaydi, F.; Albusaymi, S.; Alotaibi, S. A new mobile application of agricultural pests recognition using deep learning in cloud computing system. *Alex. Eng. J.* **2021**, *60*, 4423–4432. [CrossRef]
15. Malathi, V.; Gopinath, M. Classification of pest detection in paddy crop based on transfer learning approach. *Acta Agric. Scand. Sect. B—Soil Plant Sci.* **2021**, *71*, 552–559. [CrossRef]
16. He, J.; Chen, K.; Pan, X.; Zhai, J.; Lin, X. Advanced biosensing technologies for monitoring of agriculture pests and diseases: A review. *J. Semicond.* **2023**, *44*, 023104. [CrossRef]
17. Wang, H.; Li, Y.; Dang, L.M.; Moon, H. An efficient attention module for instance segmentation network in pest monitoring. *Comput. Electron. Agric.* **2022**, *195*, 106853. [CrossRef]
18. Wang, R.; Jiao, L.; Xie, C.; Chen, P.; Du, J.; Li, R. S-RPN: Sampling-balanced region proposal network for small crop pest detection. *Comput. Electron. Agric.* **2021**, *187*, 106290. [CrossRef]
19. Wei, D.; Chen, J.; Luo, T.; Long, T.; Wang, H. Classification of crop pests based on multi-scale feature fusion. *Comput. Electron. Agric.* **2022**, *194*, 106736. [CrossRef]
20. Bollis, E.; Maia, H.; Pedrini, H.; Avila, S. Weakly supervised attention-based models using activation maps for citrus mite and insect pest classification. *Comput. Electron. Agric.* **2022**, *195*, 106839. [CrossRef]
21. Luo, Q.; Wan, L.; Tian, L.; Li, Z. Saliency guided discriminative learning for insect pest recognition. In Proceedings of the 2021 International Joint Conference on Neural Networks (IJCNN), Shenzhen, China, 18–22 July 2021; pp. 1–8.
22. Rimal, K.; Shah, K.; Jha, A. Advanced multi-class deep learning convolution neural network approach for insect pest classification using TensorFlow. *Int. J. Environ. Sci. Technol.* **2022**, *20*, 4003–4016. [CrossRef]
23. Alsanea, M.; Habib, S.; Khan, N.F.; Alsharekh, M.F.; Islam, M.; Khan, S. A Deep-Learning Model for Real-Time Red Palm Weevil Detection and Localization. *J. Imaging* **2022**, *8*, 170. [CrossRef] [PubMed]
24. Dong, S.; Du, J.; Jiao, L.; Wang, F.; Liu, K.; Teng, Y.; Wang, R. Automatic Crop Pest Detection Oriented Multiscale Feature Fusion Approach. *Insects* **2022**, *13*, 554. [CrossRef] [PubMed]
25. Teng, Y.; Zhang, J.; Dong, S.; Zheng, S.; Liu, L. MSR-RCNN: A multi-class crop pest detection network based on a multi-scale super-resolution feature enhancement module. *Front. Plant Sci.* **2022**, *13*, 810546. [CrossRef] [PubMed]
26. Chodey, M.D.; Noorullah Shariff, C. Hybrid deep learning model for in-field pest detection on real-time field monitoring. *J. Plant Dis. Prot.* **2022**, *129*, 635–650. [CrossRef]
27. Li, Y.; Feng, Q.; Lin, J.; Hu, Z.; Lei, X.; Xiang, Y. 3D Locating System for Pests' Laser Control Based on Multi-Constraint Stereo Matching. *Agriculture* **2022**, *12*, 766. [CrossRef]
28. Ung, H.T.; Ung, H.Q.; Nguyen, B.T. An efficient insect pest classification using multiple convolutional neural network based models. *arXiv* **2021**, arXiv:2107.12189.
29. Thenmozhi, K.; Reddy, U.S. Crop pest classification based on deep convolutional neural network and transfer learning. *Comput. Electron. Agric.* **2019**, *164*, 104906. [CrossRef]
30. Ebrahimi, M.; Khoshtaghaza, M.H.; Minaei, S.; Jamshidi, B. Vision-based pest detection based on SVM classification method. *Comput. Electron. Agric.* **2017**, *137*, 52–58. [CrossRef]
31. Ren, F.; Liu, W.; Wu, G. Feature reuse residual networks for insect pest recognition. *IEEE Access* **2019**, *7*, 122758–122768. [CrossRef]
32. Liu, L.; Wang, R.; Xie, C.; Yang, P.; Wang, F.; Sudirman, S.; Liu, W. PestNet: An end-to-end deep learning approach for large-scale multi-class pest detection and classification. *IEEE Access* **2019**, *7*, 45301–45312. [CrossRef]
33. Pattnaik, G.; Shrivastava, V.K.; Parvathi, K. Transfer learning-based framework for classification of pest in tomato plants. *Appl. Artif. Intell.* **2020**, *34*, 981–993. [CrossRef]
34. Lu, C.-Y.; Rustia, D.J.A.; Lin, T.-T. Generative adversarial network based image augmentation for insect pest classification enhancement. *IFAC-Pap.* **2019**, *52*, 1–5. [CrossRef]
35. Dey, A.; Bhoumik, D.; Dey, K.N. Automatic detection of whitefly pest using statistical feature extraction and image classification methods. *Int. Res. J. Eng. Technol.* **2016**, *3*, 950–959.
36. Kasinathan, T.; Uyyala, S.R. Machine learning ensemble with image processing for pest identification and classification in field crops. *Neural Comput. Appl.* **2021**, *33*, 7491–7504. [CrossRef]

37. Haridasan, A.; Thomas, J.; Raj, E.D. Deep learning system for paddy plant disease detection and classification. *Environ. Monit. Assess.* **2023**, *195*, 120. [CrossRef] [PubMed]
38. Kuznetsova, A.; Maleva, T.; Soloviev, V. YOLOv5 versus YOLOv3 for apple detection. In *Cyber-Physical Systems: Modelling and Intelligent Control*; Springer: Berlin/Heidelberg, Germany, 2021; pp. 349–358.
39. Solawetz, J.; Nelson, J. How to Train YOLOv5 on a Custom Dataset. Volume 19. 2020. Available online: https://blog.roboflow.com/how-to-train-yolov5-on-a-custom-dataset/ (accessed on 10 December 2022).
40. Beer, K.; Bondarenko, D.; Farrelly, T.; Osborne, T.J.; Salzmann, R.; Scheiermann, D.; Wolf, R. Training deep quantum neural networks. *Nat. Commun.* **2020**, *11*, 808. [CrossRef]
41. Langenfeld, S.; Morin, O.; Körber, M.; Rempe, G. A network-ready random-access qubits memory. *Npj Quantum Inf.* **2020**, *6*, 86. [CrossRef]
42. Gyurik, C.; Dunjko, V. Structural risk minimization for quantum linear classifiers. *Quantum* **2023**, *7*, 893. [CrossRef]
43. Wu, X.; Zhan, C.; Lai, Y.-K.; Cheng, M.-M.; Yang, J. Ip102: A large-scale benchmark dataset for insect pest recognition. In Proceedings of IEEE/CVF Conference on Computer Vision and Pattern Recognition, Long Beach, CA, USA, 15–20 June 2019; pp. 8787–8796.
44. Couliably, S.; Kamsu-Foguem, B.; Kamissoko, D.; Traore, D. Explainable deep convolutional neural networks for insect pest recognition. *J. Clean. Prod.* **2022**, *371*, 133638. [CrossRef]
45. Albattah, W.; Masood, M.; Javed, A.; Nawaz, M.; Albahli, S. Custom CornerNet: A drone-based improved deep learning technique for large-scale multiclass pest localization and classification. *Complex Intell. Syst.* **2022**, 1–18. [CrossRef]
46. Feng, F.; Dong, H.; Zhang, Y.; Zhang, Y.; Li, B. MS-ALN: Multiscale Attention Learning Network for Pest Recognition. *IEEE Access* **2022**, *10*, 40888–40898. [CrossRef]
47. Zhang, W.; Sun, Y.; Huang, H.; Pei, H.; Sheng, J.; Yang, P. Pest region detection in complex backgrounds via contextual information and multi-scale mixed attention mechanism. *Agriculture* **2022**, *12*, 1104. [CrossRef]
48. Li, M.; Cheng, S.; Cui, J.; Li, C.; Li, Z.; Zhou, C.; Lv, C. High-Performance Plant Pest and Disease Detection Based on Model Ensemble with Inception Module and Cluster Algorithm. *Plants* **2023**, *12*, 200. [CrossRef] [PubMed]
49. Murugan, D. Paddy Doctor: A Visual Image Dataset for Paddy Disease Classification. *arXiv* **2022**, arXiv:2205.11108.
50. Debnath, O.; Saha, H.N. An IoT-based intelligent farming using CNN for early disease detection in rice paddy. *Microprocess. Microsyst.* **2022**, *94*, 104631. [CrossRef]
51. Anwar, Z.; Masood, S. Exploring Deep Ensemble Model for Insect and Pest Detection from Images. *Procedia Comput. Sci.* **2023**, *218*, 2328–2337. [CrossRef]
52. Gong, H.; Liu, T.; Luo, T.; Guo, J.; Feng, R.; Li, J.; Ma, X.; Mu, Y.; Hu, T.; Sun, Y. Based on FCN and DenseNet Framework for the Research of Rice Pest Identification Methods. *Agronomy* **2023**, *13*, 410. [CrossRef]

agriculture

MDPI

Article

A Hybrid Framework for Detection and Analysis of Leaf Blight Using Guava Leaves Imaging

Sidrah Mumtaz [1], Mudassar Raza [1], Ofonime Dominic Okon [2], Saeed Ur Rehman [1,*], Adham E. Ragab [3] and Hafiz Tayyab Rauf [4,*]

1 Department of Computer Science, COMSATS University Islamabad, Wah Campus, Wah Cantt 47040, Pakistan
2 Department of Electrical/Electronics & Computer Engineering, Faculty of Engineering, University of Uyo, Uyo 520103, Nigeria
3 Industrial Engineering Department, College of Engineering, King Saud University, P.O. Box 800, Riyadh 11421, Saudi Arabia
4 Independent Researcher, Bradford BD8 0HS, UK
* Correspondence: srehman@ciitwah.edu.pk (S.U.R.); hafiztayyabrauf093@gmail.com (H.T.R.)

Abstract: Fruit is an essential element of human life and a significant gain for the agriculture sector. Guava is a common fruit found in different countries. It is considered the fourth primary fruit in Pakistan. Several bacterial and fungal diseases found in guava fruit decrease production daily. Leaf Blight is a common disease found in guava fruit that affects the growth and production of fruit. Automatic detection of leaf blight disease in guava fruit can help avoid decreases in its production. In this research, we proposed a CNN-based deep model named SidNet. The proposed model contains thirty-three layers. We used a guava dataset for early recognition of leaf blight, which consists of two classes. Initially, the YCbCr color space was employed as a preprocessing step in detecting leaf blight. As the original dataset was small, data augmentation was performed. DarkNet-53, AlexNet, and the proposed SidNet were used for feature acquisition. The features were fused to get the best-desired results. Binary Gray Wolf Optimization (BGWO) was used on the fused features for feature selection. The optimized features were given to the variants of SVM and KNN classifiers for classification. The experiments were performed on 5- and 10-fold cross validation. The highest achievable outcomes were 98.9% with 5-fold and 99.2% with 10-fold cross validation, confirming the evidence that the identification of Leaf Blight is accurate, successful, and efficient.

Keywords: AlexNet; BGWO; CNN; DarkNet-53; deep learning; entropy; KNN; ROI; SVM; YCbCr

1. Introduction

Food is the fundamental requirement for the existence of human beings, and it is the notable outcome of agricultural activities. Agriculture is assumed to be the backbone of economic development, as it exhibits the cultivation of multiple crops, fruits, and vegetables. There is a large difference between the cultivation and annual production of fruits because of inappropriate advancements in technology, lack of knowledge, and diseases that negatively affect the production [1]. Disease detection in plants is a challenging task and is essential to diagnose at early stages. Diseases are mostly diagnosed through leaves because they tend to highlight contaminated parts immediately. Guava is an important fruit in agriculture; therefore, its leaves are selected for the detection and recognition of diseases [2]. Guava is nutritionally beneficial, serving calcium and iron to the human body. It is cultivated in America, especially in Mexico, Thailand, South Africa, and many other countries. Many laboratories such as the Central Institute of Subtropical Horticulture (CISH) and different institutes are continuing to work on guava production in different areas of the world [3]. Several diseases, such as bacterial and fungal diseases, attack the guava fruit, which badly affects its production [4]. There are different techniques of ML applied for disease detection. Almost 177 types of diseases are found that damage leaves, causing leaf blight and leaf spots. Known diseases include brown roots, twig drying, bacterial wilt, anthracnose, ring rots, and many others [5].

Many researchers aim for innovations in disease detection. Disease detection relies on five major steps. Usually, the first step in image processing is image acquisition. After obtaining images, preprocessing incorporates multiple steps that result in better accuracy. After preprocessing, feature extraction is performed, where the features of the images are boosted for further computation and selection. The final stage is classification. A variety of models are presented using diverse methodologies such as convolution neural network (CNN), gradient descent (GD), and many others for classification purposes [6]. Convolution neural networks play an essential part in the extraction of features through hidden layers, as manual extraction is costly and time-consuming [7]. Plant pathologists need an automatic detection system to diagnose leaf blight in plants.

The main focus of the proposed methodology is the detection and classification of leaf blight. Leaf blight affects plants as a result of a pathogenic organism infecting leaves. Therefore, an automated system is needed to detect leaf blight disease. Research of diseases in guava fruit is a challenging task, as it seeks a variety of data regarding diseases in the relevant field [8]. The forecasted production of guava is 498.95 thousand tonnes in 61.37 acres in the year 2020–2021 and the production of guava is 499.68 thousand tonnes in 61.37 acres in the year 2021–2022. Evaluation of the researched tasks becomes critical with time due to the wide range of diseases, and a great deal of effort has already been applied towards the relevant field [9].

There are several limits and difficulties in detecting and classifying guava plant diseases in the existing literature. Some major problems per the literature are the poor contrast, variation in shape, texture, and size, and illumination problems found in disease images that make them difficult to recognize and classify.

This article presents a new methodology for the detection of leaf blight. The purpose is to classify the healthy and diseased images of guava leaves with improved accuracy. The significant contributions presented in this research are as follows:

- A new deep CNN Net named SidNet is presented, which consists of 33 layers along with 35 connections. The pretraining of SidNet is performed on a plant imaging dataset. The features are extracted from the proposed SidNet, darknet53, and AlexNet, which are further fused using serial fusion. The deep features are also known as automatic features; they automatically solve the issues related to contrast, shape, texture, and illumination.
- The features are sorted using an Entropy Algorithm, and for better feature selection, Binary Gray Wolf Optimization is used. The selected features are used to make a single feature vector for classification using an SVM and KNN Classifier to achieve the best performance and results.
- Data Augmentation is performed, as the selected dataset is small; therefore, the images are flipped both horizontally and vertically to make the dataset large.

The paper consists of five sections, where Section 1 explains the introduction, motivation, contribution, and problem statements for leaf blight detection. Section 2 covers the recent existing work. Section 3 provides the details of the presented proposed framework and Section 4 describes the details of the experiments and outcomes. Lastly, Section 5 covers the conclusion.

2. Related Work

Diseases in fruit plants and leaves are a major cause of destruction and economic loss. Automated systems help greatly with the detection of diseases at early stages. While considering the field of detection of disease in plants, deep neural networks work perfectly to identify and classify diseases. These networks are mentored to conduct high-value results in detecting and classifying diseases, and to fulfill the demands of food deformation prevention.

There are different methods for collecting images under certain conditions. Images are captured by multiple appliances, such as cameras, sensors, mobile phones, and other devices. In this era, more datasets containing guava are publicly available on multiple forums, such as Kaggle, Mendeley, and many others [10]. Pre-processing of images is an important phase in image processing. The pre-processing phase entails multiple steps which help highlight the focused parts and remove irrelevant information from guava leaf images. In the real world, label noise on images is a matter of concern. Multiple techniques have attained the best results in denoising images, especially mixed noise, speckle noise, and salt and pepper noise. Low contrast and color distortion in guava leaf images make them blur. Scattering and light absorption also affect clear image visualization [11]. Images are preprocessed by using rotational filters such as horizontal and vertical flipping. Data augmentation techniques are used, such as applying rotations and zooming into images [12]. Color spacing techniques are extensively applied in image processing. RGB, CIELAB, and CMYK models are mostly used as color spacing techniques to give the best results.

Feature extraction is a process in which reatures are reduced from the raw dataset and new features for manageable processing are created. Texture analysis has a wide range of applications [13]. Pattern recognition requires feature extraction to solve problems in prediction, cluster discrimination, and representation of data in the best way [14,15]. Content-Based Image Retrieval (CBIR) converts high-level image visuals into feature vectors that contain some properties [16]. There are multiple techniques to extract the features from guava leaf images, such as handcrafted-based features, region-based features, deep CNN-based features, texture-based features, color-based features, morphological-based features, etc. Extraction of features is categorized into hand-crafted-based features and deep-based features.

The selection of features from plant leaf images is carried out after the extraction of hand engineered and deep-based features. The set of features is chosen while noisy, poor, and extra features are eliminated from the original set of features [17]. There are five main types of feature selection, which are (1) Linear Method, (2) Non-Linear Method, (3) Filter-Based Method, (4) Wrapper Method, (5) Embedded Method. Linear methods include PCA and LDA. PCA stands for Principal Component Analysis, which is used for data reduction [18]. LDA stands for Linear Discriminant Analysis and is used for the conversion of high dimension features into lower dimension features [19]. Non-linear methods include Entropy [20], Genetic Algorithm (GA) [21], Binary Gray Wolf [22], Slap Swarm [23], Atom Search [24] and many others. Filter methods includes mRMR [25], Missing Value Ratio [26], and many others. Wrapper methods include Jackstraw [27] and Boruta [28]. Finally, embedded methods include LASSO [29], Ridge [30], Elastic [31], and many others. Image fusion [32] helps greatly in improving classifier accuracy with less computational cost [33]. Different algorithms are proposed that use image fusion to get the best accuracy results.

Image classification is the last step in image processing [34]. Classification tends to dominate the feature vector to determine which object belongs to which class [35]. There are different types of techniques used for the classification of healthy and diseased images of plant leaves. Image classification is divided into three main categories, which are (1) Supervised Learning, (2) Unsupervised learning, and (3) Object-based image analysis. Supervised Learning is used to detect the new category of the object from training data [36]. Unsupervised Learning is a process in which an image is identified in an image collection without using labeled training data [37]. Object-based analysis involves the grouping of pixels on the basis of some similarities such as shape and neighborhood [38]. To get the most accurate results, the Plant Village dataset is used for testing and training purposes. A total of 80% of guava leaf images are used for testing while 20% of them are used for training purposes. The achievable accuracy is 97.22% using Alex-Net and Squeeze-Net after segmentation and classification [39]. Atila et al. [40] designed the Efficient-Net architecture, which is designed for classification purposes. Different architectures are applied using CNN for model training to get highly accurate results. The model training is performed on the dataset of 87,848 images. Images are preprocessed using different techniques such as downscaling and squaring methods, they are then classified, and an accuracy of 99.53% is achieved using AlexNet, VGG-16, and GoogLeNet [41]. Several algorithms are used for image classification. These are SVM [42], K-Nearest Neighbor (KNN) [43], Naïve Bayes [44], Shadow algorithms [45], Minimum Mean Distance (MMD) [46], Decision Trees [47], K-Means Clustering [48] and many others. The datasets are frequently classified by SVM. This involves supervised learning and comprises points that are in the sample space and different regions [49]. Segmentation is performed on the preprocessed data in three stages. In the first stage, the deep CNN is trained to learn the mapping from the space map. In the second stage, prediction-based labels are acquired. At the last stage, these acquired labeled images are sent to SVM for classification and achieve an accuracy of 86% [50]. In machine learning, KNN is a statistical classification algorithm. It gathers the objects selected by neighbors having the highest number of votes [51]. KNN is inspected for the detection of weeds from UAV images of the chili crop of Australia. In comparison with KNN, SVM and Random Forest (RF) are used. The achievable accuracies across RF, SVM, and KNN are 96%, 94%, and 63%, respectively [52]. KNN is also used for classifying facial expressions [53]. Additionally, KNN is used for the classification of grape leaves into healthy and unhealthy leaves. Texture-based and color-based features are extracted from grape leaf images and are classified by the KNN classifier, and an accuracy of 96.66% is achieved [54]. Table 1 depicts an overview of recent works related to plants diseases analysis

Table 1. An overview of recent literature regarding plants diseases analysis.

Ref.	Year	Techniques	Dataset	Diseases	Results %
[41]	2018	Downscaling and squaring method, AlexNet, VGG-16, AlexNetOWTBn.	87,848 58 classes	Apple Scab, Black Rot, Early Blight, Brown Leaf Spot	99.53
[55]	2018	ResNet-50, Deep Siamese convolutional network, TSNE method, KNN	PVD	Black Rot, Esca, Chlorosis	90
[56]	2018	Transfer learning, F-RCNN, classification.	4923	Phoma Rot, Leaf Miner, Target Spot	95.75
[57]	2019	F-CNN, S-CNN, Segmentation, annotation and labeling on region of interest (lesions), random transformation (stretch/rotation/brightness/contrast blur)	Independent dataset	Spider Mite, Target Spot	98.6
[58]	2019	VGG classification, resizing and transformation of images into grayscale,	2465	Black dot and scurf	96
[40]	2021	Efficient-Net (B5Ver), Alexnet, ResNet50, classification.	61,486	Late Blight, Bacterial Spot	99.97
[59]	2021	Resizing, normalizing and augmentation, Efficient-Net (B7Ver), Efficient-Net (B4Ver), U-net, and modified U-net segmentation model, Score-Cam visualization technique,	18,161	Target Spot	99.9

3. Materials and Methods

The proposed methodology consists of multiple phases of image processing. In the first step, as a preprocessing step, color spacing is performed on the images of the dataset. Images are converted from RGB color format to YCbCr color format. After getting preprocessed data, feature extraction is performed using two pretrained models and a newly proposed CNN deep model known as SidNet. This newly proposed model is based on 33 convolutional layers. This proposed CNN deep model is pretrained using the Plant-Village dataset, which consists of 38 classes. After pretraining, features are collected from the proposed CNN deep model. These extracted features are then fused with AlexNet and DarkNet-53 to find the best and most appropriate results. A deep CNN known as AlexNet was developed in 2012, which is an 8-layer deep model and consists of 5 convolution layers and 3 Fully Connected layers. After every convolution layer and fully connected layer,

ReLU is applied. It contains a dropout layer, which is applied after the first and second fully connected layer. The input size in AlexNet is 227 × 227. In all layers, the activation function is ReLU. Softmax is designed as an activation function in the output layer. AlexNet is the simplest deep CNN model and it is used to get highly accurate results. DarkNet-53 is another used deep model, which works as the backbone for YoloV3 in object detection. It contains 25 layers for batch normalization and Leaky ReLU. The input size is 256 × 256 in DarkNet-53. In the second step, the feature sorting entropy algorithm is used for the selection of the best features. These selected core features are fused using serial-based fusion. After fusing the core features, the binary gray wolf optimization algorithm is used. Finally, these extracted fused features are provided to SVM and KNN classifiers for the classification of Leaf Blight for the best achievable results. The complete view of the proposed model is shown in Figure 1. It covers a complete view of the flow of the diagram of the designed structure.

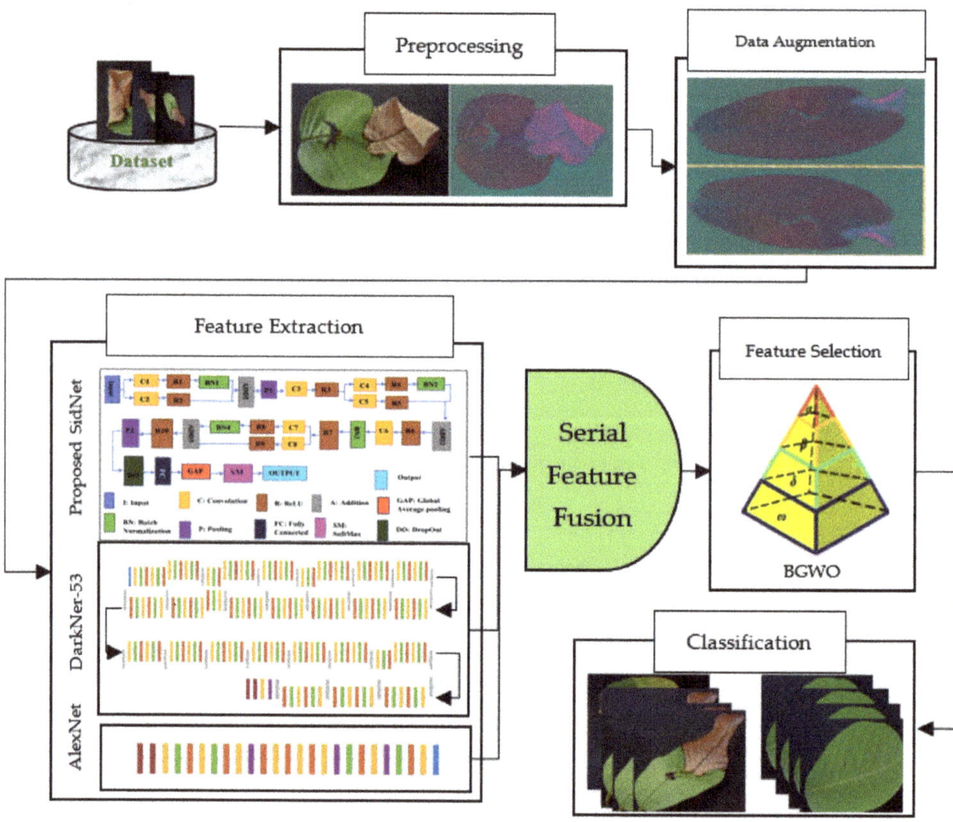

Figure 1. Proposed Framework for disease recognition in guava leaves.

3.1. Image Preprocessing

In image processing, color space conversion is an important task. *RGB* is used to store real-time images and videos because it allows for the sensitivity of color detection cells for the human visual system. *YCbCr* is beneficial for its low-resolution capability for the human visual system. Therefore, the conversion of *RGB* to *YCbCr* is mostly used in image processing. The general formula is given below:

$$YCbCr = RGB\,2YCbCr\,(RGB) \tag{1}$$

This is the general formula that is used for the conversion of *RGB* to *YCbCr* and it represents 8 bits per sample pixel in *RGB* format. The white and black colors are represented on a scale from 0 to 255. Therefore, the components of *YCbCr* are obtained from the following equations.

$$Y = 16 + 65.738/256\ R + 129.057/256\ G + 25.064/256\ B \quad (2)$$

$$Cb = 128 + 37.945/256\ R - 74.494/256\ G + 112.439/256\ B \quad (3)$$

$$Cr = 128 + 112.439/256\ R - 94.154/256\ G + 18.285/256\ B \quad (4)$$

where Y is used to represent the luma (luminance) component. Cb represents chrominance blue and Cr represents chrominance red. These numbers are the constant values that are used to adjust the value of Y. Figure 2 shows the conversion from *RGB* to *YCbCr*.

Figure 2. RGB to YCbCr conversion.

3.2. Data Augmentation

For data augmentation, horizontal and vertical flipping are used. According to the mathematical model, the horizontal flipping of the images is presented as follows:

$$H_F\ (-x,\ y) = H_O\ (x,\ y) \quad (5)$$

And the vertical flipping of the images is presented as follows:

$$H_V\ (x, -y) = H_O(x,\ y) \quad (6)$$

H_F shows the flipping function and H_O shows the original image function that is to be flipped. H_v shows the flipping function and H_O shows the original image function that is to be flipped.

3.3. Feature Extraction

After preprocessing and augmentation, the next phase is feature extraction. In this phase, with the help of pretrained models, the most optimal features are extracted. According to our proposed methodology, a newly designed model named SidNet, and other pretrained models such as DarkNet-53 and AlexNet, are used for extracting the most optimal features.

3.4. Proposed SidNet as CNN Net

This proposed CNN Net is a blockbuster CNN-based architecture used for the detection and classification of Leaf Blight. The proposed SidNet model consists of 33 layers involving 8 convolutional layers, 10 ReLU layers, 4 layers of batch normalization, one dropout layer, one softmax layer, one classification output, one fully connected, and one global average pooling layer. The Input size of the Proposed CNN Net is "227 × 227 × 3" and it contains 35 connections. The stride is "1 × 1" throughout the proposed framework's convolution layer. The number of filters is set to 96 in all convolutional layers of the framework, while the padding dimensions vary according to the convolution layer used in the different stages of architecture. The padding of the last convolution layer is 5, 5, 5, 5. Two pooling layers are used, where the stride for the two pooling layers in the architecture is "2 × 2". While the pooling size of the first pooling layer is 3, 2 and the pooling size of the second pooling layer is 3, 3. The mean decay and variance decay for all batch normalization layers are 0.1. Figure 3 shows the architecture of the proposed SidNet.

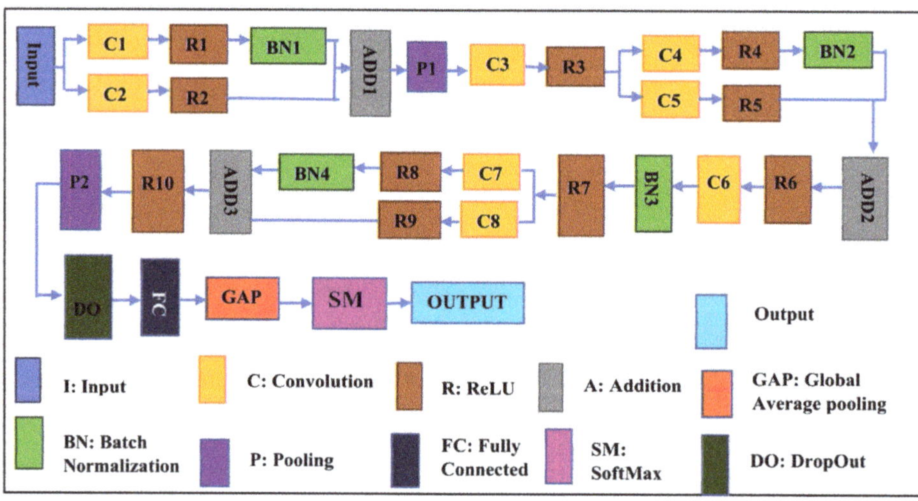

Figure 3. The architecture of the proposed CNN network, SidNet.

The proposed deep model named SidNet comprises 33 convolution layers. Its Input size is "227 × 227 × 3" and it contains 35 connections.

Due to the small number of samples of the available dataset, the proposed deep model with softmax (SM) classifier is first trained on the third-party dataset named CIFAR 100 [60]. Then, the guava leaf dataset is fed to SidNet for feature extraction. The features are extracted from the fully connected (FC) layer. These features, after feature selection, are trained and tested on various classifiers (such as SVM with its variants and KNN with its variants) for evaluation. According to our model, the features in SidNet are presented as:

$$f_{e \times f} = \{\, L1 \times 1\,,\ L1 \times 2\,,\ L1 \times 3\,,\ \ldots\ldots\ldots.L1 \times \mathrm{p}\,\} \tag{7}$$

where $L\,(1\,_{\ldots}\,p)$ represents the number of features obtained from the proposed CNN model, which is known as SidNet, and $e \times f$ is the dimension of the resultant function.

Visualization of the strongest feature maps at different convolution layers with the proposed SidNet architecture is shown in Figure 4. The visualization is performed on conv_1, conv_2, conv_3, conv_4, conv_5, conv_6, conv_7, and conv2 of the proposed architecture.

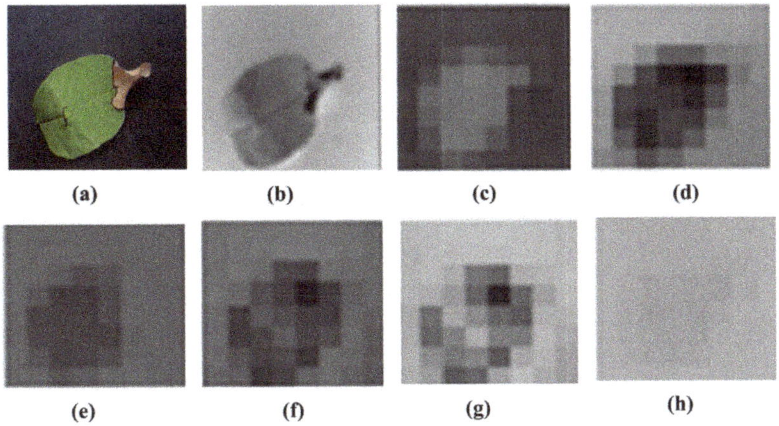

Figure 4. Visualization of the images of strong feature maps of different convolution layers. (**a**) conv_1 (**b**) conv_ 2 (**c**) conv_ 3 (**d**) conv_4 (**e**) conv_5 (**f**) conv_6 (**g**) conv_7 (**h**) conv2.

3.5. DarkNet-53

DarkNet-53 acts as the backbone of YoloV3 for object detection. DarkNet comprises 53 layers and consists of multiple convolution layers. A total of 1024 features in DarkNet are extracted with the help of the global average pooling layer. There are 25 layers in Batch normalization. The input size is 256 × 256. According to our mathematical model, the features in DarkNet are presented as:

$$f_{a\times b} = \{\, j1 \times 1\,,\ j1 \times 2\,,\ j1 \times 3\,,\ldots\ldots\ldots\ldots.j1 \times m\,\} \tag{8}$$

where $J_{(1\,\ldots\,m)}$ shows the number of features extracted from DarkNet-53 and $a \times b$ is the dimension of the resultant function. Visualization of the strongest feature maps at different convolution layers with the DarkNet-53 architecture is shown in Figure 5

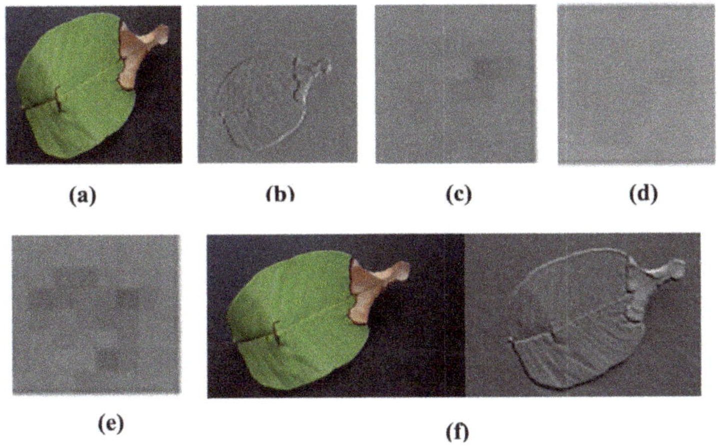

Figure 5. Visualization of the images of strong feature maps of different convolution layers of DarkNet-53. (**a**) conv3 (**b**) conv1 (**c**) conv2 (**d**) conv4 (**e**) conv5 (**f**) conv6.

3.6. AlexNet

AlexNet is the simplest model that comprises 8 layers. There are 5 convolution layers and 3 fully connected layers. Its input size is 227 × 227. The Activation function used in all layers is ReLU, which is applied after every convolution layer and fully connected layer. The Drop layer is applied after the first and second fully connected layer. The Activation function in the output layer is Softmax. The AlexNet contains 4096 features with the help of a convolution layer named a fully connected layer. According to our mathematical model, the features in AlexNet are presented as:

$$f_{c\times d} = \{ k_{1\times 1}, k_{1\times 2}, k1 \times 3, \ldots\ldots k1 \times n \} \quad (9)$$

where *k(1 … n)* shows the number of features extracted from AlexNet and $c \times d$ shows the dimensions of the resultant function. Visualization of the strongest feature maps at different convolution layers with the AlexNet architecture is shown in Figure 6

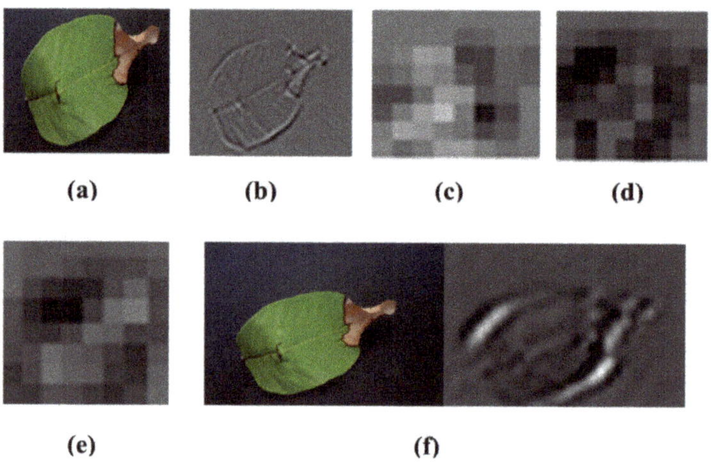

Figure 6. Visualization of the images of strong feature maps of different convolution layers of AlexNet. (**a**) conv2 (**b**) conv1 (**c**) conv3 (**d**) conv4 (**e**) conv5 (**f**) conv6.

3.7. Feature Selection

After extracting features from the proposed model and other pretrained models such as DarkNet-53, AlexNet, and SidNet, the selection of features is done with the help of feature sorting using entropy. The mathematical model for the selection of features using entropy is represented as:

$$E_{a\times b}(\alpha_1, \alpha_2, \alpha_3, \ldots\ldots \alpha_n) = -\sum f_{a\times b}\{ j_{1\times 1}, j_{1\times 2}, j_{1\times 3}, \ldots. j_{1\times m} \} log\,(j_{1\times 1}, j_{1\times 2}, j_{1\times 3}, \ldots\ldots j_{1\times m}) \quad (10)$$

$$E_{c\times d}(\beta_1, \beta_2, \beta_3, \ldots.. \beta_n - \sum f_{c\times d} = \{k_{1\times 1}, k_{1\times 2}, k_{1\times 3}, \ldots..k_{1\times n}\} loglog\,(k_{1\times 1}, k_{1\times 2}, k_{1\times 3}, \ldots\ldots\ldots k_{1\times n}) \quad (11)$$

$$E_{e\times f}(\gamma_1, \gamma_2, \gamma_3, \ldots\ldots \gamma_n = -\sum f_{e\times f} = \{L_{1\times 1}, L_{1\times 2}, L_{1\times 3}, \ldots\ldots L_{1\times p}\} loglog\,\left(L_{1\times 1}, L_{1\times 2}, L_{1\times 3}, \ldots\ldots..L_{1\times p}\right) \quad (12)$$

where $a \times b, c \times d, e \times f$ represents the dimension of features obtained after sorting features. *Log(j(1 − n)), Log(k(1 − n)),* and *Log(L(1 − n))* show the prediction of probability and *j(1 − n), k(1 − n)* and *L(1 − n)* show the selected features obtained from the extracted features. α *(1 − n), β(1 − n),* $\gamma(1 - n)$ show the features which are sorted.

3.8. Feature Fusion

Feature Fusion is performed to select the most optimal features. According to the mathematical model, the fusion of features is represented as:

$$\in f = \sum_{x=1}^{L}(E_{a\times b}) \bigcup \sum_{y=1}^{m}(E_{c\times d}) \bigcup \sum_{z=1}^{n}\left(E_{e\times f}\right) \quad (13)$$

After fusing the features using the entropy algorithm, Binary Gray Wolf Optimization is performed to obtain the most optimal results. The mathematical model representing the features selected from Binary Gray Wolf Optimization are as follows:

$$\varnothing^{\mathrm{d}} = \in 1\; if\; \gamma o < \frac{1}{3} \;\subset 2\; if\; \frac{1}{3} \leq \gamma 6 < \frac{2}{3} \;\in 3\; if\;,\; otherwise \tag{14}$$

where $\varnothing^{\mathrm{d}}$ is the function of BGWO and γo and $\gamma 6$ are the adjusting parameters that are used to set the value of the most optimal features.

3.9. Classification

Different classifiers are available for classification purposes, but SVM and KNN classifiers are chosen. These two classifiers are selected to achieve high accuracy and the most optimal results. In machine learning, features are reduced by carrying out the feature vector dimension. Different classification algorithms are available, such as Minimum Mean Distance (MMD), K means clustering, Decision Trees, Shadow algorithm, and Naïve Bayes. In this work, SVM and KNN are selected to perform the classification on the guava leaf dataset. These classifiers generate better results compared to other classifiers.

4. Results and Discussion

The purpose of this study is to classify the Leaf Blight disease with the best possible results. After the processing of the dataset using YCbCr, the extraction of features is performed using two pretrained models along with one proposed net. The selection of features is performed using BGWO. For classification purposes, SVM and KNN are chosen for the evaluation of execution. This section provides details about experiments that are performed on multiple sets of features and the results are recorded accordingly. These experiments and results are shown in two sets of test cases. Using 5 folds and 10 folds, validation experiments are performed. In comparison with other classifiers, SVM and KNN are selected, as they give the best results. The set of experiments are performed on Windows 10 (64-bit) and a Core (TM) i7-8700 CPU, 3.20 GHz (12 CPUs) 3.2 GHz processor with 16 GB RAM and an LCD and keyboard from HP. Training and testing of the designed network are performed on MATLAB R2020b.

4.1. Dataset

The assembled dataset for the classification of Leaf Blight used in this analysis [61] is small. These experiments are performed on guava leaves. The chosen dataset contains 415 guava leaf images in the original. This dataset contains a small number of images; therefore, they are augmented using horizontal and vertical flipping techniques, which increases the total number of images. The total number of images is 1000 after the augmentation technique. This dataset is publicly available on the Mendeley website. It is a binary class dataset (see Figure 7 for sample images) and many researchers have used it in their studies.

Figure 7. Sample dataset images of guava plant.

4.2. Performance Evaluation Methods

The performance measures can be used to measure and detect the performance of leaf blight disease in plants. The consequences are defined as follows: a True Positive rate as TP, True Negative rate as TN, False Positive rate as FP, and False Negative rate as FN. The Table 2 illustrates the performance measures that are used in this work.

Table 2. Performance evaluation metrics.

Measures	Mathematical Expressions
Accuracy	$\frac{\text{True Positive} + \text{True Negative}}{\text{True Positive True Negative} + \text{False Positive} + \text{False Negative}}$
Recall	$\frac{\text{True Positive}}{\text{True Positive} + \text{False Positive}}$
PRC/Positive Prediction	$\frac{\text{True Positive}}{\text{True Positive} + \text{False Positive}}$
F1 Score	$\frac{2.\text{Precision}.\text{Recall}}{\text{Precision} + \text{Recall}}$

Table 3 shows a summary of the best-achieved results performed on the guava leaf dataset, which proves that the proposed methodology is efficient and robust. Here, Quadratic SVM achieves the best results, i.e., 98.9% over 5 folds in 9.2 s and 99.2% over 10 folds in 16.2 s, with 3045 features on 5-fold cross validation.

Table 3. Summary of results.

Test Cases	Experiment #	Folds	Features	Classifier	Accuracy %	Training Time (s)
1	**1 (a)**	**5**	**3045**	**Quadratic SVM**	**98.9**	**9.2**
2	2 (a)	5	200	Fine Gaussian SVM	81.1	0.8
3	3 (a)	5	500	Quadratic SVM	84.1	2.2
4	4 (a)	5	750	Cubic SVM	85.6	2.4
5	5 (a)	5	1000	Cubic SVM	87.6	3.2
6	**1 (b)**	**10**	**3045**	**Quadratic SVM**	**99.2**	**16.2**
7	2 (b)	10	200	Fine Gaussian SVM	82.6	1.2
8	3 (b)	10	500	Fine Gaussian SVM	85.2	2.20
9	4 (b)	10	750	Cubic SVM	87.8	4.8
10	5 (b)	10	1000	Cubic SVM	87.5	6.2

The accompanying text contains some discussion on some of the experiments.

Experiment_1: Using 5-Fold and 10-Fold Validation on 3045 features

This section provides details about two test cases that were performed on 3045 features using both 5 folds and 10 folds. The selected classifiers are variants of SVM and KNN, which were chosen to get robust results. After augmentation, the chosen dataset contained 1002 images. The best results were recorded with measures such as accuracy, precision, recall, F1-score, and training time.

Experiment_1(a): Using 5 Folds and 3045 Features (1002 × 3045 features)

This test case shows the results of 3045 features upon 1002 images using SVM and KNN classifiers with 5-fold cross validation. Details are shown in Table 4.

The best results are achieved by the Quadratic SVM classifier in comparison with all KNN classifiers, which is 98.9% in 9.2 s. Here, the confusion matrix in Figure 8 and the ROC curve in Figure 9 are shown for the best results with the Quadratic SVM classifier. In the confusion matrix, A represents the diseased class, While B depicts healthy class

Table 4. Experiment_1 using 5 Folds (3045 Feature).

Classifier	Accuracy (%)	Total Cost	Precision Speed (obs/s)	Training Time (s)	Precision	F1-Score	Recall
Quadratic SVM	98.9	12	760	9.21	0.99	0.99	0.99
Linear SVM	98.0	20	780	9.65	0.98	0.98	0.98
Cubic SVM	98.6	14	770	9.43	0.985	0.985	0.99
Fine Gaussian SVM	80.1	199	490	16.3	0.78	0.785	0.86
Medium Gaussian SVM	97.3	27	770	9.76	0.97	0.975	0.97
Coarse Gaussian SVM	92.5	75	740	10.35	0.915	0.925	0.92
Weighted KNN	95.0	50	360	12.89	0.945	0.95	0.95
FINE KNN	93.5	65	360	13.62	0.935	0.935	0.94
Medium KNN	87.8	122	360	12.94	0.865	0.875	0.90
Coarse KNN	81.3	187	360	12.81	0.795	0.795	0.87
Cosine KNN	89.8	101	330	13.52	0.89	0.895	0.91
Cubic KNN	87.5	125	73	58.97	0.865	0.87	0.90

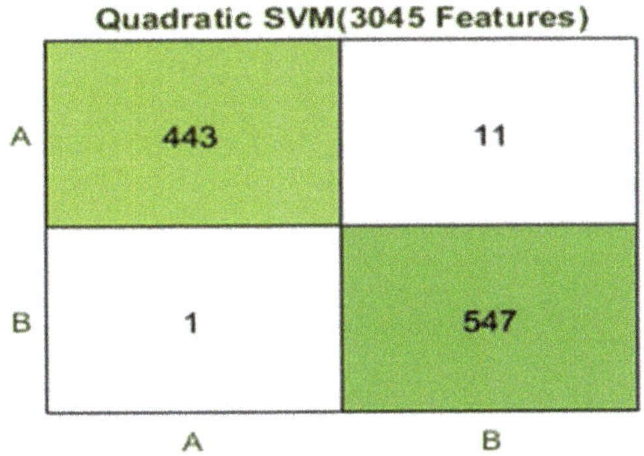

Figure 8. Confusion matrix for Q(SVM) with 3045 features and using 5-fold results.

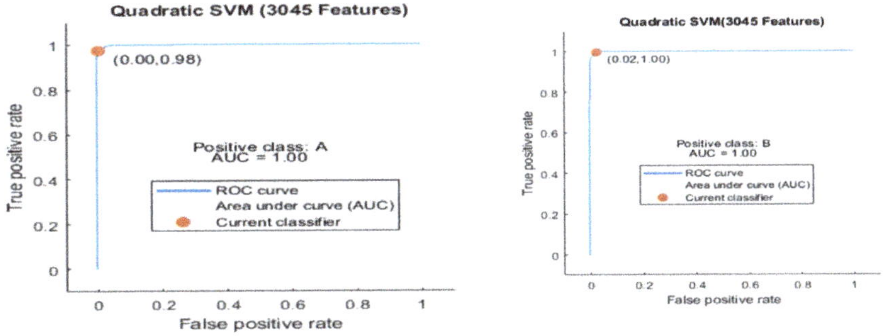

Figure 9. ROC curve for two classes A and B with 3045 features and using 5-fold results.

Here the ROC curve is shown in Figure 9 for both classes A and B, which are presented as ROC A and ROC B.

Experiment_1(b): Using 10 Folds and 3045 Features (1002 × 3045 features)

This test case shows the results of 3045 features upon 1002 images using SVM and KNN classifiers with 10 folds. Details are shown in Table 5.

Table 5. Experiment_1 using 10 Folds (3045 Feature).

Classifier	Accuracy (%)	Total Cost	Precision Speed (obs/s)	Training Time (s)	Precision	Recall	F1-Score
Quadratic SVM	99.2	8	460	16.221	0.995	0.99	0.99
Linear SVM	98.3	17	440	18.23	0.98	0.98	0.98
Cubic SVM	99.1	9	450	16.7	0.99	0.99	0.99
Fine Gaussian SVM	85.4	146	310	29.265	0.84	0.89	0.84
Medium Gaussian SVM	98.3	17	450	17.676	0.98	0.98	0.98
Coarse Gaussian SVM	93.6	64	450	18.689	0.93	0.94	0.93
Weighted KNN	94.3	57	190	23.037	0.935	0.95	0.94
FINE KNN	94.2	58	190	24.374	0.935	0.95	0.94
Medium KNN	88.2	118	190	23.12	0.87	0.91	0.87
Coarse KNN	82.7	173	190	23.278	0.81	0.88	0.81
Cosine KNN	91.0	90	180	24.438	0.9	0.92	0.90
Cubic KNN	88.3	117	57	73.713	0.87	0.91	0.87

The Quadratic SVM classifier achieved the best result in comparison with all KNN classifiers, which is 99.2%. Here, the confusion matrix as presented in Figure 10 and the ROC curve are shown for the best results against the Quadratic SVM classifier.

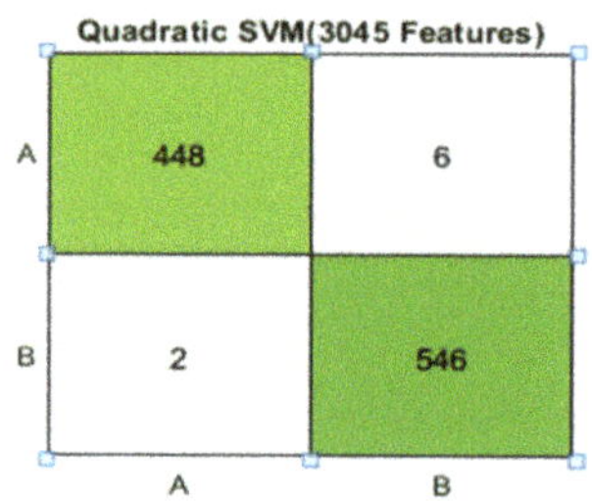

Figure 10. Confusion matrix for Q(SVM) with 3045 features and using 10-fold results.

Here, the ROC curve is shown in Figure 11 for both classes A and B, which are presented as ROC A and ROC B.

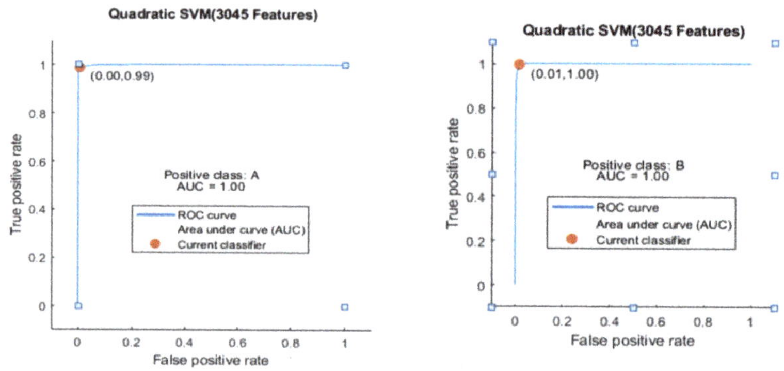

Figure 11. ROC curve for two classes A and B with 3045 features and using 10-fold results.

Experiment_2: Using 5-Fold and 10-Fold Validation on 200 features

This section provides details about two test cases that were performed on 200 features using both 5 folds and 10 folds. The efficiently selected classifiers are SVM and KNN, which were chosen to get robust results. The chosen dataset contains 1002 images. The best results are recorded with other measures, such as accuracy, precision, recall, F1-score, training time, etc. Results are shown in the Table 6 in detail.

Table 6. Experiment_2 using 5 Folds (200 Feature).

Classifier	Accuracy (%)	Total Cost	Precision Speed (obs/s)	Training Time (s)	Precision	Recall	F1-Score
Fine Gaussian SVM	81.1	189	14,000	0.81571	0.8	0.82	0.805
Linear SVM	73.7	264	16,000	1.3486	0.735	0.735	0.73
Cubic SVM	80.4	210	15,000	1.8572	0.805	0.805	0.8
Quadratic SVM	78.4	216	15,000	1.7431	0.785	0.78	0.785
Medium Gaussian SVM	72.8	273	15,000	0.7667	0.71	0.74	0.715
Coarse Gaussian SVM	60.5	396	14,000	0.671	0.565	0.71	0.5
Weighted KNN	78.2	218	7700	0.76741	0.785	0.785	0.78
FINE KNN	74.8	253	7600	1.1917	0.75	0.75	0.75
Medium KNN	66.8	333	7800	0.8367	0.66	0.665	0.66
Coarse KNN	62.4	377	7200	0.79092	0.59	0.675	0.555
Cosine KNN	67.2	329	7500	0.766606	0.67	0.665	0.67
Cubic KNN	69.0	311	1200	3.8096	0.68	0.685	0.68

Experiment_2(a): Using 5 Folds and 200 Features (1002 × 200 features)

This test case shows the results of 200 features upon 1002 images using SVM and KNN classifiers with 5 folds. Details are shown in Table 6.

The Fine Gaussian SVM classifier achieved the best result in comparison with all KNN classifiers, which is 81.1% in 0.8 s. Here, the confusion matrix in Figure 12 and the ROC in Figure 13 curve are shown for the best results with the Fine Gaussian SVM classifier.

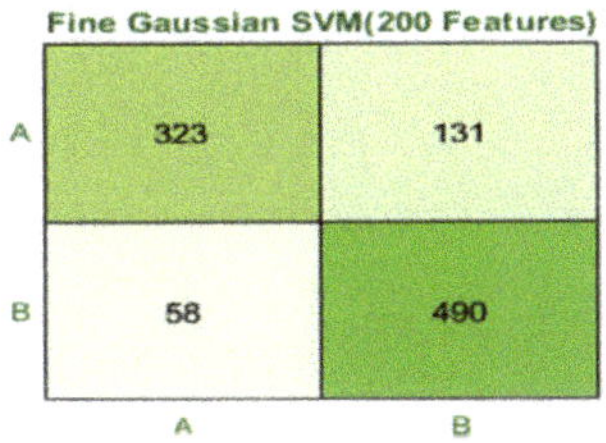

Figure 12. Confusion matrix for FG(SVM) with 200 features and using 5-fold results.

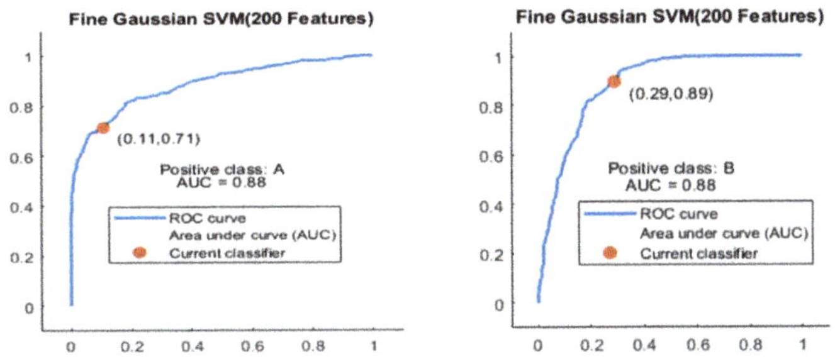

Figure 13. ROC curves for two classes A and B with 200 features and using 5-fold results.

The ROC curve is shown in Figure 13 for both classes A and B which are presented as ROC A and ROC B.

Experiment_2(b): Using 10 Folds and 200 Features (1002 × 200 features)

This test case shows the results of 200 features upon 1002 images using SVM and KNN classifiers with 10 folds. Details are shown in Table 7.

Table 7. Experiment_2 using 10 Folds (200 Feature).

Classifier	Accuracy (%)	Total Cost	Precision Speed (obs/s)	Training Time (s)	Precision	Recall	F1-Score
Fine Gaussian SVM	82.6	174	8300	1.2664	0.815	0.83	0.825
Linear SVM	73.2	269	8000	1.8576	0.725	0.73	0.73
Cubic SVM	80.8	192	8400	5.6548	0.815	0.81	0.81
Quadratic SVM	78.2	218	7800	3.4237	0.78	0.78	0.78
Medium Gaussian SVM	73.9	262	8100	1.15	0.725	0.75	0.725
Coarse Gaussian SVM	60.2	399	8100	1.1668	0.565	0.715	0.49
Weighted KNN	78.6	214	4100	1.2662	0.79	0.785	0.78
FINE KNN	75.0	250	4200	1.8358	0.755	0.75	0.75
Medium KNN	65.9	342	4000	1.8129	0.655	0.655	0.65
Coarse KNN	62.0	381	4000	1.2544	0.59	0.67	0.55
Cosine KNN	65.9	342	3900	1.274	0.665	0.665	0.655
Cubic KNN	68.5	316	960	4.57	0.675	0.68	0.68

Here, the Fine Gaussian SVM classifier achieved the best result in comparison with all KNN classifiers, which is 82.6% in 1.2 s. Here, the confusion matrix also shown in Figure 14 and the ROC curve are shown for the best results against the Fine Gaussian SVM classifier.

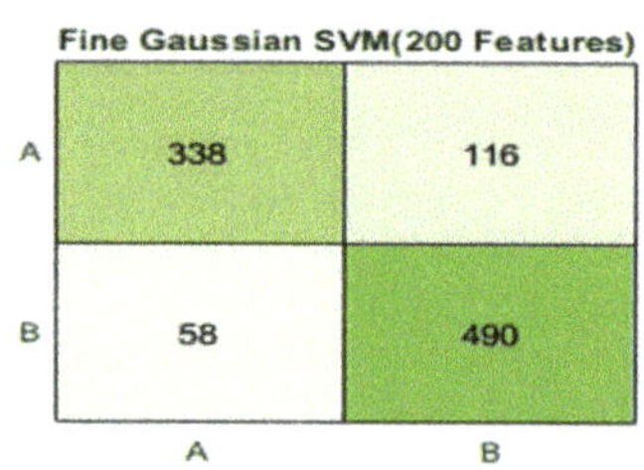

Figure 14. Confusion matrix for FG(SVM) with 200 features and using 10-fold results.

The ROC curve is shown in Figure 15 for both classes A and B, which are presented as ROC A and ROC B.

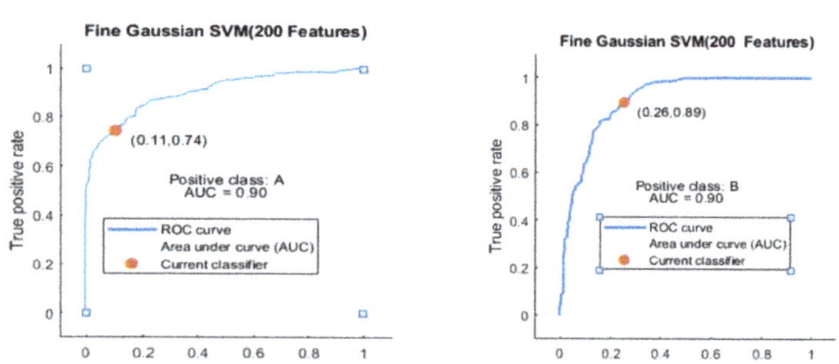

Figure 15. ROC curve for two classes A and B with 200 features and using 10-fold results.

Experiment_3: Using 5-Fold and 10-Fold Validation on 500 features.

This section provides details about two test cases that are performed on 500 features using both 5 folds and 10 folds. The efficiently selected classifiers are SVM and KNN, which were chosen to get robust results. The chosen dataset contains 1002 images. The best results are recorded with some other measures like Accuracy, precision, recall, F1-score, training time, and others. Results are shown in the Table 8 in detail.

Table 8. Experiment_3 using 5 Folds (500 Feature).

Classifier	Accuracy (%)	Total Cost	Precision Speed (obs/s)	Training Time (s)	Precision	Recall	F1-Score
Quadratic SVM	84.1	159	7000	2.2045	0.84	0.84	0.84
Linear SVM	79.0	210	6900	1.7001	0.785	0.79	0.79
Cubic SVM	82.4	176	7300	2.2117	1.64	1.65	1.65
Fine Gaussian SVM	83.6	164	6900	1.2768	1.66	1.68	1.67
Medium Gaussian SVM	75.8	242	6900	1.1461	0.745	0.765	0.75
Coarse Gaussian SVM	67.0	331	6800	1.1891	0.65	0.68	0.65
Weighted KNN	80.4	196	3100	1.5885	0.805	0.805	0.805
FINE KNN	77.6	224	3200	1.9972	0.78	0.775	0.775
Medium KNN	69.4	307	3200	1.6358	0.685	0.69	0.685
Coarse KNN	67.4	331	3100	1.584	0.65	0.695	0.64
Cosine KNN	68.5	316	3000	1.6165	0.685	0.685	0.685
Cubic KNN	73.3	268	470	9.3325	0.72	0.73	0.73

Experiment_3(a): Using 5 Folds and 500 Features (1002 × 500 features)

This test case shows the results of 500 features upon 1002 images using SVM and KNN classifiers with 5 folds. Details are shown in Table 8.

Here, the classifier Quadratic SVM achieved the best result in comparison with all KNN classifiers, which is 84.1% in 1.2 s. The confusion matrix in Figure 16 and the ROC curve in Figure 17 are shown for the best results against the Quadratic SVM classifier.

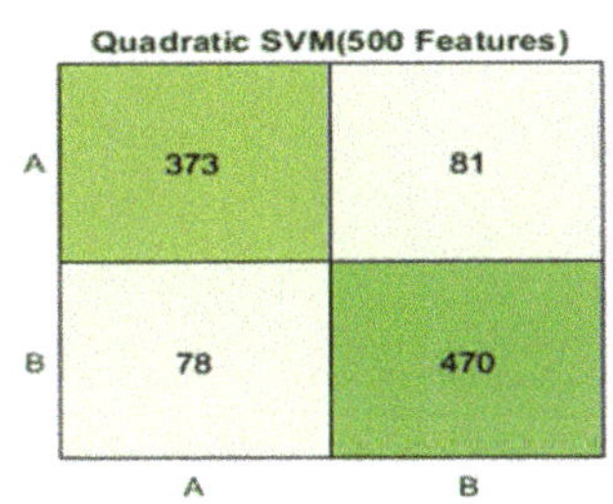

Figure 16. Confusion matrix for Q(SVM) with 500 features and using 500 fatures with 5-fold results.

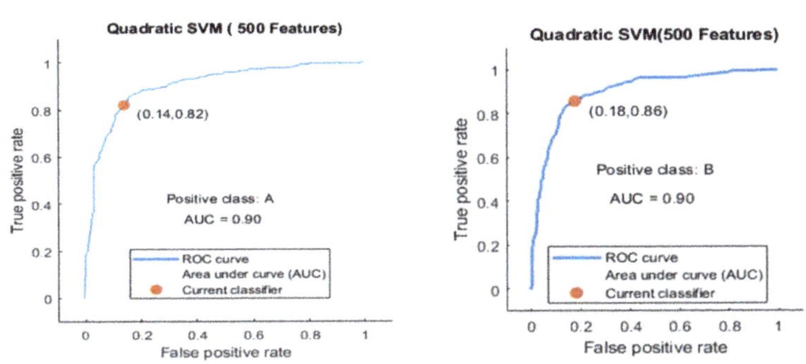

Figure 17. ROC curve for two classes A and B with 500 features and using 5-fold results.

The ROC curve is shown for both classes A and B, which are presented as ROC A and ROC B in Figure 17.

Experiment_3(b): Using 10 Folds and 500 Features (1002 × 500 features)

This test case shows the results of 500 features upon 1002 images using SVM and KNN classifiers with 10 folds. Details are shown in Table 9.

Table 9. Experiment_3 using 10 Folds (500 Feature).

Classifier	Accuracy (%)	Total Cost	Precision Speed (obs/sec)	Training Time (s)	Precision	Recall	F1-Score
Fine Gaussian SVM	85.2	148	3800	2.2088	0.845	0.86	0.85
Linear SVM	79.8	202	3600	4.5137	0.79	0.8	0.79
Cubic SVM	84.2	158	4100	4.4002	0.845	0.84	0.84
Quadratic SVM	84.7	153	4000	4.1644	0.845	0.845	0.845
Medium Gaussian SVM	75.9	241	4000	2.1057	0.75	0.765	0.755
Coarse Gaussian SVM	67.7	324	3900	2.0635	0.66	0.69	0.655
Weighted KNN	82.5	175	1800	2.6916	0.83	0.825	0.825
FINE KNN	79.4	206	1700	3.7172	0.795	0.795	0.79
Medium KNN	71.5	286	1800	2.7012	0.71	0.71	0.71
Coarse KNN	66.4	337	1800	2.667	0.645	0.685	0.63
Cosine KNN	67.9	322	1600	2.786	0.685	0.68	0.68
Cubic KNN	75.4	246	380	11.154	0.75	0.75	0.75

Here, the classifier Fine Gaussian SVM achieved the best result in comparison with all KNN classifiers, which is 85.2% in 2.2 s. Here, the confusion matrix and the ROC curve are shown for the best results against the Fine Gaussian SVM classifier. The confusion matrix for FG (SVM) is shown in Figure 18.

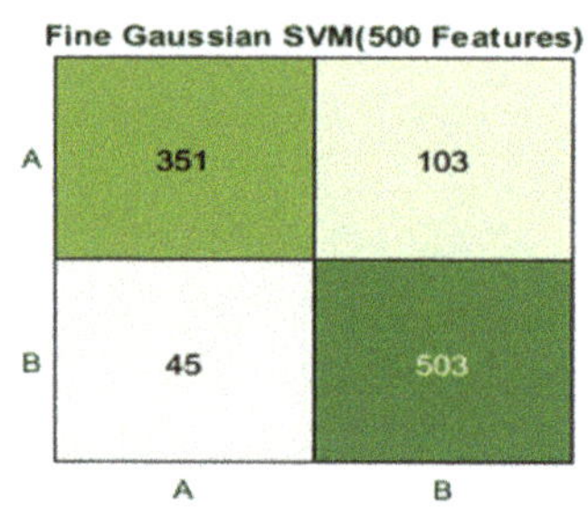

Figure 18. Confusion matrix for FG(SVM) with 500 features and using 10-fold results.

Here, the ROC curve is presented (in Figure 19) for both classes A and B, which are presented as ROC A and ROC B.

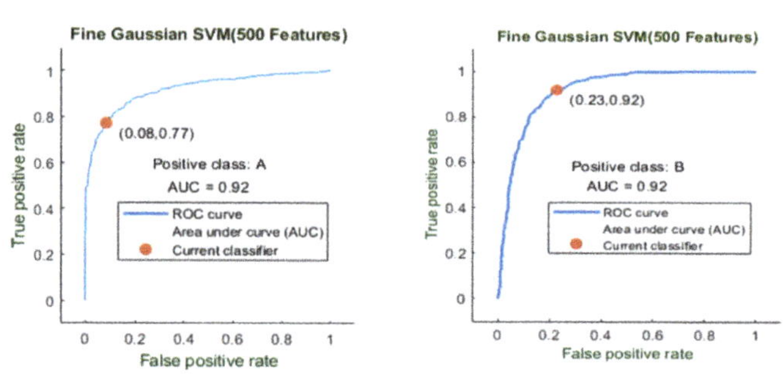

Figure 19. ROC curve for two classes A and B with 500 features and using 10-fold results.

Figure 20 shows the 5-fold best achievable results between training time and features. It shows the consumed time for training that is required for a particular set of features.

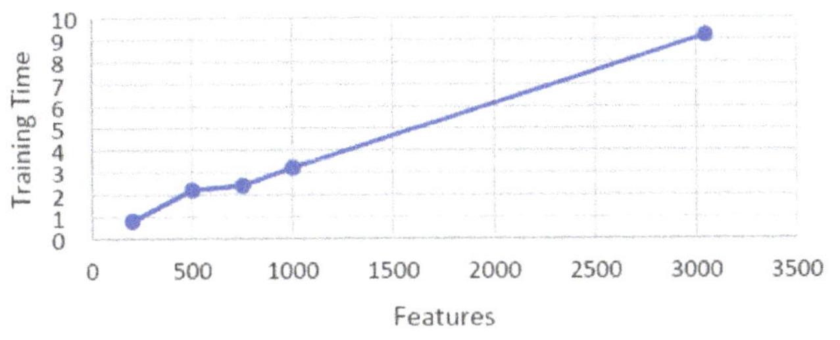

Figure 20. Graph showing training time and features on 5-fold.

Here, features are presented on the x-axis and the training time is shown on the y-axis. These results are taken on 5-folds for the best achievable results.

Similarly, the graph in Figure 21 shows the relation between accuracy and features at 5-fold results. It shows that the features are presented along the x-axis and the accuracy is presented along the y-axis, and the line presents the best achievable results.

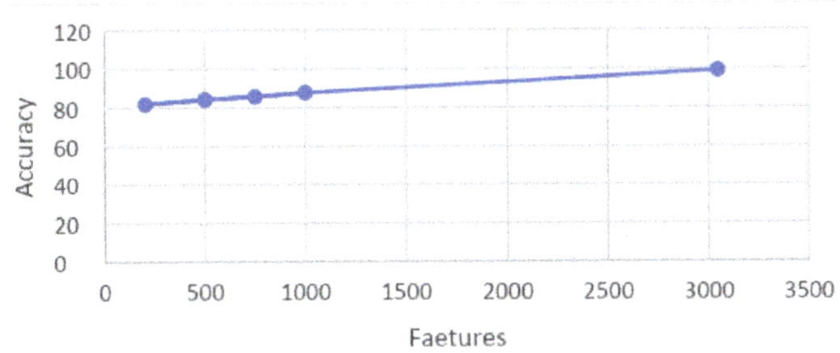

Figure 21. Graph showing accuracy and features on 5-fold.

The graph in Figure 22 shows the relation between training time and features upon results taken on 10 folds. This graph shows that features are shown along the x-axis and the training time is shown along the y-axis, and the time consumed by a particular set of features for training is illustrated.

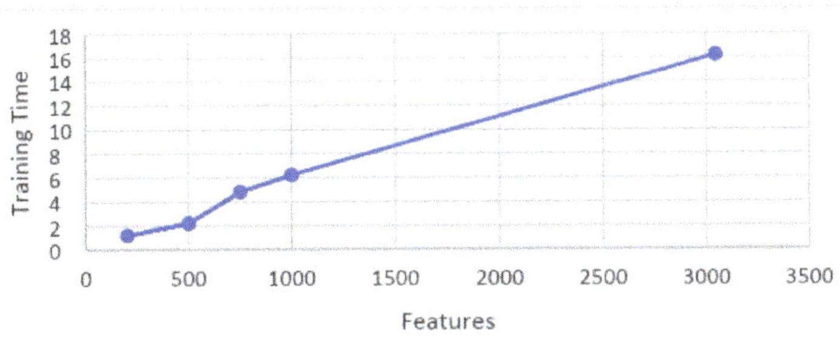

Figure 22. Graph showing features and training time.

The graph in Figure 23 shows the relation between training time and accuracy upon results taken on 10 folds. This graph shows that features are shown along the x-axis and the accuracy is shown along the y-axis, and the time consumed by a particular set of features for training is illustrated.

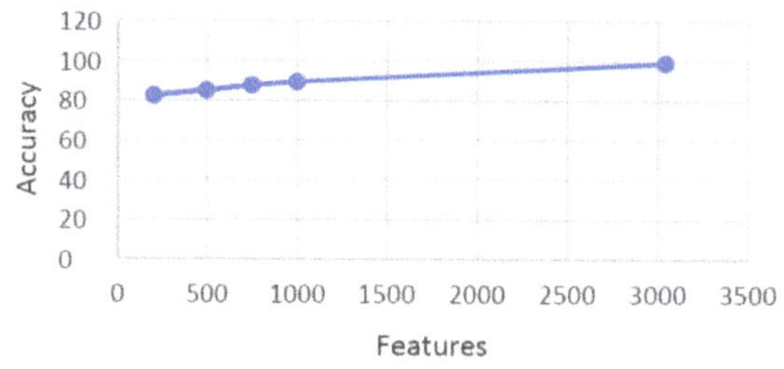

Figure 23. Graph showing features and accuracy on 10 folds.

5. Conclusions

Agriculture is the key to the development and rise of emergent nations. Diseases in plants cause crop damage. Detection of leaf blight is enormously important as it affects the annual production of guava fruit. Finally, the development of an automated system becomes indispensable. In this paper, leaf blight can be detected, analyzed, and classified through the proposed methodology. In this proposed methodology, our own deep CNN is designed, containing thirty-three layers. In the first phase of image processing, preprocessing is done by using color spacing YCbCr. The Guava dataset is chosen for the identification and analysis of leaf blight. Because the dataset is small, data augmentation is performed. Horizontal and vertical flipping were performed on images of guava leaves. After preprocessing, feature extraction was performed using Darknet-53 and AlexNet, as well as the proposed SidNet. For the selection of the best features, optimization algorithms such as Entropy and Binary Gray Wolf are used. Finally, classification is performed on guava leaf images and the best results with higher accuracy and less computational cost are achieved. Multiple experiments are performed while using the set of selected features (200, 500, 750, 1000 features using 5- and 10-fold validation). Based on the selected features, 98.9% of the results are achieved using an SVM classifier, as it proves that this proposed methodology is robust and efficient.

In the future, this work can be explored with quantum deep learning for improved performance. Quantum computing-based machine learning and deep convolutional neural networks can detect and classify leaf blight at its initial stage more precisely and meticulously, which will help save crops and fruits, save plants from destruction, and increase the production of guava fruit.

Author Contributions: Conceptualization, S.M., M.R., O.D.O., S.U.R., A.E.R. and H.T.R.; methodology, S.M. and M.R.; investigation, M.R., O.D.O., S.U.R., A.E.R. and H.T.R.; resources, A.E.R. and H.T.R.; writing—original draft preparation, S.M., M.R., O.D.O., S.U.R., A.E.R. and H.T.R.; writing—review and editing, S.M., M.R., O.D.O., S.U.R., A.E.R. and H.T.R.; supervision, M.R, A.E.R. and H.T.R.; funding acquisition, A.E.R. All authors have read and agreed to the published version of the manuscript.

Funding: The authors extend their appreciation to King Saud University for funding this work through Researchers Supporting Project number (RSPD2023R711), King Saud University, Riyadh, Saudi Arabia.

Institutional Review Board Statement: The study was conducted in accordance with the Declaration of Helsinki, and approved by the Institutional Review Board.

Conflicts of Interest: The authors declare no conflict of interest.

References

1. Rehman, A.; Jingdong, L.; Shahzad, B.; Chandio, A.A.; Hussain, I.; Nabi, G.; Iqbal, M.S. Economic perspectives of major field crops of Pakistan: An empirical study. *Pac. Sci. Rev. B Humanit. Soc. Sci.* **2015**, *1*, 145–158. [CrossRef]
2. Rai, M.K.; Asthana, P.; Jaiswal, V.; Jaiswal, U. Biotechnological advances in guava (*Psidium guajava* L.): Recent developments and prospects for further research. *Trees* **2010**, *24*, 1–12. [CrossRef]
3. Mitra, S.; Thingreingam Irenaeus, K. Guava cultivars of the world. In *International Symposia on Tropical and Temperate Horticulture—ISTTH2016*; CIRAD Publications: Cairns, Queensland, Australia, 2016; pp. 905–910.
4. Almadhor, A.; Rauf, H.T.; Lali, M.I.U.; Damaševičius, R.; Alouffi, B.; Alharbi, A. AI-Driven Framework for Recognition of Guava Plant Diseases through Machine Learning from DSLR Camera Sensor Based High Resolution Imagery. *Sensors* **2021**, *21*, 3830. [CrossRef] [PubMed]
5. Misra, A. Guava diseases—Their symptoms, causes and management. In *Diseases of Fruits and Vegetables*; Springer: Berlin/Heidelberg, Germany, 2004; Volume II, pp. 81–119.
6. Dhiman, B.; Kumar, Y.; Hu, Y.-C. A general purpose multi-fruit system for assessing the quality of fruits with the application of recurrent neural network. *Soft Comput.* **2021**, *25*, 9255–9272. [CrossRef]
7. Ashraf, R.; Habib, M.A.; Akram, M.; Latif, M.A.; Malik, M.S.A.; Awais, M.; Dar, S.H.; Mahmood, T.; Yasir, M.; Abbas, Z. Deep convolution neural network for big data medical image classification. *IEEE Access* **2020**, *8*, 105659–105670. [CrossRef]
8. Naranjo-Torres, J.; Mora, M.; Hernández-García, R.; Barrientos, R.J.; Fredes, C.; Valenzuela, A. A review of convolutional neural network applied to fruit image processing. *Appl. Sci.* **2020**, *10*, 3443. [CrossRef]
9. Lloret, E.; Plaza, L.; Aker, A. The challenging task of summary evaluation: An overview. *Lang. Resour. Eval.* **2018**, *52*, 101–148. [CrossRef]
10. Bojer, C.S.; Meldgaard, J.P. Kaggle forecasting competitions: An overlooked learning opportunity. *Int. J. Forecast.* **2021**, *37*, 587–603. [CrossRef]
11. Fu, X.; Cao, X. Underwater image enhancement with global–local networks and compressed-histogram equalization. *Signal Process. Image Commun.* **2020**, *86*, 115892. [CrossRef]
12. Tang, Z.; Gao, Y.; Karlinsky, L.; Sattigeri, P.; Feris, R.; Metaxas, D. *OnlineAugment: Online Data Augmentation with Less Domain Knowledge*; Springer: Berlin/Heidelberg, Germany, 2020; pp. 313–329.
13. Liu, Z.; Lai, Z.; Ou, W.; Zhang, K.; Zheng, R. Structured optimal graph based sparse feature extraction for semi-supervised learning. *Signal Process.* **2020**, *170*, 107456. [CrossRef]
14. Ghojogh, B.; Samad, M.N.; Mashhadi, S.A.; Kapoor, T.; Ali, W.; Karray, F.; Crowley, M. 'Feature selection and feature extraction in pattern analysis: A literature review. *arXiv* **2019**, arXiv:1905.02845.
15. Zhang, J.; Liu, B. A review on the recent developments of sequence-based protein feature extraction methods. *Curr. Bioinform.* **2019**, *14*, 190–199. [CrossRef]
16. Latif, A.; Rasheed, A.; Sajid, U.; Ahmed, J.; Ali, N.; Ratyal, N.I.; Zafar, B.; Dar, S.H.; Sajid, M., Khalil, T. Content-based image retrieval and feature extraction: A comprehensive review. *Math. Probl. Eng.* **2019**, *2019*, 9658350. [CrossRef]
17. Raj, R.J.S.; Shobana, S.J.; Pustokhina, I.V.; Pustokhin, D.A.; Gupta, D.; Shankar, K. Optimal feature selection-based medical image classification using deep learning model in internet of medical things. *IEEE Access* **2020**, *8*, 58006–58017. [CrossRef]
18. Ali, L.; Wajahat, I.; Golilarz, N.A.; Keshtkar, F.; Bukhari, S.A.C. 'LDA–GA–SVM: Improved hepatocellular carcinoma prediction through dimensionality reduction and genetically optimized support vector machine. *Neural Comput. Appl.* **2021**, *33*, 2783–2792. [CrossRef]
19. Ozyurt, B.; Akcayol, M.A. A new topic modeling based approach for aspect extraction in aspect based sentiment analysis: SS-LDA. *Expert Syst. Appl.* **2021**, *168*, 114231. [CrossRef]
20. Zhao, J.; Liang, J.-M.; Dong, Z.-N.; Tang, D.-Y.; Liu, Z. Accelerating information entropy-based feature selection using rough set theory with classified nested equivalence classes. *Pattern Recognit.* **2020**, *107*, 107517. [CrossRef]
21. Maleki, N.; Zeinali, Y.; Niaki, S.T.A. A k-NN method for lung cancer prognosis with the use of a genetic algorithm for feature selection. *Expert Syst. Appl.* **2021**, *164*, 113981. [CrossRef]
22. Hu, P.; Pan, J.-S.; Chu, S.-C. Improved binary grey wolf optimizer and its application for feature selection. *Knowl. Based Syst.* **2020**, *195*, 105746. [CrossRef]
23. Tubishat, M.; Idris, N.; Shuib, L.; Abushariah, M.A.; Mirjalili, S. Improved Salp Swarm Algorithm based on opposition based learning and novel local search algorithm for feature selection. *Expert Syst. Appl.* **2020**, *145*, 113122. [CrossRef]
24. Too, J.; Rahim Abdullah, A. Binary atom search optimisation approaches for feature selection. *Connect. Sci.* **2020**, *32*, 406–430. [CrossRef]
25. Özyurt, F. A fused CNN model for WBC detection with MRMR feature selection and extreme learning machine. *Soft Comput.* **2020**, *24*, 8163–8172. [CrossRef]
26. De Silva, K.; Jönsson, D.; Demmer, R.T. A combined strategy of feature selection and machine learning to identify predictors of prediabetes. *J. Am. Med. Inform. Assoc.* **2020**, *27*, 396–406. [CrossRef] [PubMed]
27. Germain, P.-L.; Sonrel, A.; Robinson, M.D. pipeComp, a general framework for the evaluation of computational pipelines, reveals performant single cell RNA-seq preprocessing tools. *Genome Biol.* **2020**, *21*, 1–28. [CrossRef]
28. Kaur, P.; Singh, A.; Chana, I. Computational techniques and tools for omics data analysis: State-of-the-art, challenges, and future directions. *Arch. Comput. Methods Eng.* **2021**, *28*, 4595–4631. [CrossRef]

29. Ghosh, P.; Azam, S.; Jonkman, M.; Karim, A.; Shamrat, F.J.M.; Ignatious, E.; Shultana, S.; Beeravolu, A.R.; De Boer, F. Efficient Prediction of Cardiovascular Disease Using Machine Learning Algorithms With Relief and LASSO Feature Selection Techniques. *IEEE Access* **2021**, *9*, 19304–19326. [CrossRef]
30. Deepa, N.; Prabadevi, B.; Maddikunta, P.K.; Gadekallu, T.R.; Baker, T.; Khan, M.A.; Tariq, U. An AI-based intelligent system for healthcare analysis using Ridge-Adaline Stochastic Gradient Descent Classifier. *J. Supercomput.* **2021**, *77*, 1998–2017. [CrossRef]
31. Amini, F.; Hu, G. A two-layer feature selection method using genetic algorithm and elastic net. *Expert Syst. Appl.* **2021**, *166*, 114072. [CrossRef]
32. Zhang, H.; Xu, H.; Xiao, Y.; Guo, X.; Ma, J. Rethinking the Image Fusion: A Fast Unified Image Fusion Network Based on Proportional Maintenance of Gradient and Intensity. *Proc. AAAI Conf. Artif. Intell.* **2020**, *34*, 12797–12804. [CrossRef]
33. Jung, H.; Kim, Y.; Jang, H.; Ha, N.; Sohn, K. Unsupervised deep image fusion with structure tensor representations. *IEEE Trans. Image Process.* **2020**, *29*, 3845–3858. [CrossRef]
34. Shakya, S. Analysis of artificial intelligence based image classification techniques. *J. Innov. Image Process. (JIIP)* **2020**, *2*, 44–54. [CrossRef]
35. Hong, D.; Wu, X.; Ghamisi, P.; Chanussot, J.; Yokoya, N.; Zhu, X.X. Invariant attribute profiles: A spatial-frequency joint feature extractor for hyperspectral image classification. *IEEE Trans. Geosci. Remote Sens.* **2020**, *58*, 3791–3808. [CrossRef]
36. Ke, Z.; Qiu, D.; Li, K.; Yan, Q.; Lau, R.W. *Guided Collaborative Training for Pixel-Wise Semi-Supervised Learning*; Springer: Berlin/Heidelberg, Germany, 2020; pp. 429–445.
37. Wang, F.; Liu, H.; Guo, D.; Sun, F. Unsupervised Representation Learning by InvariancePropagation. *arXiv* **2020**, *preprint*. arXiv:2010.11694.
38. Karantanellis, E.; Marinos, V.; Vassilakis, E.; Christaras, B. Object-based analysis using unmanned aerial vehicles (UAVs) for site-specific landslide assessment. *Remote Sens.* **2020**, *12*, 1711. [CrossRef]
39. Durmuş, H.; Güneş, E.O.; Kırcı, M. *Disease Detection on the Leaves of the Tomato Plants by Using Deep Learning*; IEEE: Piscataway, NJ, USA, 2017; pp. 1–5.
40. Atila, Ü.; Uçar, M.; Akyol, K.; Uçar, E. Plant leaf disease classification using EfficientNet deep learning model. *Ecol. Inform.* **2021**, *61*, 101182. [CrossRef]
41. Ferentinos, K.P. Deep learning models for plant disease detection and diagnosis. *Comput. Electron. Agric.* **2018**, *145*, 311–318. [CrossRef]
42. Azarmdel, H.; Jahanbakhshi, A.; Mohtasebi, S.S.; Muñoz, A.R. Evaluation of image processing technique as an expert system in mulberry fruit grading based on ripeness level using artificial neural networks (ANNs) and support vector machine (SVM). *Postharvest Biol. Technol.* **2020**, *166*, 111201. [CrossRef]
43. Kasinathan, T.; Singaraju, D.; Uyyala, S.R. Insect classification and detection in field crops using modern machine learning techniques. *Inf. Process. Agric.* **2021**, *8*, 446–457. [CrossRef]
44. Yao, J.; Ye, Y. The effect of image recognition traffic prediction method under deep learning and naive Bayes algorithm on freeway traffic safety. *Image Vis. Comput.* **2020**, *103*, 103971. [CrossRef]
45. Atouf, I.; Al Okaishi, W.Y.; Zaaran, A.; Slimani, I.; Benrabh, M. A real-time system for vehicle detection with shadow removal and vehicle classification based on vehicle features at urban roads. *Int. J. Power Electron. Drive Syst.* **2020**, *11*, 2091. [CrossRef]
46. Khan, M.A.; Khan, M.A.; Ahmed, F.; Mittal, M.; Goyal, L.M.; Hemanth, D.J.; Satapathy, S.C. Gastrointestinal diseases segmentation and classification based on duo-deep architectures. *Pattern Recognit. Lett.* **2020**, *131*, 193–204. [CrossRef]
47. Chen, J.; Lian, Y.; Li, Y. 'Real-time grain impurity sensing for rice combine harvesters using image processing and decision-tree algorithm. *Comput. Electron. Agric.* **2020**, *175*, 105591. [CrossRef]
48. Moubayed, A.; Injadat, M.; Shami, A.; Lutfiyya, H. Student engagement level in an e-learning environment: Clustering using k-means. *Am. J. Distance Educ.* **2020**, *34*, 137–156. [CrossRef]
49. Nanglia, P.; Kumar, S.; Mahajan, A.N.; Singh, P.; Rathee, D. A hybrid algorithm for lung cancer classification using SVM and Neural Networks. *ICT Express* **2021**, *7*, 335–341. [CrossRef]
50. Wu, W.; Li, D.; Du, J.; Gao, X.; Gu, W.; Zhao, F.; Feng, X.; Yan, H. An intelligent diagnosis method of brain MRI tumor segmentation using deep convolutional neural network and SVM algorithm. *Comput. Math. Methods Med.* **2020**, *2020*, 6789306. [CrossRef]
51. Basha, C.Z.; Rohini, G.; Jayasri, A.V.; Anuradha, S. *Enhanced and Effective Computerized Classification of X-Ray Images*; IEEE: Piscataway, NJ, USA, 2020; pp. 86–91.
52. Islam, N.; Rashid, M.M.; Wibowo, S.; Xu, C.-Y.; Morshed, A.; Wasimi, S.A.; Moore, S.; Rahman, S.M. Early Weed Detection Using Image Processing and Machine Learning Techniques in an Australian Chilli Farm. *Agriculture* **2021**, *11*, 387. [CrossRef]
53. Sasankar, P.; Kosarkar, U. A Study for Face Recognition Using Techniques PCA and KNN. EasyChair Print. 2021. Available online: https://wwww.easychair.org/publications/preprint_download/gS7Q (accessed on 30 January 2023).
54. Bharate, A.A.; Shirdhonkar, M. *Classification of Grape Leaves Using KNN and SVM Classifiers*; IEEE: Piscataway, NJ, USA, 2020; pp. 745–749.
55. Goncharov, P.; Ososkov, G.; Nechaevskiy, A.; Uzhinskiy, A.; Nestsiarenia, I. *Disease Detection on the Plant Leaves by Deep Learning*; Springer: Berlin/Heidelberg, Germany, 2018; pp. 151–159.
56. De Luna, R.G.; Dadios, E.P.; Bandala, A.A. *Automated Image Capturing System for Deep Learning-Based Tomato Plant Leaf Disease Detection and Recognition*; IEEE: Piscataway, NJ, USA, 2018; pp. 1414–1419.

57. Sharma, P.; Berwal, Y.P.S.; Ghai, W. Performance analysis of deep learning CNN models for disease detection in plants using image segmentation. *Inf. Process. Agric.* **2020**, *7*, 566–574. [CrossRef]
58. Oppenheim, D.; Shani, G.; Erlich, O.; Tsror, L. Using deep learning for image-based potato tuber disease detection. *Phytopathology* **2019**, *109*, 1083–1087. [CrossRef]
59. Chowdhury, M.E.; Rahman, T.; Khandakar, A.; Ayari, M.A.; Khan, A.U.; Khan, M.S.; Al-Emadi, N.; Reaz, M.B.I.; Islam, M.T.; Ali, S.H.M. Automatic and Reliable Leaf Disease Detection Using Deep Learning Techniques. *AgriEngineering* **2021**, *3*, 294–312. [CrossRef]
60. Krizhevsky, A.; Hinton, G. Learning Multiple Layers of Features from Tiny Images. 2009. Available online: http://www.cs.utoronto.ca/~kriz/learning-features-2009-TR.pdf (accessed on 30 January 2023).
61. Rauf, H.T.; Lali, M.I.U. A Guava Fruits and Leaves Dataset for Detection and Classification of Guava Diseases through Machine Learning. *Mendeley Data* **2021**, *1*. (Dataset).

agriculture

MDPI

Article

EffiMob-Net: A Deep Learning-Based Hybrid Model for Detection and Identification of Tomato Diseases Using Leaf Images

Zahid Ullah [1], Najah Alsubaie [2,*], Mona Jamjoom [2], Samah H. Alajmani [3] and Farrukh Saleem [1]

1 Department of Information Systems, Faculty of Computing and Information Technology, King Abdulaziz University, Jeddah 21589, Saudi Arabia
2 Department of Computer Sciences, College of Computer and Information Sciences, Princess Nourah Bint Abdulrahman University, Riyadh 84428, Saudi Arabia
3 Department of Information Technology, College of Computers and Information Technology, Taif University, Taif 21944, Saudi Arabia
* Correspondence: nmoalsubaie@pnu.edu.sa

Abstract: As tomatoes are the most consumed vegetable in the world, production should be increased to fulfill the vast demand for this vegetable. Global warming, climate changes, and other significant factors, including pests, badly affect tomato plants and cause various diseases that ultimately affect the production of this vegetable. Several strategies and techniques have been adopted for detecting and averting such diseases to ensure the survival of tomato plants. Recently, the application of artificial intelligence (AI) has significantly contributed to agronomy in the detection of tomato plant diseases through leaf images. Deep learning (DL)-based techniques have been largely utilized for detecting tomato leaf diseases. This paper proposes a hybrid DL-based approach for detecting tomato plant diseases through leaf images. To accomplish the task, this study presents the fusion of two pretrained models, namely, EfficientNetB3 and MobileNet (referred to as the EffiMob-Net model) to detect tomato leaf diseases accurately. In addition, model overfitting was handled using various techniques, such as regularization, dropout, and batch normalization (BN). Hyperparameter tuning was performed to choose the optimal parameters for building the best-fitting model. The proposed hybrid EffiMob-Net model was tested on a plant village dataset containing tomato leaf disease and healthy images. This hybrid model was evaluated based on the best classifier with respect to accuracy metrics selected for detecting the diseases. The success rate of the proposed hybrid model for accurately detecting tomato leaf diseases reached 99.92%, demonstrating the model's ability to extract features accurately. This finding shows the reliability of the proposed hybrid model as an automatic detector for tomato plant diseases that can significantly contribute to providing better solutions for detecting other crop diseases in the field of agriculture.

Keywords: tomato leaf; disease; hybrid model; detection; deep learning

1. Introduction

Tomatoes are a fast-growing crop that matures in 90 to 150 days [1]. This worldwide ever-present product has rich nutritional values [2] and can be cultivated in nearly any reasonably parched soil [3]. In recent decades, the agricultural estate has increased tomato production by above 160% [4]. Tomatoes are the most consumed vegetable, accounting for about 15% of total vegetable consumption [5], and ranking as the sixth most abundant vegetable worldwide according to the Food and Agriculture Organization (FAO) annual production statistics [6]. The key production areas of tomatoes occur in India, the USA, Iran, China, Italy, Egypt, Mexico, and Turkey [7]. However, the plant is usually infected by diseases, which could be viral or fungal, resulting in a significant reduction in both the quality and quantity of crop production [3].

Due to the large demand for tomatoes globally, there is a need to develop techniques for enhancing crop yields while allowing for the early detection of plant diseases, including viral, bacterial, and fungal diseases [8], to increase the quality and production of tomatoes to meet economic goals [9]. Accurate and timely treatment is required to prevent diseases from spreading and causing in crop losses, and ensure ideal production. In a manual scenario, human expert-based detection is required to cope with these problems [10]. Moreover, screening symptoms manually is time consuming and costly due to insufficient human infrastructure capacity [11]. An automatic detecting system can assist in identifying the symptoms of a disease through the plant leaf in a cost-efficient manner. The application of artificial intelligence (AI), particularly machine learning (ML) and deep learning (DL), has significantly contributed to efforts to detect plant diseases.

Recently, the application of DL approaches has demonstrated outstanding performance and provided solutions to real problems in a wide range of computer vision and ML jobs, including image classification, detection, recognition, and medical imaging [12]. In the literature, several techniques have been developed based on the DL approach to enhance the persistence rate of field crops through the early detection of various diseases and succeeding disease management [5]. Currently, for plant diseases, the detection and classification rate have reached 100% in laboratory-based machine vision technology [13]. DL is broadly used in agriculture for plant disease detection and classification. Moreover, a DL-based convolutional neural network (CNN) is the most commonly used method for detecting, classifying, and recognizing tomato leaf diseases because of its significant success compared with other traditional methods [14]. CNN has the capability of extracting features from objects automatically. Therefore, CNN has been extensively utilized for tomato leaf disease identification, recognition, and classification.

Based on the widespread success of DL-based CNN architectures in agriculture, particularly, the detection of plant diseases, this study proposed a hybrid DL-based model that combines two pretrained models, namely, EfficientNet and MobileNet (referred to as EffiMob-Net) for detecting tomato leaf diseases. Taking advantage of the pretrained models' architectures, the weights of both pretrained models were loaded to utilize them for feature extraction and then the outputs of both models were concatenated for the detection and classification of leaf images. The key contributions of this study are as follows:

- A deep hybrid model was proposed that combines the architectures of two pretrained models, EfficientNet and MobileNet, for extracting the significant features of tomato leaves. Their outputs were then concatenated for the detection and classification of tomato leaf diseases.
- In the proposed method, the softmax layers of both pretrained models were removed, and the output achieved from the dense layers of both models was combined. In addition, three FC layers of size 512, 256, and 128 channels were added after the concatenation process. The classification was performed using the softmax layer which was added at the end of the proposed model.
- The dataset was preprocessed and prepared for training the proposed hybrid model using various preprocessing steps.
- The proposed model was trained using the extracted features.
- The study ensured the prevention of the proposed model's overfitting by using various techniques, such as regularization, dropout, and BN.
- The proposed hybrid model was evaluated, and the classification report with descriptions is presented.

2. Related Work

This section discusses the existing work related to the application of DL approaches to the detection and classification of tomato leaf diseases. The search criteria for investigating previous work in the same domain include keywords such as tomato leaf disease detection using DL and DL approaches for detecting and classifying tomato leaf. Several well-known search engines/databases such as Google Scholar, ScienceDirect, ResearchGate, and IEEE

Explorer were explored to collect and discuss state-of-the-art methodologies used in this domain of research. The literature survey indicated that most previous related research is based on the pretrained DL models.

A study conducted by [15] utilized a plant village dataset to detect and classify tomato leaf diseases using the DL approach. For this task, several pretrained approaches such as AlexNet, GoogLeNet, SqueezeNet, Vgg16, and MobileNetv2 were applied. Vgg16 achieved higher results than the others, with an accuracy rate of 99.17%. An attempt was made by [16] to detect tomato leaf diseases using the DL method. In this regard, fuzzy-SVM, CNN, and region-based CNN (R-CNN) were applied to a dataset containing a total of 6 classes. The achieved results showed a higher performance of R-CNN, with an accuracy rate of 96.735%. Similarly, Ref. [17] utilized the mask R-CNN approach for the segmentation and identification of tomato leaf disease. The results showed a higher accuracy rate of 98%. A pretrained model and feature concatenation approach were used by [4] for tomato leaf disease classification. In this method, the features were extracted using pretrained models and concatenated, while the classification was performed using traditional ML methods. The study concluded that multinomial logistic regression (MLR) achieved the highest results, with 97% accuracy.

A multimodal hybrid DL-based approach using attention-based dilated CNN logistic regression (ADCLR) was proposed by [18] to identify tomato leaf diseases. In this approach, feature extraction was performed using attention-based dilated CNN. The processed features were combined and classified using logistic regression (LR). The classification results show a higher accuracy rate of 96.6%. A hybrid model CNN-SVM was developed by [19] to predict seven predominant diseases related to tomato leaves. The highest results were achieved with a 92.6% accuracy. Another hybrid SVM-LR model was proposed by [20] for detecting powdery mildew disease of tomato leaves. The results demonstrated that the proposed model reached 92.37% accuracy.

An optimized DL-based method was proposed by [21] to detect tomato leaf diseases. Various pretrained models were applied, and the performance of each model was tested using different optimizers. The study concluded that MobileNetv3 Large using the Adagrad optimizer outperformed other models, with an accuracy rate of 99.81%. An image-based forecast using CNN was proposed by [22], who detected the early blight disease (EBD) of tomato plants. The study reported a 98.10% accuracy rate for the model. Similarly, an optimized transfer learning approach was proposed by [23], in which two pretrained models were applied to the tomato early blight disease (TEBD) dataset. The results concluded that Vgg16 outperformed ResNet50, with an accuracy rate of 99%. A study by [24] detected nine diseases of tomato leaf using a DL approach. For this purpose, a CRNN model with GRU was implemented to classify and detect tomato leaf diseases. The model achieved 99.62% accuracy when detecting tomato leaf diseases. A classification of tomato leaves using DL methods by utilizing various optimizers and learning rates (LR) was performed by [25]. Two DL pretrained models were applied to a dataset containing tomato leaf diseases. The reported results showed that Xception with Adam optimizer and an LR of 0.0001 outperformed other combinations with Xception and the Resnet50 model. The highest accuracy achieved was 99%.

A comparative study between ML and DL methods was conducted by [13] to classify tomato leaf diseases. The results of both approaches were compared, and DL methods outperformed ML methods. Moreover, among the DL methods, ResNet34 achieved the highest accuracy rate at 97.7%. Another DL-based approach was proposed by [26] to detect tomato leaf diseases. The higher classification rates of the proposed model occurred for 5, 7, and 10 classes, which were 99.51%, 98.65%, and 97.11%, respectively. The authors of [11] proposed an image-based diagnostic system using several DL methods, which were applied to a dataset collected from a village plant database and privately collected images containing a total of 24 classes. The reported results showed a higher performance by the DenseNet121 model, which yielded a classification accuracy of 95.31%. The study by [27] classified tomato plant diseases using the Vgg16 model. The classification accuracy for multi-class classification reached 99% while binary classification (healthy and unhealthy) reached 100%, with no preprocessing of images.

A robust DL-based detector for tomato leaf and pest recognition was proposed by [28]. In this regard, 3 detectors referred to as DL meta-architecture—were combined into VggNet and ResNet. The study reported that faster R-CNN in combination with Vgg16 has a higher recognition capability. Another robust intelligent system for detecting tomato disease using the DL approach was proposed by [29]. To train the model, a dataset containing 9 diseases was utilized. The results showed that the proposed model accomplished a higher accuracy rate of 99.12% on the same dataset, compared to 71.43% on other images from a different dataset. In the study by [30], two pretrained models were trained for detecting tomato leaf diseases on a dataset acquired from a plant village database. The results indicated that AlexNet outperformed Vgg16 and accomplished 97.49% accuracy.

A study by [31] attempted to classify and visualize the symptoms of tomato leaf diseases using the DL method. The model accomplished higher accuracy, at 99.18%. A CNN approach was used by [9] to detect tomato leaf disease; several pretrained methods were trained using an open dataset acquired from plant health. The study reported better performance of the ResNet model and achieved a higher accuracy rate of 97.28%. Another CNN model was proposed by [32] to detect tomato leaf diseases. The model was trained and reported 99.84% accuracy.

3. Deep Learning Architectures

From a broad view, DL belongs to the family of ML techniques utilizing artificial neural networks (ANN) to solve real-world problems related to images (i.e., segmentation, detection, and classification of images) that are widely applied in the fields of computer vision and image processing and have shown the best performance with optimal results. DL has also recently been used in agriculture to detect plant diseases using image analysis and significantly contributed to farming with outstanding outcomes. This study presents a hybrid DL model that combines two different state-of-the-art DL models to detect tomato leaf diseases. In order to better understand the proposed hybrid model, this section highlights the core concepts of each individual model and its architectural design, followed by the proposed hybrid model.

3.1. EfficientNetB3

EfficientNetB3 belongs to the EfficientNet family [33], ranges from B0 to B7, and is regarded as one of the most computationally efficient DL models developed using ImageNet [34]. EfficientNet is a CNN architecture and scaling technique that uses a compound coefficient to consistently scale all depth, width, and resolution dimensions [33]. Furthermore, the scaling method evenly scales network width, depth, and resolution using a set of immovable scaling coefficients, in contrast to standard practice, which scales these variables arbitrarily [33]. In CNN, the kernel is a filter which is utilized to retrieve attributes from images [35], while convolution is utilized to construct a feature map. The model architecture of EfficientNetB3 consists of a convolution layer of kernel size (3×3) with BN and swish activation followed by 26 MBconvolution blocks. The MBconvolution blocks are varied with kernel sizes of (3×3) and (5×5), as shown in Figure 1. The last block of MBconvo is followed by a convolution layer. Global average pooling is utilized at the end of the convolution layers for dimensionality reduction of the feature maps. Fully connected (FC) and softmax are used at the end of the model architecture to generate the output.

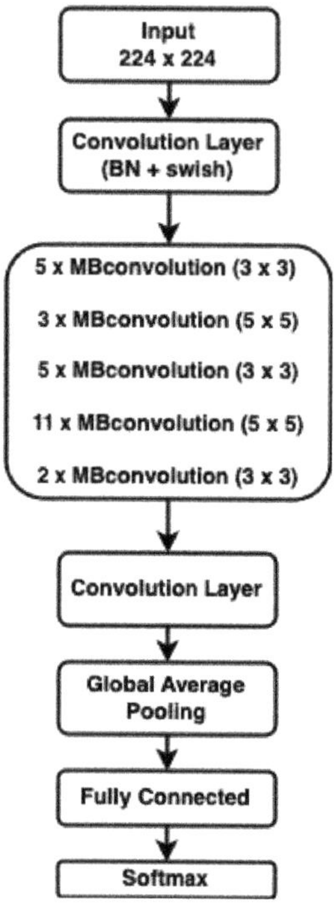

Figure 1. Basic architecture of EfficientNetB3.

3.2. MobileNet

MobileNet, a CNN-based model developed by [36], has a simplified architecture that builds lightweight deep convolutional neural networks using depth-wise separable convolutions. In the model architecture described by [36], MobileNet factorizes standard convolutions into a depth-wise convolution and a (1 × 1) pointwise convolution, as shown in Figure 2. A single convolution on every channel is performed using depth-wise filters, while the output of a depth-wise convolution is combined with the (1 × 1) pointwise convolution [37]. Due to factorization, the computation and model sizes significantly decrease, which eventually enhances the performance of the model. ReLu activation is used between the layers in order to flatten the nonlinear outputs of the preceding layer and provide it to the succeeding layer as input [12].

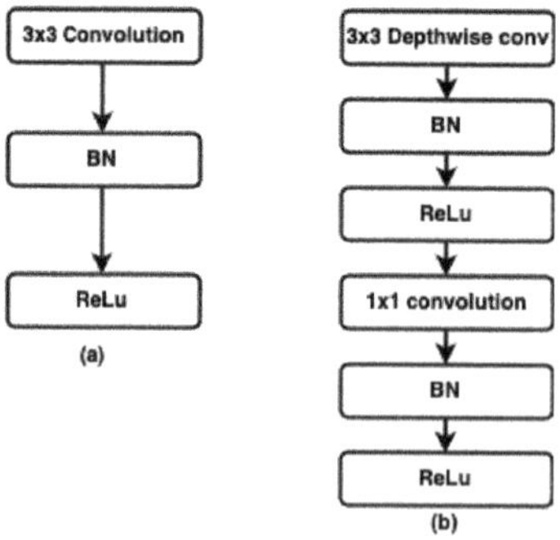

Figure 2. Difference between standard and depth-wise separable convolutions (**a**) and standard (**b**) depth-wise separable convolutions with depth-wise and pointwise layers [36].

3.3. Proposed Hybrid Model

A hybrid model can be used to improve predictive performance by running two or more relevant but distinct models and combining the results into a single score [38]. The literature review revealed that tomato leaf diseases were mostly detected and classified using individual DL models such as EfficientNet, MobileNet, and others, or a hybrid of ML and DL models. This study proposes EffiMob-Net, a hybrid DL model for detecting tomato leaf diseases that is a combination of two individual pretrained DL models, EfficientNet and MobileNet (see Table S1 in Supplementary Materials). A total of 10 diseases related to tomato leaves are recognized and classified using the hybrid EffiMob-Net. According to [39], accurate classification can be achieved by fusing diverse models with different hypotheses concerning class labels, which may not be viable with separate models. Using this approach, we took advantage of the standard architectures of both DL models in which the formerly trained weights of both DL models were loaded for the feature extraction of leaf images and combined for detection purposes, as shown in Figure 3.

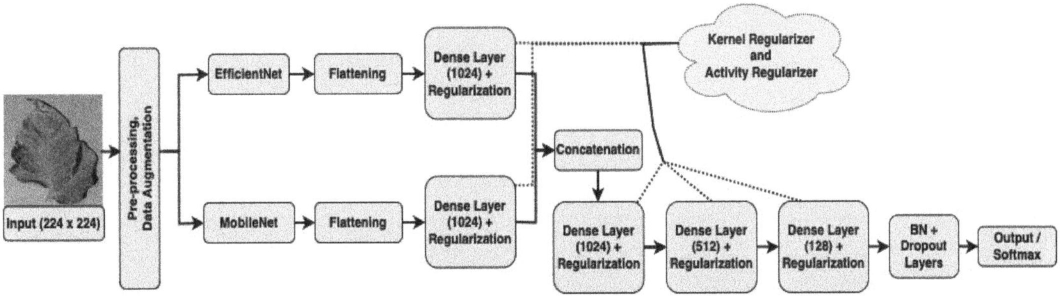

Figure 3. The hybrid EffiMob-Net model architecture proposed in this study.

The model architecture of the EffiMob-Net is simple in that the softmax layers (output layer) are removed from both individual models, the output of each model is flattened and is passed to the fully connected (FC) layer of each model. The outputs of the dense layers (layers of neurons in which each neuron in the following layer receives information from each neuron in the preceding layer) of both models are then combined using the concatenation function, and three additional FC layers containing 1024, 512, and 128 channels are added after concatenating the models, as exhibited in Figure 3. Regularization is used to fine-tune the model in order to decrease the regulated loss function and avoid overfitting and underfitting [40]. The risk of model overfitting is handled using regularization operations (i.e., kernel regularizer and activity regularizer), which are added to the last three FC layers. Moreover, in order to avoid the model overfitting issue, BN and dropout are also used after the last FC layer. The detection of tomato leaf diseases is performed using the softmax layer, which is added at the end of the hybrid model. ReLu activation is used throughout the FC layers except for the softmax layer. Figure 3 shows the detailed architecture of the proposed deep hybrid EffiMob-Net model.

4. Dataset

The proposed hybrid EffiMob-Net model was trained using an openly available dataset gathered from multiple sources, mostly from a plant village database [41] containing a total of 11 classes. Among the 11 classes, one was healthy and the remaining 10 represented different diseases of tomato leaf. The dataset consisted of a total of 32,535 images acquired from a plant village dataset and some collected images distributed into two separate folders: training and validation sets. In this study, the whole validation set is utilized for testing purposes; therefore, the validation set is changed to the test set shown in Figure 3. Thus far, this is the largest publicly available dataset of tomato leaf diseases. The training set contained 25,851 images; 6684 images were part of the test set. The images in both sets were distributed to 11 classes as described in Figure 4 Figure 5shows the number of images per class in the training set. Figure 6 shows sample images in the training set. The dataset is suitable for building a DL model that can predict a particular disease of a tomato leaf and classify them accordingly.

Class name	Training set	Testing set	Total
Bacterial_spot	2826	732	3558
Early_blight	2455	643	3098
Late_blight	3113	792	3905
Leaf_Mold	2754	739	3493
Septoria_leaf_spot	2882	746	3628
Two_Spider_mites	1747	435	2182
Target spot	1827	457	2284
Tomato_Yellow_Leaf_Curl_Virus	2039	498	2537
Tomato_mosaic_virus	2153	584	2737
Healthy	3051	806	3857
Powdery_mildew	1004	252	1256

Figure 4. Dataset description.

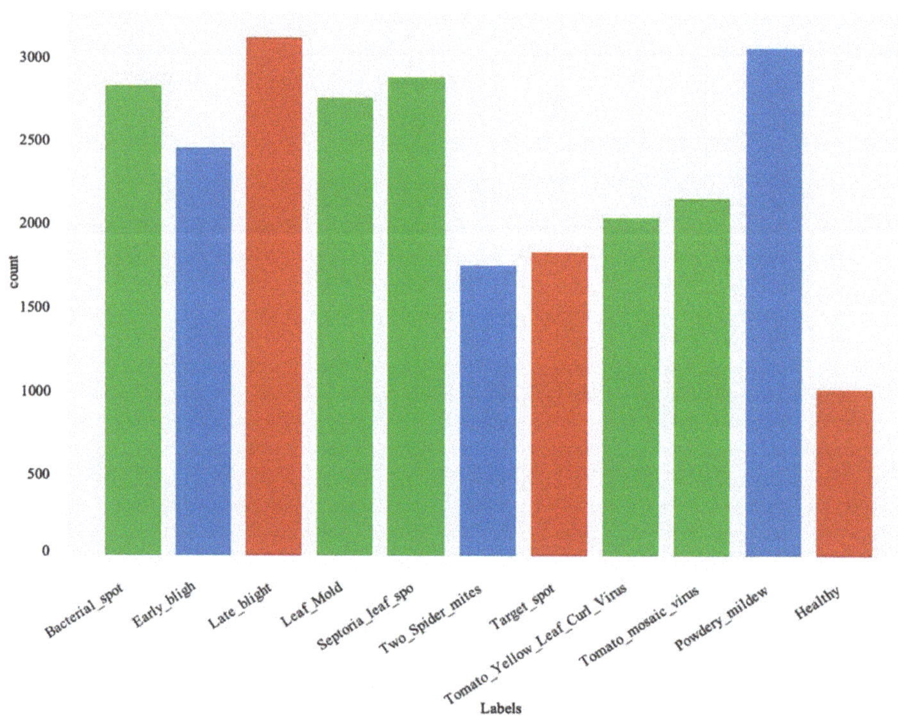

Figure 5. Image distribution per class in the training set.

Figure 6. Sample images in the training set.

Data Preprocessing

Data preprocessing is an indispensable procedure that converts data into a structure that can be easily and proficiently processed in ML and other data science tasks [42]. Removing garbage from data augments the quality of the data [43], which directly affects the performance of the trained models and ensures improved results [44]. In the first step, the images were resized to the required sizes for training the proposed model. As described in [45], CNN typically allows fixed-size images, which creates several challenges for data collection and model building. Such challenges were overcome by resizing the images to the required size of (224 × 224) when building the proposed model. TensorFlow in Python programming was used to resize images to the desired size. The images were also normalized in a pixel value of range 0 to 1 by dividing them by 255 and feeding them into the network. In the last step, the images in both sets were reshuffled to increase the predictability power of the proposed model.

5. Experimental Setup

The dataset used in this study was split into two separate sets: training and testing at a ratio of 80% to 20%, respectively. According to [46], experimental research indicates that using 20–30% of the data for testing and the remaining 70–80% of the data for training yields optimal results. In this study, 80:20 achieved optimal results and was thus chosen for data splitting. The training set was utilized to train the hybrid EffiMob-Net model on a Google Colab in a GPU environment using Python programming language. The testing set was used to validate the model performance. The experiment was executed using

MacBook Pro for 20 iterations in 40 batches. The model was compiled using the Adamax optimizer with a learning rate of 0.001. The best classifier with respect to accuracy metrics was selected to show the results for detecting tomato leaf diseases. The 20% testing set was used to verify the performance of the hybrid EffiMob-Net model using training and validation accuracies and losses. Categorical cross-entropy was used as a loss function to measure the losses. The experiment was repeated several times, and the best-fitting model with respect to accuracy metrics was finalized. The finalized trained hybrid model was then saved to the local directory for future use. Figure 7 depicts the training and validation accuracies. Normally, the curve of training accuracy is greater; however, both curves come closer to each other as the epochs advance. An epoch represents one iteration of training a model with all training data. The best epoch in which both curves coincide is epoch 20, which was one of the main reasons for executing the model for 20 epochs. Likewise, the training and validation loss shown in Figure 8 demonstrates the validity of the proposed hybrid EffiMob-Net in that both curves come closer to each other, progress simultaneously as the epochs advance, then coincide at epoch 13 and progress together in the same manner. This indicates the lack of overfitting of the hybrid EffiMob-Net model, which was avoided by using regularization, dropout, and BN techniques. The performance of the model was measured using accuracy, precision, recall, and F1-scores from the following equations.

$$Accuracy = \frac{TP + TN}{TP + TN + FP + FN} \tag{1}$$

$$Precision = \frac{TP}{TP + FP} \tag{2}$$

$$Recall\ or\ Sensitivity = \frac{TP}{TP + FN} \tag{3}$$

$$\mathrm{F1} - score = \frac{(2 * Precision * Recall)}{Precision + Recall} \tag{4}$$

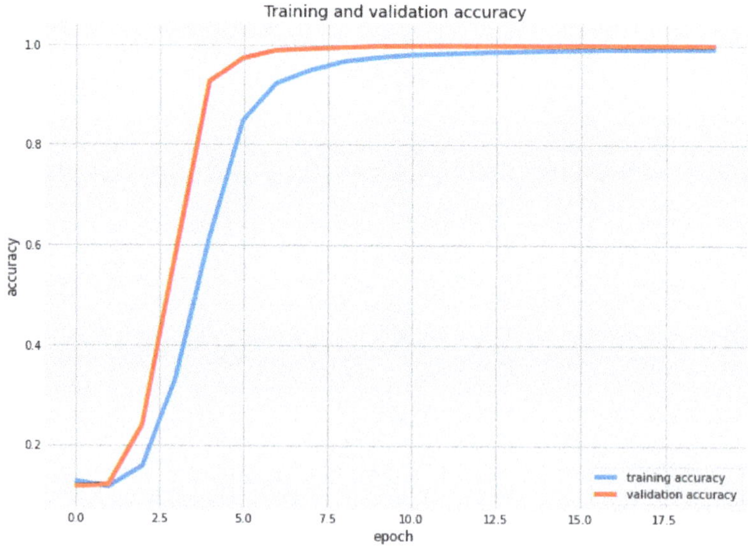

Figure 7. Training and validation accuracy.

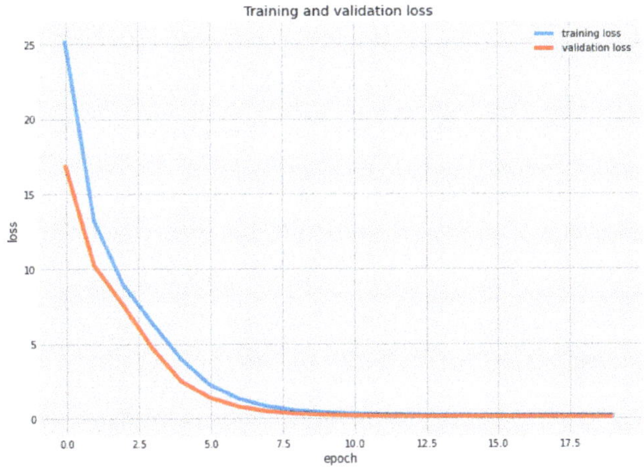

Figure 8. Training and validation loss.

6. Results and Discussion

After implementing and testing the hybrid EffiMob-Net model on the testing set, the performance of the model was measured, and the highest accuracy rate achieved was 99.92%, which is thus far the highest accuracy in the same domain. Moreover, the classification report based on Equations (1)–(4) was measured, and the outcomes are reported in Table 1.

Table 1. Classification report of EffiMob-Net model.

Class	Accuracy	Precision	Recall	F1-Score
Bacterial spot	99.84%	99.29%	99.20%	99.23%
Early blight	99.84%	98.98%	99.29%	99.14%
Late blight	99.87%	99.51%	99.36%	99.44%
Leaf mold	99.84%	99.17%	99.28%	99.23%
Septoria leaf spots	99.86%	99.31%	99.39%	99.35%
Two spider mites	99.86%	98.99%	98.86%	98.93%
Target spot	99.86%	99.04%	98.91%	98.97%
Tomato yellow leaf curl virus	99.89%	99.27%	99.39%	99.33%
Tomato mosaic virus	99.87%	99.19%	99.30%	99.25%
Powdery mildew	99.87%	99.43%	99.43%	99.43%
Healthy	99.85%	98.25%	97.76%	98.01%

The results shown in the classification report table for all 11 classes are above 99% for all measures with the exception of a few values. For example, the precisions of early blight, two spider spots, and healthy are 98.98%, 98.99%, and 98.25%, respectively. Similarly, recalls of two spider spots, target spot, and healthy are 98.86%, 98.91%, and 97.76%, respectively. Likewise, F1-scores for two spider spots, target spot, and healthy are 98.93%, 98.97%, and 98.01%, respectively. The mentioned values with respect to classes surpassed 98% except for the F1-score of healthy, which was close to 98%, showing the reliability of the proposed hybrid EffiMob-Net model when used as a smart detecting system for identifying tomato leaf diseases.

The overall accuracy of 99.92% and the classification report in Table 1 demonstrate the high performance of the proposed hybrid EffiMob-Net with a classification error of only 0.08%, which is negligible. The idea of distinct feature extraction using two separate DL models and the fusion of these features for detecting and classifying tomato leaf diseases is superior to that achieved when using an individual model, as discussed in the related work section. The conventional methods in which the feature extraction is handcrafted require high expertise; otherwise, the model efficacy can be poor. Additionally, such methods require more effort and time-consuming tasks. Therefore, DL-based methods are more useful for automatically generating features and have shown a high success rate in the identification and classification of images. Similarly, the feature extraction using multiple DL methods and the fusion features resulting from different methods show increased model accuracy. This discussion and the facts presented in the tables and figures demonstrate the reliability of the proposed hybrid EffiMob-Net model, which can be used as a reliable detector for detecting and identifying tomato leaf diseases.

7. Conclusions

The necessary precautionary measures should be taken to prevent tomato plant diseases in order to increase the cultivation of tomato crops. This study proposed a hybrid DL-based model that accurately detects and classifies 10 different tomato plant diseases through leaf images. The model architecture was designed by the fusion of two DL models in order to extract the distinct features from tomato leaf images, which were then combined to achieve the accurate identification of each disease with respect to classes. Several techniques (e.g., regularization, dropout, and BN) were used to prevent the model from being overfitted. During implementation, the optimal parameters were set in the model based on hyperparameter tuning using a random grid search technique. The proposed hybrid EffiMob-Net model was tested on processed images of tomato leaf diseases with a split ratio of 80/20 for the training/testing datasets. The results achieved demonstrate the efficacy of the proposed hybrid EffiMob-Net in accurately extracting the distinct features from tomato leaf images, with an accuracy rate of 99.92%, and a classification error of only 0.08%. Moreover, the classification report on factors such as precision, recall, and F1-score demonstrates the high performance of the proposed hybrid model in detecting tomato leaf diseases. The model is efficient in its performance based on the results achieved and, thus, can be used as an automatic detector for identifying tomato leaf diseases early in the growing process in order to increase production. The proposed hybrid model can also be used to detect other plant diseases in the agriculture field based on leaf images.

Supplementary Materials: The following supporting information can be downloaded at: https://www.mdpi.com/article/10.3390/agriculture13030737/s1, Table S1: Comparison of proposed hybrid EffiMob-Net model with existing models.

Author Contributions: Conceptualization, Z.U. and M.J.; methodology, Z.U.; software, Z.U., F.S.; validation, Z.U., N.A. and M.J.; formal analysis, F.S., N.A; investigation, M.J., Z.U.; resources, S.H.A., N.A.; data curation, Z.U., M.J.; writing—original draft preparation, Z.U., S.H.A., M.J., N.A.; writing—review and editing, Z.U., N.A., M.J., S.H.A.; supervision, S.H.A.; project administration, S.H.A.; funding acquisition, Z.U. All authors have read and agreed to the published version of the manuscript.

Funding: This research work was funded by Institutional Fund Project under grant no. (IFPIP: 310-611-1443). Therefore, the authors gratefully acknowledge technical and financial support from the Ministry of Education and King Abdulaziz University, DSR, Jeddah, Saudi Arabia.

Institutional Review Board Statement: Not applicable.

Data Availability Statement: The dataset is available on [41].

Conflicts of Interest: The authors declare no conflict of interest.

References

1. FAO Food and Agriculture Organization of United States. Available online: https://www.fao.org/land-water/databases-and-software/crop-information/tomato/en/ (accessed on 2 January 2023).
2. Bhujel, A.; Kim, N.E.; Arulmozhi, E.; Basak, J.K.; Kim, H.T. A Lightweight Attention-Based Convolutional Neural Networks for Tomato Leaf Disease Classification. *Agriculture* **2022**, *12*, 228. [CrossRef]
3. Chen, H.C.; Widodo, A.M.; Wisnujati, A.; Rahaman, M.; Lin, J.C.W.; Chen, L.; Weng, C.E. AlexNet Convolutional Neural Network for Disease Detection and Classification of Tomato Leaf. *Electronics* **2022**, *11*, 951. [CrossRef]
4. Al-Gaashani, M.S.A.M.; Shang, F.; Muthanna, M.S.A.; Khayyat, M.; Abd El-Latif, A.A. Tomato leaf disease classification by exploiting transfer learning and feature concatenation. *IET Image Process.* **2022**, *16*, 913–925. [CrossRef]
5. Zhao, S.; Peng, Y.; Liu, J.; Wu, S. Tomato leaf disease diagnosis based on improved convolution neural network by attention module. *Agriculture* **2021**, *11*, 651. [CrossRef]
6. Vadivel, T.; Suguna, R. Automatic recognition of tomato leaf disease using fast enhanced learning with image processing. *Acta Agric. Scand. Sect. B Soil Plant Sci.* **2022**, *72*, 312–324. [CrossRef]
7. Elnaggar, S.; Mohamed, A.M.; Bakeer, A.; Osman, T.A. Current status of bacterial wilt (*Ralstonia solanacearum*) disease in major tomato (*Solanum lycopersicum* L.) growing areas in Egypt. *Arch. Agric. Environ. Sci.* **2018**, *3*, 399–406. [CrossRef]
8. Chowdhury, M.E.H.; Rahman, T.; Khandakar, A.; Ayari, M.A.; Khan, A.U.; Khan, M.S.; Al-Emadi, N.; Reaz, M.B.I.; Islam, M.T.; Ali, S.H.M. Automatic and Reliable Leaf Disease Detection Using Deep Learning Techniques. *AgriEngineering* **2021**, *3*, 294–312. [CrossRef]
9. Zhang, K.; Wu, Q.; Liu, A.; Meng, X. Can deep learning identify tomato leaf disease? *Adv. Multimed.* **2018**, *2018*, 6710865. [CrossRef]
10. Alshammari, H.; Gasmi, K.; Ben Ltaifa, I.; Krichen, M.; Ben Ammar, L.; Mahmood, M.A. Olive Disease Classification Based on Vision Transformer and CNN Models. *Comput. Intell. Neurosci.* **2022**, *2022*, 3998193. [CrossRef]
11. Khatoon, S.; Hasan, M.M.; Asif, A.; Alshmari, M.; Yap, Y.K. Image-based automatic diagnostic system for tomato plants using deep learning. *Comput. Mater. Contin.* **2021**, *67*, 595–612. [CrossRef]
12. Ksibi, A.; Ayadi, M.; Soufiene, B.O.; Jamjoom, M.M.; Ullah, Z. MobiRes-Net: A Hybrid Deep Learning Model for Detecting and Classifying Olive Leaf Diseases. *Appl. Sci.* **2022**, *12*, 10278. [CrossRef]
13. Tan, L.; Lu, J.; Jiang, H. Tomato Leaf Diseases Classification Based on Leaf Images: A Comparison between Classical Machine Learning and Deep Learning Methods. *AgriEngineering* **2021**, *3*, 542–558. [CrossRef]
14. Bahhar, C.; Ksibi, A.; Ayadi, M.; Jamjoom, M.M.; Ullah, Z.; Soufiene, B.O.; Sakli, H. Wildfire and Smoke Detection Using Staged YOLO Model and Ensemble CNN. *Electronics* **2023**, *12*, 228. [CrossRef]
15. Wagle, S.A.; Harikrishnan, R. A deep learning-based approach in classification and validation of tomato leaf disease. *Trait. Signal* **2021**, *38*, 699–709. [CrossRef]
16. Nagamani, H.S.; Sarojadevi, H. Tomato Leaf Disease Detection using Deep Learning Techniques. *Int. J. Adv. Comput. Sci. Appl.* **2022**, *13*, 305–311.
17. Kaur, P.; Harnal, S.; Gautam, V.; Singh, M.P.; Singh, S.P. An approach for characterization of infected area in tomato leaf disease based on deep learning and object detection technique. *Eng. Appl. Artif. Intell.* **2022**, *115*, 105210. [CrossRef]
18. Islam, M.S.; Sultana, S.; Al Farid, F.; Islam, M.N.; Rashid, M.; Bari, B.S.; Hashim, N.; Husen, M.N. Multimodal Hybrid Deep Learning Approach to Detect Tomato Leaf Disease Using Attention Based Dilated Convolution Feature Extractor with Logistic Regression Classification. *Sensors* **2022**, *22*, 6079. [CrossRef] [PubMed]
19. Garg, N.; Gupta, R.; Kaur, M.; Kukreja, V.; Jain, A.; Tiwari, R.G. Classification of Tomato Diseases using Hybrid Model (CNN-SVM). In Proceedings of the International Conference on Reliability, Infocom Technologies and Optimization (Trends and Future Directions) (ICRITO), Noida, India, 4–5 June 2020; IEEE: Noida, India, 2020.
20. Bhatia, A.; Chug, A.; Singh, A.P. Hybrid SVM-LR classifier for powdery mildew disease prediction in tomato plant. In Proceedings of the International Conference on Signal Processing and Integrated Networks, Noida, India, 27–28 February 2020; pp. 218–223.
21. Tarek, H.; Aly, H.; Eisa, S.; Abul-Soud, M. Optimized Deep Learning Algorithms for Tomato Leaf Disease Detection with Hardware Deployment. *Electronics* **2022**, *11*, 140. [CrossRef]
22. Sareen, N.; Chug, A.; Singh, A.P. An image based prediction system for early blight disease in tomato plants using deep learning algorithm. *J. Inf. Optim. Sci.* **2022**, *43*, 761–779. [CrossRef]
23. Sareen, N.; Chug, A.; Singh, A.P. An Image-Based Tomato Early Blight Disease Prediction Using Optimized Transfer. *Vivekananda J. Res.* **2021**, *10*, 1–13.
24. Mondal, D.; Roy, K.; Pal, D.; Kole, D.K. Deep Learning-Based Approach to Detect and Classify Signs of Crop Leaf Diseases and Pest Damage. *SN Comput. Sci.* **2022**, *3*, 433. [CrossRef]
25. Patokar, A.M.; Gohokar, V.V. Classification of Tomato Leaf Diseases: A Comparison of Different Optimizers. In *Intelligent Systems and Applications, Proceedings of the ICISA 2022, Pune, India, 4–6 May 2022*; Lecture Notes in Electrical Engineering; Springer: Singapore, 2023; pp. 27–33.
26. Abbas, A.; Jain, S.; Gour, M.; Vankudothu, S. Tomato plant disease detection using transfer learning with C-GAN synthetic images. *Comput. Electron. Agric.* **2021**, *187*, 106279. [CrossRef]

27. Habiba, S.U.; Islam, M.K. Tomato Plant Diseases Classification Using Deep Learning Based Classifier From Leaves Images. In Proceedings of the International Conference on Information and Communication Technology for Sustainable Development (ICICT4SD), Dhaka, Bangladesh, 27–28 February 2021; IEEE: Dhaka, Bangladesh, 2021.
28. Fuentes, A.; Yoon, S.; Kim, S.C.; Park, D.S. A robust deep-learning-based detector for real-time tomato plant diseases and pests recognition. *Sensors* **2017**, *17*, 2022. [CrossRef]
29. Afify, M.; Loey, M.; Elsawy, A. A Robust Intelligent System for Detecting Tomato Crop Diseases Using Deep Learning. *Int. J. Softw. Sci. Comput. Intell.* **2022**, *14*, 1–21. [CrossRef]
30. Rangarajan, A.K.; Purushothaman, R.; Ramesh, A. Tomato crop disease classification using pre-trained deep learning algorithm. *Procedia Comput. Sci.* **2018**, *133*, 1040–1047. [CrossRef]
31. Brahimi, M.; Boukhalfa, K.; Moussaoui, A. Deep Learning for Tomato Diseases: Classification and Symptoms Visualization. *Appl. Artif. Intell.* **2017**, *31*, 299–315. [CrossRef]
32. Ashqar, B.A.M.; Abu-Naser, S.S. Image-Based Tomato Leaves Diseases Detection Using Deep Learning. *Int. J. Acad. Eng. Res.* **2018**, *2*, 10–16.
33. Tan, M.; Le, Q.V. EfficientNet: Rethinking Model Scaling for Convolutional Neural Networks. In Proceedings of the International Conference on Machine Learning, PMLR, Long Beach, CA, USA, 9–15 June 2019; pp. 6105–6114.
34. Salian, S.R.; Sawarkar, S.D. Melanoma skin lesion classification using improved efficientnetb3. *Jordanian J. Comput. Inf. Technol.* **2022**, *8*, 45–57. [CrossRef]
35. Intellipaat What Is Kernel in CNN? Available online: https://intellipaat.com/community/46826/what-is-kernel-in-cnn#:~:text=In%20Convolutional%20neural%20network%2C%20the,the%20matrix%20of%20dot%20products (accessed on 14 March 2023).
36. Howard, A.G.; Zhu, M.; Chen, B.; Kalenichenko, D.; Wang, W.; Weyand, T.; Andreetto, M.; Adam, H. MobileNets: Efficient Convolutional Neural Networks for Mobile Vision Applications. *arXiv* **2017**, arXiv:1704.04861.
37. Wang, W.; Li, Y.; Zou, T.; Wang, X.; You, J.; Luo, Y. A novel image classification approach via dense-mobilenet models. *Mob. Inf. Syst.* **2020**, *2020*, 7602384. [CrossRef]
38. Burns, E. Ensemble Modeling. Available online: https://www.techtarget.com/searchbusinessanalytics/definition/Ensemble-modeling (accessed on 4 February 2023).
39. Ahmed, S.; Choi, K.Y.; Lee, J.J.; Kim, B.C.; Kwon, G.R.; Lee, K.H.; Jung, H.Y. Ensembles of Patch-Based Classifiers for Diagnosis of Alzheimer Diseases. *IEEE Access* **2019**, *7*, 73373–73383. [CrossRef]
40. Simplilearn. The Best Guide to Regularization in Machine Learning. Available online: https://www.simplilearn.com/tutorials/machine-learning-tutorial/regularization-in-machine-learning#:~:text=Regularizationreferstotechniquesthat,andpreventoverfittingorunderfitting.&text=UsingRegularization%2Cwecanfit,re (accessed on 14 March 2023).
41. Khan, Q. Tomato Disease Multiple Sources. Available online: https://www.kaggle.com/datasets/cookiefinder/tomato-disease-multiple-sources?resource=download-directory (accessed on 13 December 2022).
42. Lawton, G. Data Preprocessing. Available online: https://www.techtarget.com/searchdatamanagement/definition/data-preprocessing (accessed on 28 December 2022).
43. Al-Mudimigh, A.S.; Ullah, Z. Prevention of Dirty Data and the Role of MADAR Project. In Proceedings of the UKSim 5th European Symposium on Computer Modeling and Simulation, Madrid, Spain, 16–18 November 2011.
44. Ullah, Z.; Al-Mudimigh, A.S. Integration and Communication to Prevent Dirty Data: The Role of MADAR Project. *Information* **2012**, *15*, 3459.
45. Ullah, Z.; Jamjoom, M. An intelligent approach for Arabic handwritten letter recognition using convolutional neural network. *PeerJ Comput. Sci.* **2022**, *8*, e995. [CrossRef]
46. Gholamy, A.; Kreinovich, V.; Kosheleva, O. Why 70/30 or 80/20 Relation between Training and Testing Sets: A Pedagogical Explanation. *Dep. Tech. Rep.* **2018**, 1–6.

agriculture

MDPI

Article

EfficientPNet—An Optimized and Efficient Deep Learning Approach for Classifying Disease of Potato Plant Leaves

Tahira Nazir [1], Muhammad Munwar Iqbal [2], Sohail Jabbar [3], Ayyaz Hussain [4] and Mubarak Albathan [3,*]

1 Faculty of Computing, Riphah International University, Gulberg Greens Campus Islamabad, Islamabad 45320, Pakistan; tahira.nazir@riphah.edu.pk
2 Department of Computer Science, University of Engineering and Technology, Taxila 47080, Pakistan; munwar.iq@uettaxila.edu.pk
3 College of Computer and Information Sciences, Imam Mohammad Ibn Saud Islamic University (IMSIU), Riyadh 11432, Saudi Arabia; sjjabar@imamu.edu.sa
4 Department of Computer Science, Quaid-i-Azam University, Islamabad 44000, Pakistan; ayyaz.hussain@qau.edu.pk
* Correspondence: mmalbathan@imamu.edu.sa; Tel.: +966-503-451-575

Abstract: The potato plant is amongst the most significant vegetable crops farmed worldwide. The output of potato crop production is significantly reduced by various leaf diseases, which poses a danger to the world's agricultural production in terms of both volume and quality. The two most destructive foliar infections for potato plants are early and late blight triggered by *Alternaria solani* and *Phytophthora infestans*. In actuality, farm owners predict these problems by focusing primarily on the alteration in the color of the potato leaves, which is typically problematic owing to uncertainty and significant time commitment. In these circumstances, it is vital to develop computer-aided techniques that automatically identify these disorders quickly and reliably, even in their early stages. This paper aims to provide an effective solution to recognize the various types of potato diseases by presenting a deep learning (DL) approach called EfficientPNet. More specifically, we introduce an end-to-end training-oriented approach by using the EfficientNet-V2 network to recognize various potato leaf disorders. A spatial-channel attention method is introduced to concentrate on the damaged areas and enhance the approach's recognition ability to effectively identify numerous infections. To address the problem of class-imbalanced samples and to improve network generalization ability, the EANet model is tuned using transfer learning, and dense layers are added at the end of the model structure to enhance the feature selection power of the model. The model is tested on an open and challenging dataset called PlantVillage, containing images taken in diverse and complicated background conditions, including various lightning conditions and the different color changes in leaves. The model obtains an accuracy of 98.12% on the task of classifying various potato plant leaf diseases such as late blight, early blight, and healthy leaves in 10,800 images. We have confirmed through the performed experiments that our approach is effective for potato plant leaf disease classification and can robustly tackle distorted samples. Hence, farmers can save money and harvest by using the EfficientPNet tool.

Keywords: agriculture; classification; deep learning; transfer learning; convolutional neural networks; EfficientNet; potato diseases

1. Introduction

According to the UN Food and Agriculture Organization (FAO), the global population could reach 9.1 billion by 2050. Due to the rising population, food consumption will increase [1]. In the meantime, the lack of farmland and access to clean water makes it hard for nutrient levels to rise. In order to meet human needs, there is an immediate need to boost food security while using the least amount of growing area. As opposed to this, a number of crop anomalies cause a significant decrease in meal productivity and quality. Therefore, immediate detection of these plant pathogens is necessary, as they have the potential to reduce agricultural profits and increase inflation rates. Such outcomes may cause market-wide economic uncertainty. Additionally, agricultural crop disorders at their most severe stages can wipe out harvests and cause hunger in a country, especially in developing nations with poor incomes. Typically, plant assessments are performed with the aid of domain specialists; however, this is a laborious and time-consuming task that depends on the participation of local professionals. Additionally, such methods of crop evaluation are not regarded as highly trustworthy, and it is difficult for people to individually evaluate each crop [2]. Therefore, it is critical to accurately and promptly identify the numerous plant illnesses that might prevent growers from deploying pricey treatment techniques while improving the food growth rate. The science world is concentrating its effort on the creation of computerized plant illness diagnosis and recognition systems to address the aforementioned issues with manual approaches [3].

Despite the existence of numerous different crops, such as tomatoes, onions, strawberries, and cherries, among others, the potato plant is a highly consumed crop around the globe. The potato crop is regarded as the major staple by more than a billion people globally, and is considered the third largest food crop on the planet after rice and wheat. More than 300,000 tons are produced globally each year, providing both nutrients and an essential source of calories for people [4]. In addition to providing a sizeable share of the world's nutrition, potatoes are a common source of raw ingredients for industry. Potatoes are produced all over the world, with the top three exporters being China, India, and Russia [5].

Following a survey performed by the UN Food and Agriculture Organization (FOA), the prevalence of many illnesses, the majority of those which originate from the leaves of the potato crop and cause a reduction in output amount from 9% to 11% annually [6], is the main obstacle to the pace of potato growth. To examine potato crop leaf disorders, the scientific world initially used methods from the fields of biological sciences and cell biology [7,8]. These methods, however, exhibit high processing complexity and demand a significant need for expert skills [9]. The majority of agricultural production is done by low-income individuals; hence, such pricey methods are not practical for farmers [10]. The rapid advancement of machine vision and object classification algorithms is being used in existing works to design automated methods for identifying crop pathogens. Image processing and machine learning (ML) studies are receiving more focus, and these methods are emerging as appealing alternatives to ongoing crop infection surveillance. Several conventional ML predictors, such as K-Nearest Neighbors (KNN), Random Forest Tree (RFT) [11], and Support Vector Machine (SVM), are highly employed in existing works for accomplishing classification tasks related to various plant-related diseases. Although these ML techniques are simpler to comprehend and only need a minimal quantity of samples to build models, they take time and rely heavily on expert human capital. Additionally, the classic ML information computation methods consistently necessitate a compromise between processing effort and classification results [12].

Deep learning (DL) techniques are currently being evaluated to address the shortcomings of ML algorithms. Different DL methodologies, including CNN [13], RNN [14], and long short-term memory (LSTM) [15], are currently widely praised in the field of food security. DL methods are capable of accurately estimating the informative collection of sample feature characteristics without the assistance of domain experts. Both these strategies for object recognition and deep learning (DL) imitate how the human brain functions when a person locates and recognizes a variety of items by looking at examples of them. DL approaches provide reliable results in the field of modern agriculture research, and are effectively suited to a variety of jobs, whereas different kinds of deep neural networks (DNNs) exhibit greater precision than multispectral evaluation. The agricultural production field is intensively investigating methods such as GoogLeNet [16], DenseNet [17], Inception, VGG [18], and Residual Net [19] for problems including quantifying grain volume, detecting plant heads, quantifying fruits, crop disorder diagnosis and categorization, etc. Because of their capacity to utilize the structural and morphological information afrom the investigated images, these approaches are able to demonstrate excellent recognition accuracy while minimizing processing effort [20].

Even though experts have carried out a significant amount of work to classify potato crop leaf infections, it remains difficult to identify illness in the initial stages, as infected and healthy plant sections share many similar characteristics [20]. Recognition is made more difficult by varied plant leaf shapes, fluctuations in lighting and luminosity, the inclusion of distortion, and blurring in the processed images. Thus, there remains an opportunity for potential improvement in terms of computing power as well as correctness in identifying potato plant diseases. In the presented work, we attempted to tackle the existing problem of potato plant leaf disease classification by proposing an effective DL approach, namely, EfficientPNet. We have modified the existing EfficientNet-v2 model by introducing an attention mechanism (AM) and additional dense layers at the end of the framework structure. The presented EfficientPNet approach robustly extracts high-level signs of infected regions and associates them with related groups via employing an end-to-end training mechanism. In addition, the AM strategy improves the recall power of the proposed solution by passing relevant details of noticeable attributes such as diseased areas of plant leaves. The distinctive contributions of this work can be elaborated as follows:

(1) An effective light DL approach called EfficientPNet is suggested that is proficient in calculating relevant and distinctive sample characteristics and shows improved potato plant leaf disease classification results with little computational effort.
(2) The model includes the pixel and channel attention approach in the feature computation phase, which improves its ability to comprehend crosslinks and spacewise orientation properties to accelerate the diagnosis of potato leaf disorders in realistic scenarios.
(3) Transfer learning and multi-class focal loss are adopted to cope with the problem of class imbalance and network overfitting, which improves the precision of classifying potato leaf infected regions.
(4) In order to demonstrate the effectiveness of the suggested EfficientPNet model, we performed huge comparison evaluations to check the classification results by utilizing a collection of images of potato crop disease taken from a standard sample repository called PlantVillage. The suggested method successfully categorizes potato crop illnesses, even in the context of challenging external factors such as noise, distortion, unbalanced lighting, and variations in the shape, color, and placement of infection marks.
(5) To increase the size and ensure balance between the training and testing datasets, we have applied data augmentation techniques. Using these data augmentation techniques, the classifier become more able to generalize.

The rest of this paper is organized as follows: related works is presented in Section 2, and the proposed method in Section 3; we discuss the obtained results in Section 4; finally, the work is concluded, and our future research plans are elaborated in Section 5.

2. Related Works

There have been several approaches presented for potato leaf disease detection from leaf images. In [21], the authors proposed a pre-trained ResNet50 CNN model for the classification and detection of plant diseases. This method was applied to potato leaves taken from the PlantVillage dataset. The presented approach included augmentation and segmentation, which were then passed to ResNet-50 for classification, achieving 98% accuracy. The method performed well; however, its accuracy depends on augmentation and needs further improvements. Bhagat et al. [22] presented bag-of-words (BoWs) and SURF-based techniques for the identification of potato leaf diseases. The bag of words approach was utilized for feature extraction in the initial phase. After that, the SURF method was selected to extract the strongest features, which were then passed to an SVM for classification. Experiments were performed on potato leaves taken from the PlantVillage dataset, and the model attained 97% accuracy. The method in [22] performed well; however, the model did not consider unseen or real-world samples.

Pal et al. [23] proposed the AgriDet (Agriculture Detection) approach. Their method utilized the conventional Inception-Visual Geometry Group Network and Kohonen for the detection and identification of potato leaf disease. The multi-variate Grabcut was applied to reduce the occlusion problem. This method was applied to the PlantVillage dataset to detect and segment potato leaf disease classification. The model achieved good results, with 92.12 % accuracy. The presented approach can tackle the overfitting problem through the dropout layer. However, it is unable to recognize multiple instances of the same disease in one image. Yu, H. et al. described an improved deep learning model for classifying potato plant leaf diseases in their paper [24]. They used a convolutional neural network (CNN) and a transfer learning approach to train their model on a large dataset of potato leaf images. The model achieves high accuracy rates in classifying different types of diseases, and outperforms several other deep-learning models in terms of accuracy and training time [24,25].

Chen, X. et al. presented a study on potato leaf disease classification using an improved deep learning model. The authors used a modified Inception-V3 model and a transfer learning approach to train the model on a dataset of potato leaf images. The model achieves high accuracy rates in classifying different types of diseases and outperforms several other deep-learning models in terms of accuracy and training time [26]. In [27], Kang et al. proposed a lightweight CNN-based approach for the recognition of potato leaf diseases. The authors utilized multi-scale pyramid fusion technology for efficient feature selection. This fusion of features was achieved using the improved backbone model and optimized features. This lightweight technique recognized and identified plant leaf diseases, achieving 93% accuracy. However, the presented model needs further improvements in accuracy.

To detect and classify potato leaves, Kumar et al. [28] presented an automated method based on Gaussian filtering and Fuzzy c-means clustering. This method extracted different types of features, including textual, geometrical, and statistical features. The extracted features were then passed to a PCA for efficient feature selection. At last, several classifiers were employed for the classification of potato leaves. The unbalanced data makes [29] machine learning models more biased and leads to overfitting issues. This study shows a way to add more information to data based on an image-to-image translation model. This helps eliminate the bias from adding these bad potato leaf images. To produce pictures representing more obvious disease textures, the authors suggested that the augmentation approach translates healthy and unhealthy leaf images and uses attention processes.

Rashid et al. [30] proposed a multi-level DL-based model for the recognition of potato leaf diseases. In the initial stage, the YOLOv5 technique was employed for the segmentation of images. Second, the Deep CNN model was utilized for potato leaf identification from images. Experiments were performed on a proprietary dataset and achieved good results. However, the presented model is unable to detect multiple diseases from a single image. Tiwari et al. [31] proposed a deep learning technique for the detection of potato leaf diseases. Their model was based on numerous approaches. In the first step, features were extracted through a VGG19 model. The extracted features were then classified using different classifiers, in which logistic regression performed well compared to the others, achieving 97.8% accuracy on the PlantVillage dataset. The presented model needs further improvements to efficiently detect unseen examples. Similarly, a CNN approach was utilized in [32] to recognize potato leaf diseases. The technique was based on the Adam optimizer and cross-entropy for model analysis. The final classification was performed using a softmax layer. Another CNN-based approach was employed in [33] for the detection of potato leaf diseases. Experimentation was performed on the Kaggle dataset, and the model attained 97% accuracy. However, the presented model tackles only binary classification. Iqbal et al. [34] proposed a method for the segmentation and classification of potato leaf diseases. The PlantVillage dataset was utilized for evaluation of the proposed technique. The random forest approach was employed for classification of leaves into two types, diseased or healthy, with an accuracy of 97%. A deep learning technique was proposed to efficiently detect potato leaf diseases using PlantVillage dataset in [35]. The model was based on the lightweight MobileNet-V2, which was then modified using the addition of a layer in the model. The model achieved 97.33% accuracy on the classification of potato leaf disease. However, this model is computationally light in terms of time. In [36], the authors presented a deep learning-based approach for the classification of potato leaf diseases. The proposed technique was based on four types of models: MobileNet, VGG16, VGG19, and ResNet. Fine-tuning of parameters was performed to enhance the accuracy of the proposed model. Experiments were performed on the PlantVillage dataset, achieving 97.8% accuracy. However, the presented approach did not tackle real-world samples.

3. Materials and Methods

Our proposed work is based on the EfficientNet approach called improved EfficientNetV2 for the recognition and classification of potato leaf diseases. To test and validate the performance of the proposed system, the PlantVillage dataset, with a total number of 54,306 images of potato plants, was utilized. To balance this dataset in each class, we applied data augmentation techniques. The proposed work is focused on improving the EfficientNet approach for potato leaf disease recognition and classification by introducing additional layers at the bottom of the model. These additional layers were designed to enhance the model's performance by allowing it to identify more complex patterns and features in the images. The improved model, called improved EfficientNetV2, was trained on a large dataset of potato leaf images consisting of both healthy and diseased leaves. The model was trained using a supervised learning approach in which it was provided with labelled examples of healthy and diseased leaves and learned to classify new images based on the patterns identified in the training data. The additional layers at the bottom of the improved EfficientNetV2 model allow it to capture more low-level features and patterns in the images, which in turn can improve the accuracy and robustness of the model. Techniques such as transfer learning, data augmentation, and regularization can be employed to further improve the model's performance.

The proposed work has the potential to contribute to the development of more accurate and reliable models for potato leaf disease recognition and classification, which can help farmers and agricultural researchers in their efforts to improve crop yields and reduce losses. The improved model has additional layers at the bottom of the model, which help to enhance the performance. The complete flow of our improved model is shown in Figure 1. The overall process is explained in Algorithm 1.

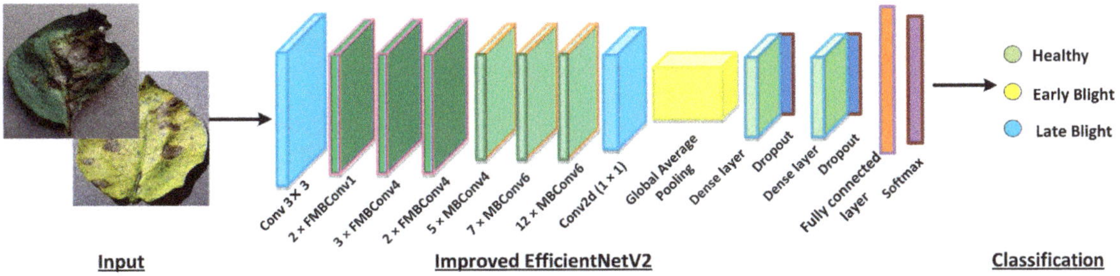

Figure 1. Flow of the proposed EfficientNetV2 framework.

Algorithm 1: Steps followed by EfficientPNet for potato plant leaf abnormality categorization

START
INPUT: TP, Labels
OUTPUT: The category of potato plant leaf diseased region, EfficientPNet
    TP: Total potato images with various abnormalities.
    Labels: Class of each potato sample
    EfficientPNet: improved EfficientNet-V2 model.
//Data preparation and augmentation to balance dataset
    Data Augmentation (x)
    SampleDimension ← [j h]
    // Labels associated with each input sample
    Ä← ReadClassLabel (TP, Class)
// training phase //Functions

1. EffiNetV2(): employed to measure the keypoints with EfficientNet-V2 network
2. EvaluatFramework(): employed to accomplish the model training

    // Improved EfficientNet-V2 model
EfficientPNET ← EffiNetV2 (SampleDimension, *Ä*)
    [TrainingPart, TestPart ← Database distribution
    For each sample c in → TrainingPart
        Compute *improved-EfficientNet-V2* features →*tm*
    End
    Utilize *tm* images EfficientPNet training, and calculate time
    £abelA ← IdentifyPotatoLeafAffectedAreaLabel (*tm*)
    Ap← EvaluatFramework (*improved-EfficientNet-V2, LoclizeA*)
// test phase
    For each image C in → TestPart
        (a) βC ← Compute features via employing the trained model EfficientPNet
        (b) [*ConfidenceScore, ClassLabel*] ←Predict (βC)
        (c) show samples *ClassLabel*
        End
Exit

3.1. *EfficientPNet Framework*

A robust set of image features is essential to obtaining superior classification results, as it directly helps to distinguish the numerous image data groupings. The use of dense DL networks can help in calculating a collection of more effective characteristics, which in turn causes the recall rate of methods to increase [16]. The deployment of these CNN techniques depends heavily on the availability of processing power and memory needs, which places a computational constraint on the models when deep networks are used. Consequently, the cost of computing and the results of the evaluation are always tradeoffs. For this reason, it is necessary to provide a system for identifying leaf diseases that can demonstrate improved classification accuracy while maintaining computing costs [37]. In this study, we introduce a simple and reliable computational strategy to improve model performance for categorizing various anomalies.

An enhanced EfficientNetV2-B4 model is introduced for the identification of potato plant diseases and given the name of EfficientPNet. EfficientNetV2is an expanded version of EfficientNet [38]. Essentially, the improved EfficientNetV2 model is presented to increase available resources while maintaining a high recall rate. The improved EfficientNetV2 model was created using a quick and effective composite scaling method that enables a regular ConvNet to be scaled to any resource limitations while maintaining the method capability. Therefore, the proposed approach offers an ideal choice for network design, i.e., network layers or feature vector size, as well as an optimal solution for computing cost. The EfficientNetV2 technique conducts classification operations robustly and only uses a limited number of model parameters. Furthermore, it performs well in terms of efficiency compared to other methods such as GoogleNet [16], AlexNet [39], DenseNet [40], ResNet [41], and MobileNet [42].

The motivation behind EfficientNet-V2 with dense layers for recognition of potato leaf diseases is that it is an efficient and lightweight approach that requires less training time and contains fewer parameters. The EfficientNetV2 approach makes use of neural architecture search to increase classification accuracy while reducing the size of feature vectors and training time (NAS). Additionally, by including the Fused-MBConv (FMBConv) blocks [43] in the EfficiceiNetV2 architecture, the operative power is optimized and mobile or server accelerators are employed effectively, whereas the conventional EfficientNet technique, which only uses depth-wise convolutions, uses MBConv blocks [44] as its primary building block. Despite the fact that depth-wise convolutions reduce the number of operations required, they do not fully utilize new hardware accelerators. The EfficientNetV2 technique fully utilizes both MBConv and FMBConv blocks to achieve computational gains. The depth-wise 3*3 convolution is replaced in the FMBConv by conventional 3*3 convolution layers. The main objective is to boost the implementation speed of the model while keeping the classification results [45] as shown in Figure 2.

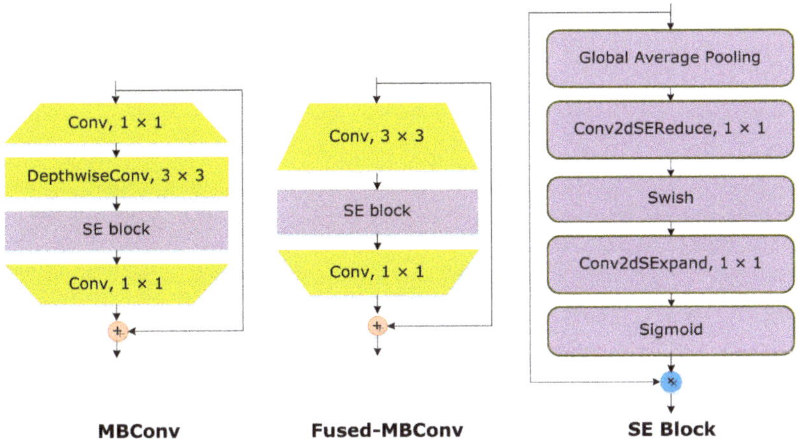

Figure 2. Graphic form of MBConv4, Fused-MBConv4, and SE block.

We used EfficientNetV2 with the B4 architecture to complete the classification task. The B4 base was chosen primarily because it shows a good trade-off between time complexity and model classification performance. Table 1 provides a thorough overview of the enhanced EfficientNetV2 model. The revised EfficientNet-V2 model uses FMBConv blocks at the bottom layers while using MBConv blocks with 3*3 and 5*5 convolutions, squeeze-and-excitation block (SEB) [46], and swish activation at the advanced level. The MBConv blocks preserve an up-set residual link through the SEB to produce robust classification results.

ReLU activation (ReLUAF) is replaced in the framework by the swish activation function (*SAF*) [47], as ReLU excludes values lower than zero and loses an essential component of the used ECG signal. The following equation can be used to calculate the *SAF* (1):

$$SAF(x) = X.Sigmoid(x) \quad (1)$$

Additionally, a Batch normalization layer was added at the beginning of a framework to down-sample the input image sizes. Only three FMBConv blocks were used, as they include many parameters for large values of O. After the MBConv, a global average pooling layer was introduced to reduce the model parameters in order to prevent the issue of model overfitting. Together with the ReLUAF and dropout layers, we included two additional inner-dense layers that help to compute the more effective collection of image characteristics by effectively presenting them. A dropout rate of 30% was chosen arbitrarily in order to progress the model's performance. At the end, a softmax layer was utilized for the classification of potato leaf diseases.

Table 1. Details of blocks and layers used in the proposed model.

Sr No.	Layers
1	BatchNormalization
2	ConvL (3 × 3)
3	2 × FMBConv1 Block
4	3 × FMBConv4 Block
5	2 × FMBConv4 Block
6	5 × MBConv4 Block
7	7 × MBConv6 Block
8	12 × MBConv6 Block
9	Conv2d (1 × 1) Block
10	Global average pooling
11	Dense Layer
12	Dropout
13	Dense Layer
14	Dropout
15	FC Layer
16	Softmax Layer

3.2. Loss Function (LF)

The loss function (*LF*) is a task employed by models to assess their performance. Networks use automated learning to find rules and provide predictions for enormous amounts of data. The primary goal of the *LF* is to determine how much the real and anticipated values have changed. Throughout the model training process, the *LF* is adjusted regularly until a robust fitting value is obtained to reduce error.

We removed the final layer of the EfficientPNet model by introducing an output neuron to accomplish the categorization task for high-quality and distorted samples. For this reason, the hyperparameters of the framework were nominated using an empirical strategy. In our proposed approach, we have adopted the Adadelta optimizer during the model training phase, along with a learning score of 0.1. Moreover, we used twenty epochs for model training. The cross-entropy *LF* uses the Softmax function for classification tasks to assess the variance between calculated and real values. Calculating the cross-entropy *LF* is done as follows:

$$LF = \frac{1}{N}\sum_{k=1}^{n} \log(\frac{e^{s_j}}{\sum_i e^{s_k}}) \quad (2)$$

Here, N represents the total neurons, s_k indicates the input vector, and sj is the estimated label. The model permits the fine-tuning of only 20% of the entire framework parameters without adjusting the remaining 80%. A validation set was utilized to ensure the avoidance of model overfitting issues. Adaptive Moment Estimation [48] was adopted to compute the value of the learning rate against each parameter. This method works by storing the exponential decay of the previous gradient by adopting the impulse approach, as shown in Equations (3) and (4), respectively.

$$M_t = b1M_t - 1 + (1 - b1)G_t \quad (3)$$

$$V_t = b2V_t - 1 + (1 - b2){G^2}_t \quad (4)$$

Here, $b1$, and $b2$ are constants with scores of 0.9, and 0.999, respectively, G indicates the gradient score, and M_t and V_t represent the first-moment and second-moment vectors. The values of these two factors show the link between the updated and previous gradient values. Higher scores of these parameters show a close link between the previous and new gradient values. Initially, the values of both moments are initialized to zero, which requires the bias correction factors *b*1 and *b*2 to avoid the 0 biases. Such biases can be removed by computing the bias-corrected Mt, as elaborated in Equations (5) and (6), respectively:

$$M_t = M_t - (bt1) \quad (5)$$

$$V_t = V_t - (bt2) \quad (6)$$

The optimization approach in our model uses Equation (7) to update the gradient value.

$$W_{t+1} = W_t - \eta/(V_t + \epsilon M)^{0.5}t \quad (7)$$

Here, ϵ is a constant, η is a learning rate with a score of 0.00001, which is employed to avoid the denominator from becoming zero, and $W(t + 1)$ shows the framework parameters at a given time $(t + 1)$.

4. Experimental Results

This section briefly describes the dataset used to train and evaluate the classification results of the proposed technique for classifying various types of potato plant leaf diseases. In addition, it illustrates used performance measures. Finally, we carried out a huge comparison with various other models to show the effectiveness of our approach.

4.1. Dataset Acquisition

To check the recognition ability of our framework, a standard dataset called the PlantVillage repository [49] is utilized in this work. This data sample is free and available online for model simulation. The PlantVillage dataset is a large collection of plant leaf images with a total of 54,306 images. As the presented approach is associated with classifying plant leaf diseases only in potato crops, only a sample of the mentioned category was used for the performance evaluation. Table 2 demonstrates the list of categories included in the PlantVillage dataset. The reason for nominate this data sample for performance testing is that it comprises samples that vary in mass, structure, size, and orientation of both leaves and infected regions. Moreover, samples suffer from several distortions, including clutter, blur, intensity variations, and color variations. A few samples of this dataset are shown in Figure 3.

Table 2. List of categories included in the PlantVillage dataset without data augmentation.

Class	Images in Dataset	Training Set	Test Set
Healthy Leaves	600	480	120
Early Blight	1200	960	240
Late Blight	1200	960	240
Total	3000	2400	600

4.2. Data Augmentation

We used the PlantVillage dataset to obtain pictures of potato leaf diseases that we used to train, validate, and test the proposed DL model. The collection featured images of late blight, early blight, and healthy potato leaf conditions. The resolution of each image in the group was (256 × 256) pixels. The images of healthy potato leaves portrayed leaves in a normal, healthy state. In contrast, the early and late blight photos illustrated the two stages of shattering potato leaf disease. For the three classes in the dataset, we used the indices 0, 1, and 2. The distribution of the dataset's total number of pictures among its many categories is shown in Table 2. In contrast to images of the other two groups of potato blight, the dataset contained far fewer pictures of healthy potato leaves. The dataset's photos were all randomly chosen to create a training and test set with an 80/20 ratio.

Figure 3. A visual figure of sample dataset.

By randomly picking ten healthy potato leaf photographs and making ten duplicates of each, we increased the quantity of healthy potato leaf images to balance the dataset. This procedure was repeated five times to balance the dataset in terms of photos of healthy potato leaves. Table 3 lists the total number of pictures in each class in the dataset after balancing. Originally, each category had 1200 images for early and late blight and 600 images of healthy potato leaves. After data augmentation, each category had 3600 images for early and late blight and 3600 images of healthy potato leaves.

Table 3. List of categories included in the PlantVillage dataset after data augmentation.

Class	Images in Dataset	Training Set	Test Set
Healthy Leaves	3600	2880	720
Early Blight	3600	2880	720
Late Blight	3600	2880	720
Total	10,800	8640	2160

We normalized the data and increased the size of the training set to train the model and ensure that it would not overfit. The photos were rotated between 20 and +20, sheared between 40 and +40, and moved by width and height within a range of 0.2 for augmentation. Figure 4 displays a visualization of the augmentation process.

Figure 4. Visualized augmented images: (**a**) sheared, (**b**) rotated, (**c**) shifted, (**d**) vertically flipped, (**e**) horizontally flipped, (**f**) height-shifted.

4.3. Performance Metrics

To quantitatively estimate the categorization results of our approach for recognizing diseases of potato plant leaves, we used the standard measures of accuracy, F1 measure, precision (p), and recall (r). The mathematical formulation of the accuracy measure, p, r, and F1 is provided in Equations (8) to (11).

$$Accuracy = \frac{TP + TN}{TP + FP + TN + FN} \tag{8}$$

$$p = \frac{TP}{TP + FP} \tag{9}$$

$$r = \frac{TP}{TP + FN} \tag{10}$$

$$F1 = \frac{2 \times p \times r}{p + r} \tag{11}$$

4.4. Experimental Results

In the first phase of model evaluation, we tested the performance of the proposed strategy in terms of class-wise results to check how much well approach is able to recognize various types of potato plant leaf abnormalities. For this, we measured the performance of our approach using different performance metrics. The results are discussed below. In addition, the experimental results were verified by an expert who is currently working as a Plant Pathologist. Figure 5 shows the results for their training/validation loss and accuracy of the proposed model.

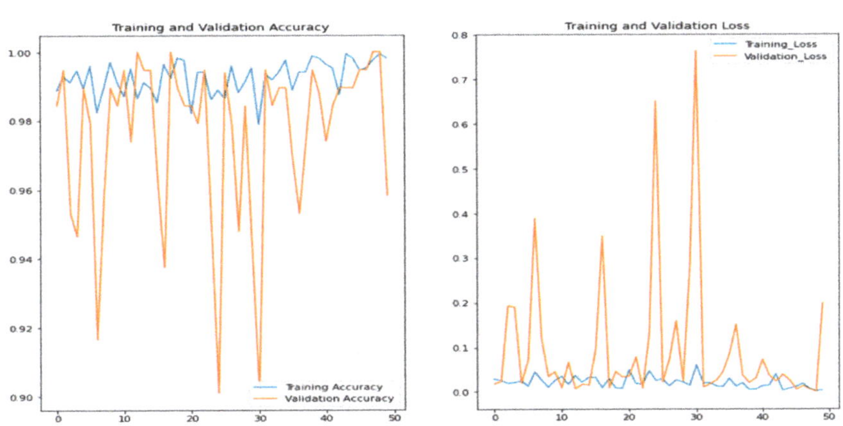

Figure 5. Graph of training and validation accuracy vs. loss for the proposed model.

First, the classification results of this approach are discussed in terms of precision and recall measures, as these are the standard way of elaborating model categorization results. The attained values are provided in Figure 6 for all three classes, showing healthy, early blight, and late blight, respectively. The scores attained in Figure 6 clearly indicate that our approach is able to effectively recognize all three classes in the employed dataset. For the precision metric our approach attained results of 98.26%, 98.03%, and 97.99% for healthy, early blight, and late blight, respectively, while for recall our solution showed values of 97.41%, 97.15%, and 97.10% for healthy, early blight, and late blight, respectively.

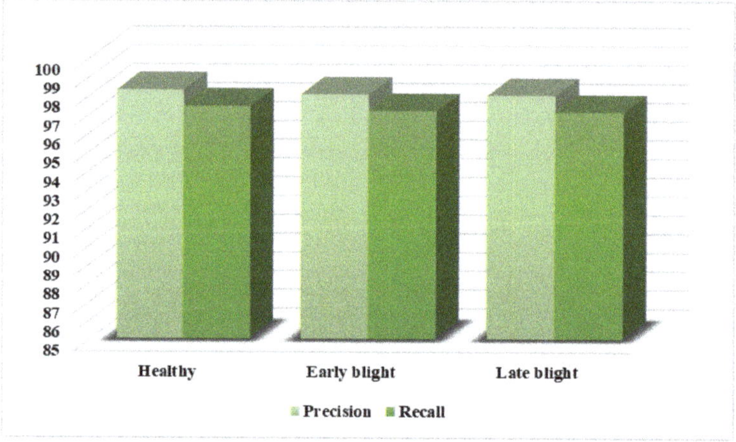

Figure 6. A graphical depiction of the attained precision and recall results.

Next, the model behavior is assessed from the perspective of the F1-score and error rate, as the precision and recall metrics are unable to fully capture the classification behavior of a model. This is because certain approaches are unable to attain a better value of recall for a high value of precision, and vice-versa. Hence, employing the F1-score measure can provide an overall performance assessment of a classification approach by employing both the precisions and recall measures. The attained results for all three classes of the employed dataset are provided in Figure 7. The suggested method reaches an average F1-score value of 97.65%, as depicted in Figure 7. Moreover, we the highest and lowest error scores are 2.46%, and 2.17%, for the late blight and healthy classes of potato plant leaves, respectively.

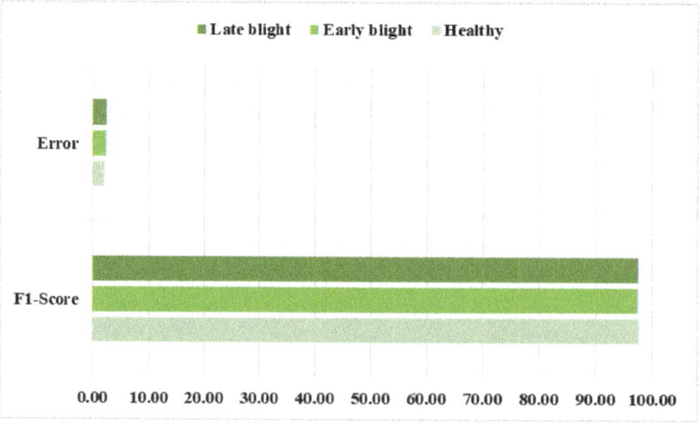

Figure 7. Graphical depiction of attained F1-score and error rate results.

Further, class-wise accuracy value was computed for all three groups of potato plant leaves; the obtained evaluation is shown with the help of box plots in Figure 8. Box plots are proficient in providing a thorough understanding of attained performance results by plotting the maximum, mean, and minimum values. The class-wise accuracy values shown in Figure 8 clearly prove the effectiveness of our approach for categorizing the infected areas of potato plant leaves. More descriptively, for the healthy, early, and late blight classes, the proposed solution acquires average values of 98.24%, 98.11%, and 98.01%, respectively.

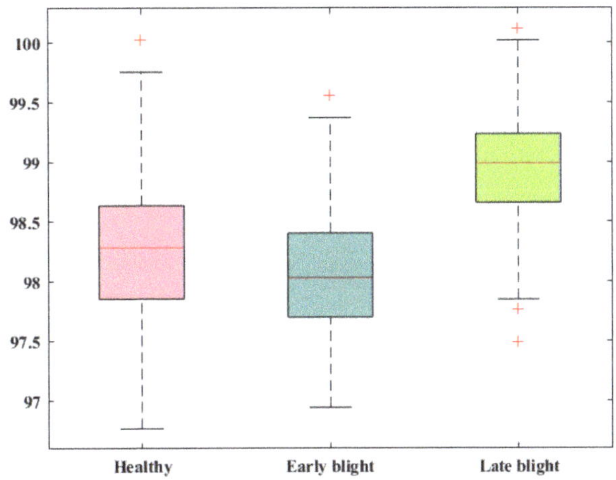

Figure 8. Graphical depiction of attained accuracy results.

Finally, we further depict the class-wise results of our approach by reporting the confusion matrix, which is a powerful plot for showing the recognition ability of a framework by reporting the values in terms of the true positive rate. The confusion matrix for our proposed strategy is shown in Figure 9, demonstrating that our model achieves better values on all three classes of potato plant leaves. Clearly, our approach attains an average TPR of %, which shows its better recall behavior. Moreover, we attain a minimum error of 97.22%, while the highest error rate of 1.61% is reported for the late and healthy blight classes, which could be due to the structural resemblance of the infected regions in these classes. However, both classes are highly differentiable.

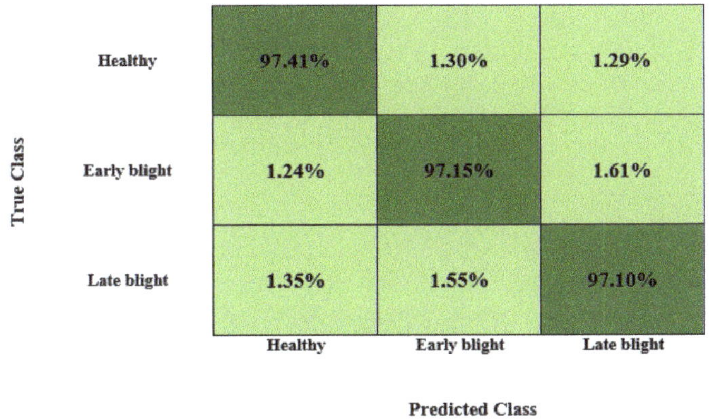

Figure 9. Attained results in the form of a confusion matrix.

All class-wise performance evaluations of the proposed solution with the help of the standard measures confirm the better recognition ability of our approach, which enables it to better classify all the samples in all three classes. The major reason for the improved classification behavior of our approach is due to the relevant and distinctive sample characteristics computation of our approach, which assists and enhances its recall rate and increases its classification performance.

4.5. Comparison with DL Models

In this section of the paper, a comparative analysis of the proposed work with other DL approaches is accomplished to show the efficacy of our work in comparison. For this purpose, a series of well-known DL frameworks, including VGG16 [50], VGG19 [51], MobileNet [52], ResNet50 [53], and DenseNet-101 [54], were nominated. We compared these DL architectures from the perspectives of model structure and performance by comparing the total number of model parameters and accuracy. The results of the evaluation are presented in Table 4. The values clearly depict our approach as being both effective and efficient in comparison to the other DL frameworks. Clearly, the presented work comprises the lowest number of model parameters, with 11 million. Comparatively, the VGG19 model is more expensive in terms of model structure, with a total of 1.96 million parameters. In terms of model accuracy, the lowest performance result is attained by ResNet50, with a score of 73.75%. The second lowest performance score is reached by MobileNet, at 78.84%. The DenseNet approach shows better performance outcomes, with an accuracy value of 93.93%; however, this approach is complex in terms of network structure, with a total of 40 million parameters. In comparison, our approach performs well with an accuracy score of 98.12% and has a total of 11 million model parameters. Clearly, the comparison of these approaches shows an average score of 83.92%, and is 98.12% for our model. Thus, we have achieved a performance gain of 14.20%, that clearly showing the efficacy of our model.

Table 4. Assessment of the suggested approach compared to other DL models.

Sr No.	Model	Parameters (million)	Accuracy (%)
1.	VGG16	138	92.69
2.	VGG19	196	80.39
3.	MobileNet	13	78.84
4.	ResNet50	23	73.75
5.	DenseNet	40	93.93
6.	**Proposed**	**11**	**98.12**

The main cause of these better model classification results is that the other techniques are quite complex in terms of their model structure, which causes issues with model overfitting. Comparatively, our approach is lighter in structure and better able to tackle the overfitting issue. Moreover, our technique adopts the pixel and channel attention approach during the feature computation phase and introduces layers at the end of the network structure, which assists in better nominating the effective set of image characteristics and enhances the cataloguing score. Thus, it can be said that we have presented both an efficient and effective approach to recalling the various groups of potato plant leaf illnesses.

4.6. Proposed Approach in Comparison with the Latest Techniques

Next, we performed another experiment to check the potato plant disease classification results of our model against other new techniques from history. Numerous latest approaches [36,55–57] were nominated for this reason, and performance results in terms of classification results are evaluated. The attained performance comparison is shown in Table 5, from which it is quite clearly confirmed that our model is more robust for classifying the abnormalities of potato plants as compared to the other approaches shown in Figure 10.

Table 5. Comparison of the proposed approach with new methods.

Sr. No	Reference	Accuracy (%)
1.	Chen et al. [55]	97.73
2.	Barman et al. [56]	96.98
3.	Mahum et al. [57]	97.20
4.	Chakraborty et al. [36]	97.89
5.	**Proposed Technique**	**98.12**

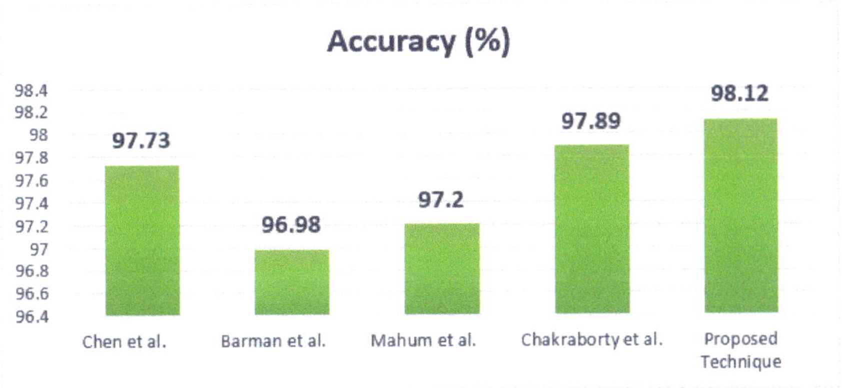

Figure 10. A comparison with the latest works developed by Chen et al. [55], Barman et al. [56], Mahum et al. [57] and Chakraborty et al. [36].

Chen et al. [55] used a DL approach called MobOca_Net to recognize different potato plant leaves by introducing pixel and channel-wise attention units in the base network. This approach attained an accuracy rate of 97.73%. Barman et al. [56] used a self-introduced CNN model to classify various infections found on the leaf areas of the potato crop, and achieved an accuracy of 96.98%. Another model, discussed in [57], used the concept of transfer learning to perform potato plant leaf diseases categorization, and attained a classification score of 97.20%, and the approach in [36] showed an accuracy value of 97.89%.

In comparison with these techniques, the proposed approach attains the highest accuracy rate at 98.12%. The compared techniques exhibit an average accuracy rate of 97.45%, compared to 98.12% for the presented strategy. Consequently, we have provided a performance gain of 0.67% in terms of the accuracy metric. The major cause of this effective performance result is that the approach in [55] is unable to tackle the distorted samples, while the technique in [56] lacks the ability to handle noisy data. On the other hand, the approaches in [36,57] suffer from issues with model overfitting. Comparatively, our approach is better able to handle these issues than existing approaches by presenting an effective model that adopts the pixel and channel AM in the feature computation phase and introduces dense layers at the end of the network structure, which results in nominating a reliable set of sample features even in the presence of various image distortions, thereby enhancing the classification score.

5. Conclusions

Farmers lose money and harvest due to potato plant diseases. Potato leaves are mostly affected by early and late blight. According to estimates, these illnesses are the cause of the majority of yield loss in potatoes. We divided photos of potato leaves into three categories: healthy leaves, late blight leaves, and early blight leaves. To recognize these classes, a solution called EfficientPNet is implemented in this paper. EfficientPNet is a DL approach that classifies various types of potato plant leaves. We improved the EfficientNet-v2 approach by adding the AM strategy and extra layers at the end of the model structure. The presented EfficientPNet approach robustly extracts high-level signs of infected regions and associates them with related groups by employing an end-to-end learning mechanism. In addition, the AM strategy improves the recall power of the proposed solution by passing relevant information on noticeable attributes such as diseased areas of plant leaves. We accomplished rigorous experimentation on a complex data sample designated as PlantVillage to show the effectiveness of our framework, and proved through the attained performance scores that our model is proficient in recognizing potato diseases even from distorted images. As a future goal, we intend to develop another ensemble model by integrating explainable AI [58] and EfficientPNet DL architectures on other challenging datasets.

Author Contributions: Conceptualization, T.N., M.M.I., S.J. and A.H.; Data curation, T.N., M.M.I. and S.J.; Formal analysis, A.H. and M.A.; Funding acquisition, M.A.; Investigation, M.M.I.; Methodology, T.N. and M.A.; Project administration, M.A.; Resources, S.J. and A.H.; Software, S.J., A.H. and M.A.; Supervision, M.A.; Validation, T.N., S.J. and A.H.; Visualization, T.N. and M.A.; Writing—original draft, T.N., S.J., A.H. and M.A.; Writing—review and editing, M.M.I. and S.J. All authors have read and agreed to the published version of the manuscript.

Funding: The authors extend their appreciation to the Deanship of Scientific Research at Imam Mohammad Ibn Saud Islamic University (IMSIU) for funding and supporting this work through Research Partnership Program no. RP-21-07-11.

Institutional Review Board Statement: Not applicable.

Informed Consent Statement: Not applicable.

Data Availability Statement: A standard online dataset PlantVillage [49] is utilized in this paper to evaluate EfficientPNet model. It can be downloaded from https://data.mendeley.com/datasets/tywbtsjrjv/1 [accessed on 12 January 2023].

Acknowledgments: The authors extend their appreciation to the Deanship of Scientific Research at Imam Mohammad Ibn Saud Islamic University (IMSIU) for funding and supporting this work through Research Partnership Program no. RP-21-07-11. The authors would also like to thank our master student (Wasif Ali) for providing us with the code of EfficientPNet model from the Department of Computer Science, University of Engineering and Technology Taxila. This code was provided without data augmentation techniques.

Conflicts of Interest: The authors declare no conflict of interest.

References

1. Bruinsma, J. *The Resource Outlook to 2050: By How Much Do Land, Water and Crop Yields Need To Increase by 2050*; Food and Agriculture Organization of the United Nations: Rome, Italy, 2009; pp. 24–26.
2. Pantazi, X.; Moshou, D.; Tamouridou, A. Automated leaf disease detection in different crop species through image features analysis and One Class Classifiers. *Comput. Electron. Agric.* **2019**, *156*, 96–104. [CrossRef]
3. Wolfenson, K.D.M. *Coping with the Food and Agriculture Challenge: Smallholders' Agenda*; Food Agriculture Organisation of the United Nations: Rome, Italy, 2013.
4. Kumar, A.; Patel, V.K. Classification and identification of disease in potato leaf using hierarchical based deep learning convolutional neural network. *Multimedia Tools Appl.* **2023**, *81*, 1–27. [CrossRef]
5. Elnaggar, S.; Mohamed, A.M.; Bakeer, A.; Osman, T.A. Current status of bacterial wilt (*Ralstonia solanacearum*) disease in major tomato (*Solanum lycopersicum* L.) growing areas in Egypt. *Arch. Agric. Environ. Sci.* **2018**, *3*, 399–406. [CrossRef]
6. Sardogan, M.; Tuncer, A.; Ozen, Y. Plant leaf disease detection and classification based on CNN with LVQ algorithm. In Proceedings of the 2018 3rd International Conference on Computer Science and Engineering (UBMK), Sarajevo, Herzegovina, 20–23 September 2018; pp. 382–385.
7. Sankaran, S.; Mishra, A.; Ehsani, R.; Davis, C. A review of advanced techniques for detecting plant diseases. *Comput. Electron. Agric.* **2010**, *72*, 1–13. [CrossRef]
8. Dinh, H.X.; Singh, D.; Periyannan, S.; Park, R.F.; Pourkheirandish, M. Molecular genetics of leaf rust resistance in wheat and barley. *Theor. Appl. Genet.* **2020**, *133*, 2035–2050. [CrossRef] [PubMed]
9. Ferentinos, K.P. Deep learning models for plant disease detection and diagnosis. *Comput. Electron. Agric.* **2018**, *145*, 311–318. [CrossRef]
10. Patil, S.; Chandavale, A. A survey on methods of plant disease detection. *Int. J. Sci. Res.* **2015**, *4*, 1392–1396.
11. Birgé, L.; Massart, P. Gaussian model selection. *J. Eur. Math. Soc.* **2001**, *3*, 203–268. [CrossRef]

12. Bello-Cerezo, R.; Bianconi, F.; Di Maria, F.; Napoletano, P.; Smeraldi, F. Comparative Evaluation of Hand-Crafted Image Descriptors vs. Off-the-Shelf CNN-Based Features for Colour Texture Classification under Ideal and Realistic Conditions. *Appl. Sci.* **2019**, *9*, 738. [CrossRef]
13. Roska, T.; Chua, L.O. The CNN universal machine: An analogic array computer. *IEEE Trans. Circuits Syst. II Analog. Digit. Signal Process.* **1993**, *40*, 163–173. [CrossRef]
14. Zaremba, W.; Sutskever, I.; Vinyals, O. Recurrent neural network regularization. *arXiv* **2014**, arXiv:1409.2329.
15. Salakhutdinov, R.; Hinton, G. Deep Boltzmann Machines. Available online: https://www.utstat.toronto.edu/~rsalakhu/papers/dbm.pdf (accessed on 8 January 2023).
16. Szegedy, C.; Liu, W.; Jia, Y.; Sermanet, P.; Reed, S.; Anguelov, D.; Erhan, D.; Vanhoucke, V.; Rabinovich, A.; Liu, W.; et al. Going deeper with convolutions. In Proceedings of the 2015 IEEE Conference on Computer Vision and Pattern Recognition (CVPR), Boston, MA, USA, 7–12 June 2015; pp. 1–9.
17. Yuan, Z.W.; Zhang, J. Feature extraction and image retrieval based on AlexNet. In Proceedings of the Eighth International Conference on Digital Image Processing (ICDIP 2016), Chengu, China, 20–22 May 2016; p. 100330E.
18. Vedaldi, A.; Zisserman, A. *Vgg Convolutional Neural Networks Practical*; Department of Engineering Science, University of Oxford: Oxford, UK, 2016; Volume 2016, p. 66.
19. Thenmozhi, K.; Reddy, U.S. Crop pest classification based on deep convolutional neural network and transfer learning. *Comput. Electron. Agric.* **2019**, *164*, 104906. [CrossRef]
20. Paul, A.; Ghosh, S.; Das, A.K.; Goswami, S.; Choudhury, S.D.; Sen, S. A review on agricultural advancement based on computer vision and machine learning. In *Emerging Technology in Modelling and Graphics*; Springer: Berlin/Heidelberg, Germany, 2020; pp. 567–581.
21. Olawuyi, O.; Viriri, S. *Plant Diseases Detection and Classification Using Deep Transfer Learning*; Springer: Berlin/Heidelberg, Germany, 2023; pp. 270–288.
22. Bhagat, M.; Kumar, D. Efficient feature selection using BoWs and SURF method for leaf disease identification. *Multimed. Tools Appl.* **2023**, 1–25. [CrossRef]
23. Pal, A.; Kumar, V. AgriDet: Plant Leaf Disease severity classification using agriculture detection framework. *Eng. Appl. Artif. Intell.* **2023**, *119*, 105754. [CrossRef]
24. Jiang, P.; Chen, Y.; Liu, B.; He, D.; Liang, C. Real-Time Detection of Apple Leaf Diseases Using Deep Learning Approach Based on Improved Convolutional Neural Networks. *IEEE Access* **2019**, *7*, 59069–59080. [CrossRef]
25. Li, L.; Zhang, S.; Wang, B. Plant Disease Detection and Classification by Deep Learning—A Review. *IEEE Access* **2021**, *9*, 56683–56698. [CrossRef]
26. Xian, T.S.; Ngadiran, R. Plant Diseases Classification using Machine Learning. *J. Physics Conf. Ser.* **2021**, *1962*, 012024. [CrossRef]
27. Kang, F.; Li, J.; Wang, C.; Wang, F. A Lightweight Neural Network-Based Method for Identifying Early-Blight and Late-Blight Leaves of Potato. *Appl. Sci.* **2023**, *13*, 1487. [CrossRef]
28. Kumar, S.; Shukla, A. Automatic Grading of Potato Leaf using Machine learning & Computer Vision. 2022. Available online: https://assets.researchsquare.com/files/rs-2102065/v1/bf9e85cf-18fe-4287-a82d-16c252e06b4a.pdf?c=1667213983 (accessed on 1 January 2023).
29. Min, B.; Kim, T.; Shin, D.; Shin, D. Data Augmentation Method for Plant Leaf Disease Recognition. *Appl. Sci.* **2023**, *13*, 1465. [CrossRef]
30. Rashid, J.; Khan, I.; Ali, G.; Almotiri, S.H.; AlGhamdi, M.A.; Masood, K. Multi-Level Deep Learning Model for Potato Leaf Disease Recognition. *Electronics* **2021**, *10*, 2064. [CrossRef]
31. Tiwari, D.; Ashish, M.; Gangwar, N.; Sharma, A.; Patel, S.; Bhardwaj, S. Potato leaf diseases detection using deep learning. In Proceedings of the 2020 4th International Conference on Intelligent Computing and Control Systems (ICICCS), Madurai, India, 13–15 May 2020; pp. 461–466.
32. Lee, T.-Y.; Yu, J.-Y.; Chang, Y.-C.; Yang, J.-M. Health detection for potato leaf with convolutional neural network. In Proceedings of the 2020 Indo–Taiwan 2nd International Conference on Computing, Analytics and Networks (Indo-Taiwan ICAN), Rajpura, India, 14–15 February 2020; pp. 289–293.
33. Asif, M.K.R.; Rahman, M.A.; Hena, M.H. CNN based disease detection approach on potato leaves. In Proceedings of the 2020 3rd International Conference on Intelligent Sustainable Systems (ICISS), Thoothukudi, India, 3–5 December 2020; pp. 428–432.
34. Iqbal, M.A.; Talukder, K.H. Detection of potato disease using image segmentation and machine learning. In Proceedings of the 2020 International Conference on Wireless Communications Signal Processing and Networking (WiSPNET), Chennai, India, 4–6 August 2020; pp. 43–47.
35. Chen, J.; Deng, X.; Wen, Y.; Chen, W.; Zeb, A.; Zhang, D. Weakly-supervised learning method for the recognition of potato leaf diseases. *Artif. Intell. Rev.* **2022**, 1–18. [CrossRef]
36. Chakraborty, K.K.; Mukherjee, R.; Chakroborty, C.; Bora, K. Automated recognition of optical image based potato leaf blight diseases using deep learning. *Physiol. Mol. Plant Pathol.* **2022**, *117*, 101781. [CrossRef]
37. Ngugi, L.C.; Abelwahab, M.; Abo-Zahhad, M. Recent advances in image processing techniques for automated leaf pest and disease recognition—A review. *Inf. Process. Agric.* **2020**, *8*, 27–51. [CrossRef]
38. Tan, M.; Le, Q. Efficientnet: Rethinking model scaling for convolutional neural networks. In Proceedings of the International conference on machine learning, Long Beach, CA, USA, 10–15 June 2019; pp. 6105–6114.

39. Krizhevsky, A.; Sutskever, I.; Hinton, G. Imagenet classification with deep convolutional neural networks. *Adv. Neural Inf. Process. Syst.* **2012**, *25*, 1097–1105. [CrossRef]
40. Huang, G.; Liu, Z.; Van Der Maaten, L.; Weinberger, K.Q. Densely connected convolutional networks. In Proceedings of the IEEE Conference on Computer Vision and Pattern Recognition, Honolulu, HI, USA, 21–27 July 2017; pp. 4700–4708.
41. He, K.; Zhang, X.; Ren, S.; Sun, J. Deep residual learning for image recognition. In Proceedings of the IEEE Computer Society Conference on Computer Vision and Pattern Recognition (CVPR), Las Vegas, NV, USA, 27–30 June 2016; pp. 770–778.
42. Howard, A.G.; Zhu, M.; Chen, B.; Kalenichenko, D.; Wang, W.; Weyand, T.; Andreetto, M.; Adam, H. Mobilenets: Efficient convolutional neural networks for mobile vision applications. *arXiv* **2017**, arXiv:1704.04861.
43. Gupta, S.; Akin, B. Accelerator-aware neural network design using automl. *arXiv* **2020**, arXiv:2003.02838.
44. Sandler, M.; Howard, A.; Zhu, M.; Zhmoginov, A.; Chen, L. MobileNetV2: Inverted residuals and linear bottlenecks. In Proceedings of the IEEE/CVF Conference on Computer Vision and Pattern Recognition (CVPR), Salt Lake City, UT, USA, 18–23 June 2018; pp. 4510–4520.
45. Tan, M.; Le, Q.V. Efficientnetv2: Smaller models and faster training. *arXiv* **2021**, arXiv:2104.00298.
46. Hu, J.; Shen, L.; Sun, G. Squeeze-and-Excitation Networks. In Proceedings of the IEEE Conference on Computer Vision and Pattern Recognition (CVPR), Salt Lake City, UT, USA, 18–23 June 2018; pp. 7132–7141.
47. Ramachandran, P.; Zoph, B.; Le, Q.V. Searching for activation functions. *arXiv* **2017**, arXiv:1710.05941.
48. Khaire, U.M.; Dhanalakshmi, R. High-dimensional microarray dataset classification using an improved adam optimizer (iAdam). *J. Ambient. Intell. Humaniz. Comput.* **2020**, *11*, 5187–5204. [CrossRef]
49. Hughes, D.; Salathé, M. An open access repository of images on plant health to enable the development of mobile disease diagnostics. *arXiv* **2015**, arXiv:1511.08060.
50. Nawaz, M.; Masood, M.; Javed, A.; Iqbal, J.; Nazir, T.; Mehmood, A.; Ashraf, R. Melanoma localization and classification through faster region-based convolutional neural network and SVM. *Multimedia Tools Appl.* **2021**, *80*, 28953–28974. [CrossRef]
51. Carvalho, T.; De Rezende, E.R.; Alves, M.T.; Balieiro, F.K.; Sovat, R.B. Exposing computer generated images by eye's region classification via transfer learning of VGG19 CNN. In Proceedings of the 2017 16th IEEE international conference on machine learning and applications (ICMLA), Cancun, Mexico, 18–21 December 2017; pp. 866–870.
52. Qin, Z.; Zhang, Z.; Chen, X.; Wang, C.; Peng, Y. Fd-Mobilenet: Improved Mobilenet with a Fast Downsampling Strategy. In Proceedings of the 2018 25th IEEE International Conference on Image Processing (ICIP), Athens, Greece, 7–10 October 2018; pp. 1363–1367.
53. Nawaz, M.; Javed, A.; Irtaza, A. ResNet-Swish-Dense54: A deep learning approach for deepfakes detection. *Vis. Comput.* **2022**, 1–22. [CrossRef]
54. Albahli, S.; Nawaz, M. DCNet: DenseNet-77-based CornerNet model for the tomato plant leaf disease detection and classification. *Front. Plant Sci.* **2022**, *13*, 1–22. [CrossRef]
55. Chen, W.; Chen, J.; Zeb, A.; Yang, S.; Zhang, D. Mobile convolution neural network for the recognition of potato leaf disease images. *Multimedia Tools Appl.* **2022**, *81*, 20797–20816. [CrossRef]
56. Barman, U.; Sahu, D.; Barman, G.G.; Das, J. Comparative assessment of deep learning to detect the leaf diseases of potato based on data augmentation. In Proceedings of the 2020 International Conference on Computational Performance Evaluation (ComPE), Shillong, India, 2–4 July 2020; pp. 682–687.
57. Mahum, R.; Munir, H.; Mughal, Z.-U.; Awais, M.; Khan, F.S.; Saqlain, M.; Mahamad, S.; Tlili, I. A novel framework for potato leaf disease detection using an efficient deep learning model. *Hum. Ecol. Risk Assessment Int. J.* **2022**, *29*, 303–326. [CrossRef]
58. Ullah, F.; Moon, J.; Naeem, H.; Jabbar, S. Explainable artificial intelligence approach in combating real-time surveillance of COVID19 pandemic from CT scan and X-ray images using ensemble model. *J. Supercomput.* **2022**, *78*, 19246–19271. [CrossRef] [PubMed]

agriculture

MDPI

Article

Agent-Based Modelling to Improve Beef Production from Dairy Cattle: Young Beef Production

Addisu H. Addis [1,2,*], Hugh T. Blair [1], Paul R. Kenyon [1], Stephen T. Morris [1], Nicola M. Schreurs [1] and Dorian J. Garrick [3]

[1] Animal Science, School of Agriculture and Environment, Massey University, Palmerston North 4442, New Zealand
[2] Applied Biology, College of Natural and Computational Sciences, University of Gondar, Gondar P.O. Box 196, Ethiopia
[3] AL Rae Centre for Genetics and Breeding, Massey University, Hamilton 3214, New Zealand
* Correspondence: a.hailu@massey.ac.nz

Abstract: Approximately 42% of the total calves born in New Zealand's dairy industry are either euthanized on farms or commercially slaughtered as so-called bobby calves within 2 weeks of age. These practices have perceived ethical issues and are considered a waste of resources because these calves could be grown on and processed for beef. Young beef cattle harvested between 8 and 12 months of age would represent a new class of beef production for New Zealand and would allow for a greater number of calves to be utilized for beef production, reducing bobby calf numbers in New Zealand. However, the acceptance of such a system in competition with existing sheep and beef cattle production systems is unknown. Therefore, the current study employed an agent-based model (ABM) developed for dairy-origin beef cattle production systems to understand price levers that might influence the acceptance of young beef production systems on sheep and beef cattle farms in New Zealand. The agents of the model were the rearer, finisher, and processor. Rearers bought in 4-days old dairy-origin calves and weaned them at approximately 100 kg live weight before selling them to finishers. Finishers managed the young beef cattle until they were between 8 and 12 months of age in contrast to 20 to 30 months for traditional beef cattle. Processing young beef cattle in existing beef production systems without any price premium only led to an additional 5% of cattle being utilized compared to the traditional beef cattle production system in New Zealand. This increased another 2% when both weaner cattle and young beef were sold at a price premium of 10%. In this scenario, Holstein Friesian young bull contributed more than 65% of total young beef cattle. Further premium prices for young beef cattle production systems increased the proportion of young beef cattle (mainly as young bull beef), however, there was a decrease in the total number of dairy-origin cattle processed, for the given feed supply, compared to the 10% premium price. Further studies are required to identify price levers and other alternative young beef production systems to increase the number of young beef cattle as well the total number of dairy-origin beef cattle for beef on sheep and beef cattle farms. Some potential options for investigation are meat quality, retailer and consumer perspectives, and whether dairy farmers may have to pay calf rearers to utilize calves with lower growth potential.

Keywords: agent-based model; dairy cattle; young beef; price lever

1. Introduction

Dairy-origin calves contribute significantly to beef finishing systems [1–4], accounting for more than 58% of the beef cattle finished annually on sheep and beef cattle farms in New Zealand [4,5], and over 60, 80 and 87% total beef processed from Ireland [6], Finland [7], and Russia [8] respectively. Holstein–Friesian (33.1%), Jersey (8.6%), and Holstein–Friesian–Jersey crosses (48.5%) represent most dairy cows in New Zealand [9,10].

From these cows, approximately 20% of calves born on New Zealand dairy farms are beef–dairy cross-bred, which are subsequently finished for beef on sheep and beef cattle farms [11]. Early-born and heavier beef-dairy cross calves are preferred for finishing as prime heifer or steer as they grow faster and attain better conformation than their dairy breed counterparts [12–14]. Well-marked, and therefore supposedly predominantly Holstein–Friesian bull calves, are favored in New Zealand for bull-beef production [15–19].

Dairy-origin calves which are not required for dairy heifer replacements nor beef finishing have traditionally been disposed of as bobby calves [20,21]. These include calves born to Jersey cows, calves from cows not suitable for breeding dairy herd replacements, calves born to first-calving heifers, or calves born to late-calving cows. Bobby calves in New Zealand are defined as calves that are commercially slaughtered within 2 weeks of age [21–23]. In 2020, New Zealand processed approximately 1.9 million bobby calves from the dairy industry [2]. Commercial slaughtering of excess calves from the dairy industry is also common in EU countries [24] and Australia [14]. Transporting and slaughtering these calves is fraught with welfare and ethical issues which can be considered a potential threat to New Zealand dairy and beef trading in the form of non-tariff barriers [23,25,26]. Furthermore, there are concerns due to there being a high prevalence of E.coli with the processing of calves [27], and concerns of dehydration in bobby calves prior to processing [20]. Bobby calf production is also considered a waste of animal resources, as these animals could be utilized for beef production if slaughtered at an older age [20,21,28]. To provide options for the utilization of surplus calves born on dairy farms, systems of young beef cattle production have been proposed to increase the number of dairy-origin cattle finished for beef [28–32] while accounting for a fixed quantity of grazing land.

Some studies on carcass and meat qualities of young beef [29,31,32] and profitability and pasture utilization of young beef cattle production at the farm level [28,30] have been conducted in New Zealand. However, as a new potential class of beef finishing, its acceptance level in the existing New Zealand beef cattle finishing system is unknown. Therefore, this study utilized Agent-Based Modelling (ABM) to represent young beef cattle production systems that would finish dairy-origin calves for beef before their first winter (i.e., 8 to 12 months of age) in a New Zealand context. Agent-based modeling allows for repetitive and competitive interactions between agents which enables the exploration of dynamics over time and captures the adaptive and emergent phenomenon from the interaction [33–36]. A base ABM model for dairy-origin beef finishing using rearer, finisher, and processor agents, accounting for the specifics of dairy-origin beef cattle has been developed [37]. This was modified in the current study to understand the influence of cattle sale prices on the uptake of young beef cattle on sheep and beef cattle farms. The present study modeled unselected dairy-origin calves as identified in the previous study [37] for beef production slaughtered at either 8, 10, or 12 months with a weaner cattle and manufacturing beef price of NZ$4.50 per kg carcass. It was hypothesized that an increase in price would increase the number of calves selected for young beef cattle finishing systems, enabling a greater number of dairy-origin beef cattle to be finished for a given feed supply. Premiums of 10 or 20% both for weaner cattle and young beef were modeled in comparison to the current calf price and also in comparison to a scenario where rearers were provided with calves for free from dairy farms which is a possible scenario if there was a mandate for calves from dairy farms to be reared.

2. Materials and Methods

2.1. Agent-Based Modeling Development

In New Zealand, sheep and beef cattle finishing farms' pasture provides up to 95% of the diet [17,38,39]. In dairy-origin beef cattle finishing, calf producers, rearers, finishers, and processors influence each other in determining the type and number of cattle that move along the supply chain from the dairy industry to the beef industry. The interactions between rearers, finishers, and processors for weaning and finishing dairy-origin beef cattle on New Zealand beef cattle and sheep farms were modeled using "Agents.jl" [37,40] which is a Julia framework for ABM. A base ABM model for dairy-origin beef cattle finishing had previously been developed [37] and is briefly described below.

2.2. The Base Model

The number of 4-day-old, spring-born calves, available on a daily basis was assumed to follow a Poisson distribution based on the date of birth [41] over a three-month, Spring-calving period. A multivariate, normal distribution function applied to the Cholesky decomposition of the assumed variance-covariance matrix [42] was employed to simulate a positively correlated birth weight, growth rate, and price for each calf [43]. Calves with a likely higher marginal return (due to being heavier and likely to be faster growing, i.e., Holstein–Friesians and Holstein–Friesian–Jersey crossbreds were finished via the existing beef cattle finishing systems on sheep and beef cattle farms, and the remainder were processed as bobby calves.

The rearer, finisher, and processing agents simultaneously and repetitively, interacted with each other to determine the number and type of dairy-origin cattle moving along the supply chain. Rearers preferentially brought 4-day-old calves that were heavier and had the potential for faster growth and managed them until weaning at approximately 100 kg live weight, before on-selling to a finisher. If the rearing capability of the rearers (i.e., the number of calves they could successfully rear) was higher than the demand for weaned calves by finishers, they subsequently reduced their rearing capability to balance the demand for weaners by the finishers. Finishers primarily bought weaners from rearers, however, if the weaner supply from rearers was insufficient relative to their finishing capability, they sourced more weaners directly from dairy farms. Increased demand for weaner calves encouraged dairy farmers to rear more calves along with their own replacement heifer calves [44].

2.3. Agent-Based Modelling for Young-Beef Cattle Production: Price Levers on Adoption

The model was parameterized with 45,000 spring-born calves, representing 1% of the total calves produced annually on dairy farms in New Zealand, the same as was used in the previously published base model for traditional dairy-origin beef cattle finishing systems [37]. Unlike the base model, the current study fitted the growth curve of beef cattle for young and traditional beef production using seasonally adjusted von Bertalanffy growth equations [45]. This allows animals to grow faster during spring and gain less live weight change during winter to match the feed supply in pasture-based systems [28].

Unselected calves, which were comparatively slower-growing and lighter, were modeled to determine whether they could be finished at the ages of 8, 10, or 12 months for young beef. In New Zealand, pasture supply is typically highest in spring and lowest in winter. Young beef cattle finishing would allow finishers to start in Spring with a higher number of beef cattle including traditional and young beef cattle and then progressively harvest animals prior to the winter from 8 months of age to ensure feed demand equaled feed supply as pasture growth declines over the winter period. Slaughter started in May, with those approximately 8-month-old young beef cattle that were heaviest, and was completed with the harvesting of young cattle at approximately 12 months of age by August, freeing up pasture demand for the next crop of animals.

Energy requirements for maintenance and live weight gain for traditional and young beef cattle were estimated using values from [46,47]. Calves were sold from dairy farms to rearers at 4-days of age from NZ$70 to 120 per head and weaner cattle were sold to finishers from NZ$3.00 to 4.50 per kg live weight, based on values reported in [43,44] (Table 1). Carcass weights from 8–10-month old beef cattle were estimated as 0.48 times the live weights, increasing to 0.5 times the live weight for 12-month-old beef cattle [29–32]. Young beef was valued at the manufacturing beef price of NZ$4.50 per kg carcass weight [48] plus an additional 10% or 20% premiums applied for both weaner cattle and young beef. Each scenario was also simulated with the calf sale price from the dairy farm set to zero, reflecting dairy farmers giving excess calves to rearers for free which simulates a scenario where mandates for no calf wastage are in place and all excess calves need to be directed to beef production. Calves were provided to commercial rearers to avoid dairy farms bearing the cost burden of additional calf rearing. A total of 30 ABM simulations for each price scenario were conducted [37].

Table 1. Birth weight, minimum weight at slaughter, slaughter age, price per head 4-day old calves, and price per kg live weight weaner cattle parameters of various classes of dairy-origin beef cattle for traditional beef cattle finishing.

Attributes	Holstein–Friesian			Holstein–Friesian–Jersey			[a] Jersey			References
	Heifer	Steer	Bull	Heifer	Steer	Bull	Heifer	Steer	Bull	
[b] Birth weight, kg	36.1	38.2	38.2	31.7	33.9	33.9	27.6	29.8	29.8	[13]
[c] Minimum weight at slaughter (kg)	500	-	550	500	580	550	500	580	550	[49–55]
[d] Adjusted average age at slaughter (d)	610	-	600	679 *	896	805	700 *	920	880 *	[50–52,56]
4-day-old calf price/head (NZ$)	90	-	110	80	100	100	70	90	90	[57,58]
Weaner price/kg live weight (NZ$)	3.70	-	4.50	3.60	3.70	4.00	3.00	3.20	3.20	[57,58]
	Beef–Holstein–Friesian cross			Beef–Holstein–Friesian–Jersey cross			Beef–Jersey cross			
[a] Birth weight, kg	38.3	40.2	40.2	37 *	39 *	39 *	35 *	37 *	37 *	[43,53]
[c] Minimum weight at slaughter (kg)	500	580	550	500	580	550	500	580	550	[49–55]
[d] Adjusted average age at slaughter (d)	561	663	625	579 *	689	640	600 *	750 *	703 *	[50,59]
4-day-old calf price/head (NZ$)	95	120	120	90	110	110	75	95	95	[57,58]
Weaner price/kg live weight (NZ$)	3.90	4.00	4.70	3.60	3.70	4.00	3.00	3.20	3.20	[57,58]

[a] includes the "other breed" category; [b] Male calves' birth weight was 2.2 kg heavier [13]; [c] minimum slaughter weight for young beef cattle was 250 kg; [d] young beef cattle were slaughtered at ages of either 8, 10, or 12 months. * estimated based on the value of other classes and breeds; heifers', steers', and bulls' carcass weights from traditional beef cattle were estimated as 50, 54, and 52% of live weight, respectively [41,42,45].

3. Results

Allowing for the harvest of young beef cattle in the existing beef production systems, without any price premium, led to the finishing of an additional 5% of cattle compared to the traditional beef cattle production system only (Figure 1: existing calf price at 0% premium scenario vs. base model scenario). Of the total beef cattle finished for young beef, young bull beef cattle contributed 79% followed by young steers (12%) and heifers (9%). This modified farming system meets the feed demands of the young cattle by farming 20% fewer traditional beef cattle in total, with the most pronounced reduction being in the numbers of traditional heifer cattle finished for beef (Figure 1: 44% lower than that in the base model).

A price premium of 10% for weaner cattle and for the beef schedule price (Figure 1: existing calf price at 10% premium scenario) resulted in an additional 7% of calves of dairy origin being utilized for beef production compared to the traditional beef cattle finishing system (Figure 1: base model scenario). Further premium prices for young weaner cattle and beef (i.e., 20%), and the provision of free calves from the dairy farm to the rearer, increased the proportion of cattle used for young beef production, in particular, young bull beef cattle. However, in order to source pasture for the young cattle, the system decreased the total number of dairy-origin calves utilized compared to the 10% premium scenario (Figure 1).

Calves born from Holstein–Friesian dams accounted for 60% of total harvested cattle (young and traditional beef cattle). However, calves from Jersey cows contributed less than 2% of total dairy-origin beef breed cattle processed in this study (Table 2). Young bull beef cattle from Holstein–Friesian dams contributed approximately 65% of total young beef cattle processed when a 10% increased price scenario was utilized for weaner cattle and schedule prices for young beef.

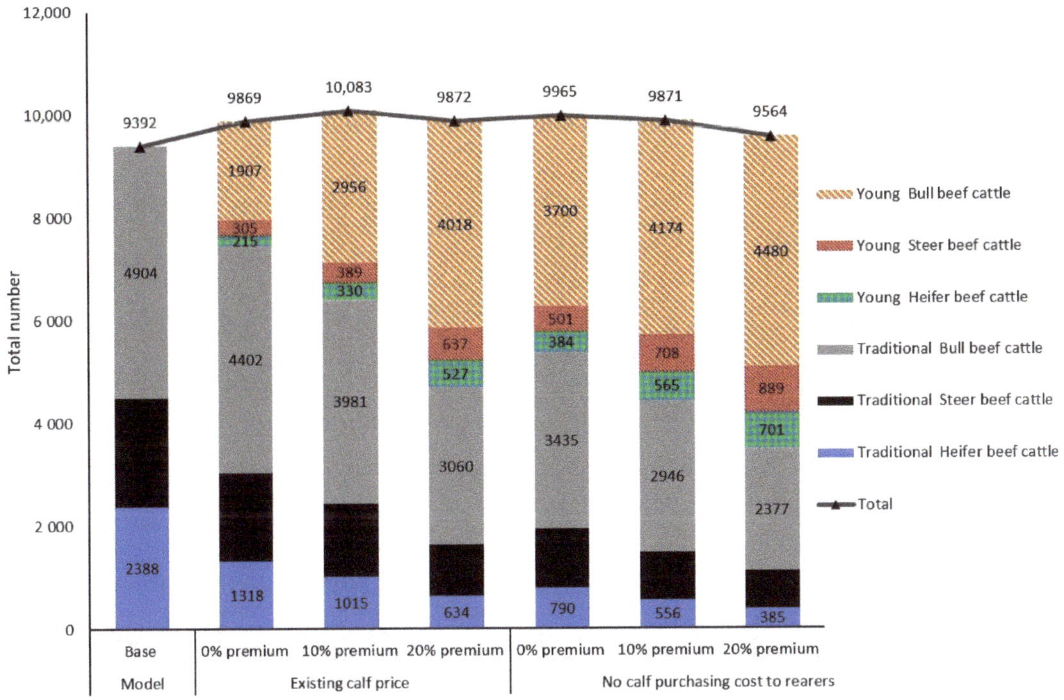

Figure 1. The mean number of traditional beef cattle (i.e., finished 20 to 30 months old) and young beef cattle sold at NZ$4.50 per kg carcass and premiums of 10% and 20% for both weaner cattle and young beef (i.e., finished at 8, 10 or 12-months old) with or without the current calf price for 45,000 modeled calves across 30 ABM simulation runs. 0%, 10%, and 20% premiums were for both weaner cattle and young beef values.

Table 2. Mean number (sd) of traditional beef cattle (finished 20 to 30 months old) and young beef cattle (finished at 8, 10, or 12 months old) at a 10% premium scenario (an existing calf price at 10% premium for both weaner and young beef) per sex and dam breed out of 45,000 modeled calves across 30 ABM simulation runs.

	Heifer				Steer				Bull			
	HF	HJ	Jr	Other	HF	HJ	Jr	Other	HF	HJ	Jr	Other
Traditional beef cattle	619 (43)	351 (92)	26 (5)	19 (8)	-	1308 (169)	60 (10)	44 (12)	2869 (57)	1046 (103)	33 (4)	33 (10)
Young beef cattle	228 (15)	101 (11)	1 (1)	0	-	371 (17)	16 (4)	2 (1)	2378 (101)	552 (61)	12 (5)	13 (4)

HF: Holstein–Friesian; HJ: Holstein–Friesian–Jersey cross, Jr: Jersey cattle, and others: other dairy breed cattle.

4. Discussion

Dairy-beef animals harvested at a young age (8 to 12 months) represent a new beef production system being considered in New Zealand. It aims to finish as many calves as possible, to reduce the bobby calf slaughter. Understanding the level of acceptance of these systems by existing sheep and beef cattle production farmers who are used to finishing animals at older ages and heavier weights, and recognition of the main constraints associated with their use, would allow farmers and processors to make informed decisions regarding the utility of such systems. The current study simulated the use of dairy-origin heifer, steer, and bull beef cattle slaughtered from 8 to 12 months of age using a previously reported ABM model [37]. It utilized historical average weaner and manufacture beef prices and 10% and 20% premiums on these prices by utilizing current sale prices for calves sold to rearers. The provision of calves to the rearer at no cost was also considered.

Young beef cattle are slaughtered at lighter weights compared to the traditional classes of cattle used for beef production and so, under the current carcass classification and payment system used for beef in New Zealand, the carcasses would be categorized into a manufacturing price of NZ$4.50 per kg carcass weight [3,22]. At that price and with a 4-day-old calf purchasing cost included, a mix of young and traditional beef cattle finishing systems processed an extra 5% of cattle compared to the traditional beef production system (i.e., without young beef cattle). A farm optimization study [28] identified that including young beef cattle in the existing beef finishing system and processing them at NZ$4.50 per kg, the carcass would enable the processing of 5% more beef cattle per farm than the traditional beef cattle finishing system. These relatively low percentage increases in the numbers of cattle indicate that young beef cattle would need to be incentivized to increase the uptake of dairy-origin calves into beef production systems. One such incentive is the price obtained for the carcass. A study conducted by [30] identified that young steers slaughtered at 8 to 12 months would require more than NZ$6.00 to break even with a traditional bull finishing system. This 33% increase in price would require a major change by the processing companies and would be unlikely to happen unless new high-value markets were identified.

Young beef is more tender due to being finished at an earlier age than beef from traditional beef cattle [29] which might enable a premium price over either traditional beef (heifer or steer beef) or processed beef (bull beef). This requires the identification of high-value markets and selling valuable cuts at a higher price to lift the value of the whole carcass. This would increase the profitability of young beef cattle production [60–62] and make the system more attractive for rearers and finishers thereby enabling the processing of more dairy-origin beef cattle for beef to reduce bobby calves numbers.

A 10% increase in the sale price per kg live weight for weaners and per carcass weight for young beef resulted in young beef cattle contributing 7% of the total dairy-origin beef cattle processed. In this scenario, Holstein Friesian young bulls contributed more than 65% of the total young beef cattle. This could be due to Holstein Friesian bull beef cattle across both systems growing faster compared to other sexes and classes [48,63]. Further, a premium of 20% both for weaner cattle and processed young-beef cattle increased the proportion of young beef cattle and the uptake of young bull beef cattle. However, this decreased the total number of cattle processed for beef compared with the 10% premium scenario. This is likely explained by a greater per-head feed requirement for bulls to achieve the target weight which decreases the total number of cattle processed for beef for the given feed supply. Thus, alternative finishing systems that would increase the uptake of slower-growing beef cattle classes would be required to increase the uptake of dairy-origin beef cattle for traditional as well young beef cattle. Alternatively, assisted reproduction techniques or better selection of beef bulls for mating with dairy cows could be used to produce better quality calves from dairy cows [64] which would increase growth rates and on-farm efficiency and the quality of meat from dairy-origin calves [65,66].

The current carcass weight payment system encourages the harvest of those fast-growing cattle that can achieve the highest carcass weights. As more fast-growing young beef cattle entered the modeled system, the per-head demand increased and thus a smaller number of traditional beef cattle would be farmed for the given feed supply. Considering value-based market beef production for slower-growing cattle including Jersey calves would make them more competitive and allow a higher number of beef cattle for the given feed resource. A study by [67] identified that beef-Jersey cross-breed cattle had higher marbling scores which increased carcass value resulting in higher value carcasses than from pure Jersey cattle.

Dairy-origin beef cattle produce 29% less greenhouse gas emissions (GHG) compared to cow-calf beef production system per kg carcass [4]. This is due to them being a byproduct of cows that provide milk for sale and calves for beef production for the same amount of dry matter consumed. Further, at a younger age (less than 12 months of age), growth is faster and there is less fat in the gain compared to older cattle [63], which would make younger cattle more efficient in terms of feed converting to saleable product [68]. Young beef cattle increased pasture utilization efficiency and reduced silage preparation and utilization [28]. It also processed higher gross carcass output per farm for a given feed supply, which meant they produced lower GHG emissions per kg carcass [68]. This implies young dairy-origin beef cattle production should be considered as a mitigation strategy to reduce GHG emissions from livestock production [69]. Finishing dairy-origin cattle for beef at a young age would have also less impact on the soil compared to heavier animals in wet seasons [4,70]. These positive environmental impacts might allow young beef to attract a higher per kg carcass value driven by consumer demand which would potentially increase the uptake of young beef cattle systems. Identifying markets that would pay extra for ethical and welfare-friendly beef and/or reducing calf selling price at 4-day old would also allow young beef to earn higher value per kg live/carcass weight [71,72].

The current study did not include beef retailer or consumer perspectives. Consumer perspectives associated with meat quality (meat color, tenderness, juiciness, flavor) and extrinsic characteristics (brand, price, labeling, package, and outlet) [73–76] and origin of beef are important factors in determining breed, sex, and class of beef cattle required for beef production and the likely premium that could be achieved for the meat product [77,78]. Considering these parameters in future studies would allow the model to provide full insight into the uses of young beef cattle in New Zealand and other countries where bobby calf production needs to be discontinued.

5. Conclusions

Young beef cattle production systems represent a new beef cattle finishing option in New Zealand, aiming to process a greater number of dairy cattle for beef. This would reduce bobby calf numbers and the associated potential ethical issues and could increase the profitability of both the beef and dairy industries. Utilizing young beef cattle production systems enabled a greater number of beef cattle to be managed on-farm and greater throughput of beef cattle from weaning to slaughter per hectare for a given feed supply. Processing young beef cattle at NZ$4.50 per kg carcass with a 10% premium allowed the system to finish only 5% and 7%, respectively, greater numbers of beef cattle compared to the traditional beef cattle production system. This small increase in uptake suggests further research is required before firm conclusions on the uses of young beef cattle can be made. Some examples include higher premium prices, lower 4-day-old calf costs, alternative finishing systems which encourage the uptake of slower-growing dairy-origin beef cattle, meat quality traits, and retailer and consumer perspectives. Given the minimal use of young beef production of dairy origin in New Zealand, agent-based modeling is a useful tool to examine the efficacy of these options.

Author Contributions: A.H.A., D.J.G., H.T.B., S.T.M., P.R.K. and N.M.S. contributed to the conceptualization, methodology, software, validation, formal analysis, investigation, data curation, original draft preparation, and review and editing. D.J.G., S.T.M., H.T.B., P.R.K. and N.M.S. contributed to the supervision, project administration, and funding acquisition. All authors have read and agreed to the published version of the manuscript.

Funding: This research and the APC were funded by Massey University (Massey Foundation). Addisu Addis was sponsored by Massey University doctoral scholarship, Pūtea Tautoko-Doctoral Financial Support Grant and Helen E. Akers Postgraduate Scholarship. We would like to thank all the financial sources for their support.

Institutional Review Board Statement: Not applicable.

Data Availability Statement: Not applicable.

Conflicts of Interest: The authors declare no conflict of interest.

References

1. Berry, D. Invited review: Beef-on-dairy—The generation of crossbred beef × dairy cattle. *J. Dairy Sci.* **2021**, *104*, 3789–3819. [CrossRef] [PubMed]
2. New Zealand Statistics Agricultural Production Statistics. 2022. Available online: http://infoshare.stats.govt.nz/ (accessed on 8 March 2022).
3. MPI (Ministry for Primary Industries): Livestock Slaughter Statistics. Record Created 10 April 2017, Last Updated 24 July 2017; 2022. Available online: https://catalogue.data.govt.nz (accessed on 10 June 2022).
4. Van Selm, B.; de Boer, I.; Ledgard, S.; van Middelaar, C. Reducing greenhouse gas emissions of New Zealand beef through better integration of dairy and beef production. *Agric. Syst.* **2021**, *186*, 102936. [CrossRef]
5. Davison, R. *Export Cattle Slaughter by Beef and Dairy Origin*; Morris, S., Ed.; Personal communication; Executive Director for Beef+ Lamb New Zealand Economic Service & Insights: Wellington, New Zealand, 2020.
6. Morrison, D. It's Not DAIRY versus Beef, It's Integration versus Isolation'. 2020. Available online: https://www.independent.ie/business/farming/beef/its-not-dairy-versus-beef-its-integration-versus-isolation-39763642.html (accessed on 8 March 2023).
7. Hietala, S.; Heusala, H.; Katajajuuri, J.-M.; Järvenranta, K.; Virkajärvi, P.; Huuskonen, A.; Nousiainen, J. Environmental life cycle assessment of Finnish beef—Cradle-to-farm gate analysis of dairy and beef breed beef production. *Agric. Syst.* **2021**, *194*, 103250. [CrossRef]
8. Schierhorn, F.; Meyfroidt, P.; Kastner, T.; Kuemmerle, T.; Prishchepov, A.V.; Müller, D. The dynamics of beef trade between Brazil and Russia and their environmental implications. *Global Food Sec.* **2016**, *11*, 84–92. [CrossRef]
9. LIC & DairyNZ (Livestock Imporvement Corporation and New Zealand Dairy Statistics). 2021. Available online: https://www.lic.co.nz (accessed on 17 July 2020).
10. Coleman, L.; Blair, H.; Lopez-Villalobos, N.; Back, P.; Hickson, R. Breed variation in tongue colour of dairy and beef-cross-dairy calves. In Proceedings of the 22nd Conference of the Association for the Advancement of Animal Breeding and Genetics (AAABG), Townsville, QLD, Australia, 2–5 July 2017; pp. 473–476.
11. Burggraaf, V. Beef and Lamb New Zealand Dairy Beef Integration Project: Final Report. 2016. Available online: https://beeflambnz.com/sites/default/files/factsheets/pdfs/dairy-beef-integration-project-final-report.pdf (accessed on 23 June 2020).
12. Twomey, A.; Ring, S.; McHugh, N.; Berry, D. Carcass and efficiency metrics of beef cattle differ by whether the calf was born in a dairy or a beef herd. *J. Anim. Sci.* **2020**, *11*, skaa321. [CrossRef]
13. Hickson, R.; Zhang, I.; McNaughton, L. BRIEF COMMUNICATION: Birth weight of calves born to dairy cows in New Zealand. In Proceedings of the New Zealand Society of Animal Production, Dunedin, New Zealand, 1 July 2015; pp. 257–259.
14. Vicic, V.; Saliba, A.; Campbell, M.; Quinn, J. Barriers to Utilizing Non-replacement Male Calves in the Australian Dairy Industry: A Qualitative Study. *Front. Vet. Sci.* **2022**, *8*, 1671. [CrossRef] [PubMed]
15. Martin, N.; Schreurs, N.; Morris, S.; McDade, J.; Craigie, C.; Hickson, R. Is it possible to produce good quality meat from Holstein-Friesian bulls? *N. Z. J. Anim. Sci. Prod.* **2018**, *78*, 194–199.
16. Pettigrew, E.; Morris, S.; Back, P.; Kenyon, P.; Berry, J.; Donald, A.; Lane, A.; Hickson, R. Growth of weaned Friesian bull calves on a herb sward or with concentrate supplementation during late summer and early autumn. *N. Z. J. Agric. Res.* **2017**, *60*, 70–79. [CrossRef]
17. Morris, S.; Kenyon, P. Intensive sheep and beef production from pasture-a New Zealand perspective of concerns, opportunities and challenges. *Meat Sci.* **2014**, *98*, 330–335. [CrossRef]
18. Muir, P.; Thomson, B.; Smith, N. Factors affecting early concentrate uptake by calves. *N. Z. J. Anim. Sci. Prod.* **2020**, *80*, 95–100.
19. Bown, M.; Muir, P.; Thomson, B. Dairy and beef breed effects on beef yield, beef quality and profitability: A review. *N. Z. J. Agric. Res.* **2016**, *59*, 174–184. [CrossRef]
20. Palmer, A.; Beausoleil, N.; Boulton, A.; Cogger, N. Prevalence of Potential Indicators of Welfare Status in Young Calves at Meat Processing Premises in New Zealand. *Animals* **2021**, *11*, 2467. [CrossRef]
21. Cook, A. The Hunt for the Missing Billion: New Zealand's Dairy Beef Opportunity. 2014. Available online: http://researcharchive.lincoln.ac.nz/handle/10182/6713 (accessed on 14 June 2019).
22. B+LNZ (Beef and Lamb New Zealand) 2022. Available online: https://beeflambnz.com/data-tools (accessed on 10 March 2022).
23. Andrew, J. Bobby Calves: The Game Changers within New Zealand's Supply Chain. 2016. Available online: https://ruralleaders.co.nz/bobby-calves-the-game-changers-within-new-zealands-supply-chain-andrew-jolly (accessed on 16 June 2019).
24. Rutherford, N.; Lively, F.; Arnott, G. A Review of Beef Production Systems for the Sustainable Use of Surplus Male Dairy-Origin Calves Within the UK. *Front. Vet. Sci.* **2021**, *8*, 635497. [CrossRef] [PubMed]
25. Ferguson, D.; Schreurs, N.; Kenyon, P.; Jacob, R. Balancing consumer and societal requirements for sheep meat production: An Australasian perspective. *Meat Sci.* **2014**, *98*, 477–483. [CrossRef]
26. Boulton, A.; Kells, N.; Beausoleil, N.; Cogger, N.; Johnson, C.; Palmer, A.; O'Connor, C. Bobby Calf Welfare Across the Supply Chain-Final Report for Year 1. 2018. Available online: http://www.mpi.govt.nz/news-and-resources/publications/ (accessed on 1 September 2021).
27. Browne, S.; Midwinter, C.; Withers, H.; Cookson, L.; Biggs, J.; Marshall, C.; Benschop, J.; Hathaway, S.; Haack, A.; Akhter, N. Molecular epidemiology of Shiga toxin-producing *Escherichia coli* (STEC) on New Zealand dairy farms: Application of a culture-independent assay and whole-genome sequencing. *Appl. Environ. Microbiol.* **2018**, *84*, e00481-18. [CrossRef]

28. Addis, A.; Blair, H.; Kenyon, P.; Morris, S.; Schreurs, N. Optimization of Profit for Pasture-Based Beef Cattle and Sheep Farming Using Linear Programming: Young Beef Cattle Production in New Zealand. *Agriculture* **2021**, *11*, 849. [CrossRef]
29. Pike, S.; Schreurs, N.; Hickson, R.; Hunt, J.; Kenyon, P.; Garrick, D.; Blair, H.; Morris, S. BRIEF COMMUNICATION: Meat quality of light-weight, yearling steers of dairy origin. *N. Z. J. Anim. Sci. Prod.* **2019**, *79*, 156–158.
30. Hunt, J.; Tozer, P.; Kenyon, P.; Schreurs, N.; Pike, S.; Hickson, R.; Blair, H.; Garrick, D. BRIEF COMMUNICATION: Relative cost of producing carcasses from dairy-origin steers slaughtered at 8–12 months of age in New Zealand. *N. Z. J. Anim. Sci. Prod.* **2019**, *79*, 153–155.
31. Nakitari, J. *Growth, Carcass and Meat Quality Atrributtes of Dairy-Beef Bulls and Steers Slaugthered at Eleven Months of Age*; Massey University: Palmerston North, New Zealand, 2021.
32. Addis, A.; Blair, H.; Morris, S.; Kenyon, P.; Schreurs, N. Prediction of the Hind-Leg Muscles Weight of Yearling Dairy-Beef Steers Using Carcass Weight, Wither Height and Ultrasound Carcass Measurements. *Animals* **2020**, *10*, 651. [CrossRef] [PubMed]
33. Epstein, J.; Axteijl, R.; Tesfatsion, L. Growing artificial societies: Social science from the bottom up. *J. Econ. Literat.* **1998**, *36*, 233–234.
34. Axelrod, R. *The Complexity of Cooperation: Agent-Based Models of Competition and Collaboration*; Princeton University Press: Princeton, New Jersey, USA, 1997.
35. Macal, C.; North, M. Tutorial on agent-based modeling and simulation. In Proceedings of the Winter Simulation Conference, Orlando, FL, USA, 4–5 December 2005; p. 14.
36. Bonabeau, E. Agent-based modeling: Methods and techniques for simulating human systems. *Proc. Natl. Acad. Sci. USA* **2002**, *99*, 7280–7287. [CrossRef]
37. Addis, H.; Blair, T.; Kenyon, R.; Morris, T.; Schreurs, M.; Garrick, J. Agent-Based Modeling to Improve Beef Production from Dairy Cattle: Model Description and Evaluation. *Agriculture* **2022**, *12*, 1615. [CrossRef]
38. Morris, S. Sheep and Beef Cattle Production Systems. In *Ecosystems Services in New Zealand*; Manaaki Whenua Press: Lincoln, New Zealand, 2013; pp. 79–84.
39. Morris, S. The New Zealand beef cattle industry. In Proceedings of the New Zealand Society of Animal Production, Hamilton, New Zealand, 2–4 June 2013; pp. 1–4.
40. Vahdati, A. Agents. jl: Agent-based modeling framework in Julia. *J. Open Sour. Softw.* **2019**, *4*, 1611. [CrossRef]
41. Poisson, S.-D.; Costabel, P.; Dugac, P. *Siméon-Denis Poisson et la Science de son Temps*; Ecole Polytechnique: Palaiseau, France, 1981.
42. Dai, M.; Guo, W. Multivariate spectral analysis using Cholesky decomposition. *Biometrika* **2004**, *91*, 629–643. [CrossRef]
43. Coleman, L.; Back, P.; Blair, H.; López-Villalobos, N.; Hickson, R. Sire effects on birth weight, gestation length, and pre-weaning growth of beef-cross-dairy calves: A case study in New Zealand. *Dairy* **2021**, *2*, 385–395. [CrossRef]
44. Nor, N.; Steeneveld, W.; Mourits, M.; Hogeveen, H. The optimal number of heifer calves to be reared as dairy replacements. *J. Dairy Sci.* **2015**, *98*, 861–871.
45. Fontoura, F.; Agostinho, A. Growth with seasonally varying temperatures: An expansion of the von Bertalanffy growth model. *J. Fish Biol.* **1996**, *48*, 569–584. [CrossRef]
46. Brookes, I.; Nicol, A. The Metabolisable energy requirments of grazing livestock. In *Pasture and Supplements for Grazing Animals*; Occassional Publication No. 14; Rattray, P., Brookes, I., Nicol, A., Eds.; New Zealand Society of Animal Production: Rotorua, New Zealand, 2017.
47. Trafford, G.; Trafford, S. *Farm Technical Manual*; Lincon Universty: Christchurch, New Zealand, 2011.
48. B+LNZ (Beef and Lamb New Zealand): Economic Service Export Cattle Slaughter. 2021. Available online: https://beeflambnz.com/data-tools/benchmark-your-farm (accessed on 22 December 2019).
49. Schreurs, N.; Hickson, R.; Coleman, L.; Kenyon, P.; Martin, N.; Morris, S. BRIEF COMMUNICATION: Quality of meat from steers born to beef-cross-dairy cows and sired by Hereford bulls. In Proceedings of the New Zealand Society of Animal Production, Napier, New Zealand, 30 June–1 July 2014; pp. 229–232.
50. Coleman, L.; Hickson, R.; Schreurs, N.; Martin, N.; Kenyon, P.; Lopez-Villalobos, N.; Morris, S. Carcass characteristics and meat quality of Hereford sired steers born to beef-cross-dairy and Angus breeding cows. *Meat Sci.* **2016**, *121*, 403–408. [CrossRef]
51. Robert, P. Optimum early-maturing calf-to-beef systems. In *Sustainable Grass-Based Production*; Teagasc, Johnstown Castle Research Center: Waxford, Ireland, 2019.
52. Martín, N.; Schreurs, N.; Morris, S.; López-Villalobos, N.; McDade, J.; Hickson, R. Sire Effects on Post-Weaning Growth of Beef-Cross-Dairy Cattle: A Case Study in New Zealand. *Animals* **2020**, *10*, 2313. [CrossRef]
53. Coleman, L.; Martín, N.; Back, P.; Blair, H.; López-Villalobos, N.; Hickson, R. Low Birthweight Beef Bulls Compared with Jersey Bulls Do Not Impact First Lactation and Rebreeding of First-Calving Dairy Heifers—A Case Study in New Zealand. *Dairy* **2022**, *3*, 87–97. [CrossRef]
54. Hopkins, D.; Roberts, A. The value of carcass weight, fat depth measures and eye muscle area for predicting the percentage yield of saleable meat in Australian grass-fed beef carcasses for Japan. *Meat Sci.* **1995**, *41*, 137–145. [CrossRef] [PubMed]
55. Perry, D.; Yeates, A.; McKiernan, W. Meat yield and subjective muscle scores in medium weight steers. *Austr. J. Exp. Agric.* **1993**, *33*, 825–831. [CrossRef]
56. Lynch, R.; French, P. Profitable dairy-beef production systems. In *Sustainable Grass-Based Production Dairy-Beef*; Teagasc Crop, Environmental and Land Use Research Center: Waxford, Ireland, 2019; pp. 12–15.
57. Mylivestock Market Reports for Manawatu, New Zealand. 2020. Available online: https://mylivestock.co.nz (accessed on 28 November 2020).

58. Interest New Zealand Rural Schedule Price of Beef from Bulls, Heifers/Steers amd Cows per kg Carcass Weight. 2022. Available online: https://www.interest.co.nz (accessed on 9 March 2022).
59. Muir, D.; Fugle, J.; Smith, B.; Ormond, A. A comparison of bull beef production from Friesian type and selected Jersey type calves. *Proc. N. Zealand Grassl. Assoc.* **2001**, *63*, 203–207. [CrossRef]
60. Mandolesi, S.; Naspetti, S.; Arsenos, G.; Caramelle-Holtz, E.; Latvala, T.; Martin-Collado, D.; Orsini, S.; Ozturk, E.; Zanoli, R. Motivations and barriers for sheep and goat meat consumption in europe: A means–end chain study. *Animals* **2020**, *10*, 1105. [CrossRef]
61. Garmyn, A.; Polkinghorne, R.; Brooks, J.; Miller, M. Consumer assessment of New Zealand forage finished beef compared to US grain fed beef. *Meat Muscle Biol.* **2018**, *3*, 22–32. [CrossRef]
62. Monteils, V.; Sibra, C. Identification of combinations of influential rearing practices applied during the heifers' whole life on the carcass quality by the decision tree method. *Livest. Sci.* **2019**, *230*, 103823. [CrossRef]
63. Berg, R.; Butterfield, R. *New Concepts of Cattle Growth*; Sydney University Press: Sydney, Australia, 1978.
64. Crowe, A.; Lonergan, P.; Butler, S. Invited review: Use of assisted reproduction techniques to accelerate genetic gain and increase value of beef production in dairy herds. *J. Dairy Sci.* **2021**, *104*, 12189–12206. [CrossRef] [PubMed]
65. Martín, N.P.; Schreurs, N.M.; Morris, S.T.; López-Villalobos, N.; McDade, J.; Hickson, R.E. Meat quality of beef-cross-dairy cattle from Angus or Hereford sires: A case study in a pasture-based system in New Zealand. *Meat Sci.* **2022**, *190*, 108840. [CrossRef]
66. Martín, N.; Schreurs, N.; Morris, S.; López-Villalobos, N.; McDade, J.; Hickson, R. Sire effects on carcass of beef-cross-dairy cattle: A case study in New Zealand. *Animals* **2021**, *11*, 636. [CrossRef]
67. Jaborek, J.R. *Use of Diverse Cattle Breeds to Understand Marbling Development and Growth for the Production of High-Quality Beef*; The Ohio State University: Columbus, OH, USA, 2019.
68. Cooke, A.; Le Grice, P.; McAuliffe, G.; Lee, M.; Rivero, J. Rethinking Efficiency: Area Under the Curve (AUC) as a Low-Input Proxy for Improving the Accuracy of Finishing within Beef Systems. *SSRN Electron. J.* **2022**. [CrossRef]
69. Molano, G.; Clark, H.; Knight, T.; Cavanagh, A. Methane emissions from growing beef cattle grazing hill country pasture. In Proceedings of the New Zealand Society of Animal Production, Napier, New Zealand, 9 January 2006; pp. 172–175.
70. Richard, J.; Gemma, M. Policy Brief: How Do Different Livestock Types, Sizes and Breeds Differ in Their Greenhouse Gas Emissions? 2019. Available online: https://www.climatexchange.org.uk/media/3651/how-do-different-livestock-types-sizes-and-breeds-differ-in-their-greenhouse-gas-emissions.pdf (accessed on 25 March 2022).
71. Krystallis, A.; Arvanitoyannis, I.; Chryssochoidis, G. Is There a Real Difference Between Conventional and Organic Meat? Investigating Consumers' Attitudes Towards Both Meat Types as an Indicator of Organic Meat's Market Potential. *J. Food Prod. Market.* **2006**, *12*, 47–78. [CrossRef]
72. Xue, H.; Mainville, D.; You, W.; Nayga, R. Consumer preferences and willingness to pay for grass-fed beef: Empirical evidence from in-store experiments. *Food Qual. Pref.* **2010**, *21*, 857–866. [CrossRef]
73. Strydom, P.; Burrow, H.; Polkinghorne, R.; Thompson, J. Do demographic and beef eating preferences impact on South African consumers' willingness to pay (WTP) for graded beef? *Meat Sci.* **2019**, *150*, 122–130. [CrossRef]
74. Banović, M.; Grunert, K.G.; Barreira, M.M.; Fontes, M.A. Beef quality perception at the point of purchase: A study from Portugal. *Food Qual. Pref.* **2009**, *20*, 335–342. [CrossRef]
75. Bredahl, L. Cue utilisation and quality perception with regard to branded beef. *Food Qual. Pref.* **2004**, *15*, 65–75. [CrossRef]
76. Bello Acebrón, L.; Calvo Dopico, D. The importance of intrinsic and extrinsic cues to expected and experienced quality: An empirical application for beef. *Food Qual. Pref.* **2000**, *11*, 229–238. [CrossRef]
77. Scalco, A.; Macdiarmid, J.; Craig, T.; Whybrow, S.; Horgan, G. An Agent-Based Model to Simulate Meat Consumption Behaviour of Consumers in Britain. *J. Artific. Soc. Soc. Simulat.* **2019**, *22*. [CrossRef]
78. Foraker, B.; Frink, J.; Woerner, D. Invited review: A carcass and meat perspective of crossbred beef × dairy cattle. *Translat. Anim. Sci.* **2022**, *6*, txac027. [CrossRef]

agriculture

MDPI

Article

Identification of Characteristic Parameters in Seed Yielding of Selected Varieties of Industrial Hemp (*Cannabis sativa* L.) Using Artificial Intelligence Methods

Dominika Sieracka [1], Maciej Zaborowicz [2] and Jakub Frankowski [1,*]

[1] Department of Bioeconomy, Institute of Natural Fibers and Medicinal Plants—National Research Institute, Wojska Polskiego 71B, 60-630 Poznan, Poland; dominika.sieracka@iwnirz.pl

[2] Department of Biosystems Engineering, Poznan University of Life Sciences, Wojska Polskiego 28, 60-637 Poznan, Poland; maciej.zaborowicz@up.poznan.pl

[*] Correspondence: jakub.frankowski@iwnirz.pl

Abstract: Currently, there is a significant increase in interest in hemp cultivation and hemp products around the world. The hemp industry is a strongly developing branch of the economies of many countries. Short-term forecasting of the hemp seed and grain yield will provide growers and processors with information useful to plan the demand for employees, technical facilities (including appropriately sized drying houses and crop cleaning lines) and means of transport. This will help to optimize inputs and, as a result, increase the income from cultivation. One of the methods of yield prediction is the use of artificial intelligence (AI) methods. Neural modeling proved to be useful in predicting the yield of many plants, which is why work was undertaken to use it also to predict hemp yield. The research was carried out on selected, popular hemp varieties—Białobrzeskie and Henola. Their aim was to identify characteristic factors: climatic, cultivation and agrotechnical, affecting the size and quality of the yield. The collected data allowed the generation of Artificial Neural Network (ANN) models. It has been shown that based on a set of characteristics obtained during the cultivation process, it is possible to create a predictive neural model. Modeling using one output variable, which is seed yield, can be used in short-time prediction of industrial crops, which are gaining more and more importance.

Keywords: neural modeling; artificial neural networks; sensitivity analysis; hemp cultivation; seed material

1. Introduction

Due to their properties, Artificial Neural Networks (ANN) perform identification and prediction tasks similarly to the human brain; however, the use of computer methods eliminates subjective analysis and evaluation, which cannot be ruled out when performing similar analyses by a human. ANN are increasingly used in many fields of science, including mechanical and agricultural engineering, and connected problems, especially related to the scope of classification and prediction [1–3]. In the broadly understood agricultural industry, they were used, e.g., in research on starch content in potatoes [4], optimization of methods and parameters of drying willow [5], determination of the moment when crop irrigation should start [6], or the possibility of using unsold cut flowers of the most popular species for energy purposes [7].

For the efficient operation of farms and agricultural enterprises, in addition to the highest possible yield, it is also important to reduce losses associated with the storage of manufactured products. This maximizes the production volume of agricultural crops. This topic also interested scientists using Artificial Intelligence (AI) methods in their research [8,9]. The research that can be used in practice is the work on the possibility of using computer image analysis methods and neural modeling in the process of qualitative assessment of greenhouse tomatoes. Works were carried out by Zaborowicz and his research team. Generated neural model can do this by implementing into a computer system [10].

The hemp industry has been developing very strongly in recent years, and the area of hemp cultivation is constantly growing [11]. In Europe, the number of hectares of hemp plantations increased by 70% between 2013 and 2018 [12]. Hence, in recent years there has also been a lot of interesting research in the field of AI conducted on hemp. The use of ANN was undertaken to assess the effect of different types and concentrations of carbohydrate sources and the potency of nutrients on seed germination rates and morphological features of hemp seedlings grown in vitro [13]. AI methods have been used to detect and classify hemp diseases [14,15]. Mathematical models have also been developed to predict the dry density, compressive strength and thermal conductivity of hemp-based biocomposites using the AI-based gene expression programming (GEP) technique [16].

There have also been many works successfully using ANN in research on forecasting the yield of agricultural plants [17–19]. However, there is currently no objective and easy-to-use system for predicting the yield of industrial hemp seeds. So far, yields have been predicted using average amounts of seeds harvested in previous years, taking into account, e.g., variety or form of harvest. Cultivation of hemp, especially for seed purposes, is difficult, time-consuming and labor intensive. It is also burdened with a high risk of failure, but the market demand for hemp products, and thus for high-quality seed material, is growing dynamically. This indicates the need to undertake scientific research aimed at developing a new, effective method of seed yield prediction of selected industrial hemp varieties.

This topic was raised by Frankowski's team. Data collected from experimental plots were used to study the effect of sowing density and fertilization on the yield of Henola hemp seeds and straw. ANN studies were a supplement to standard research and statistical methods [20]. The results achieved are reported in the discussion section of this article. The achieved results proved to be promising. This prompted researchers to continue to develop them. Analyzing a much larger amount of data coming not from experimental plots but from hemp seed plantations, an attempt was made to identify the cultivation parameters characteristic of the seed yield of selected industrial hemp varieties. The aim of the research was to answer the question whether the ANN model can effectively predict the yield of industrial hemp seeds, based on the information obtained during the cultivation process. The result of the research was the generation of six ANN models. Thanks to the sensitivity analysis of the variables of the created neural models, it was possible to determine the indicators that are most important for their operation. The conducted research allowed the formulation of the main conclusion: based on the set of characteristics obtained during the agrotechnical process, it is possible to create a predictive neural model for assessing the yield of industrial hemp seeds.

2. Materials and Methods

2.1. Research Material

In order to generate training sets, information collected from seed plantations managed in Poland in 2019 and 2020 for Białobrzeskie and Henola varieties was used. The varieties to be tested were selected due to their very high popularity both in Europe and in the world, and the largest number of plantations and batches from them, among other varieties contracted by the Institute of Natural Fibers and Medicinal Plants—National Research Institute (INF&MP-NRI). The Białobrzeskie variety is a monoecious, stabilized variety with a high fiber content, cultivated for textile purposes since the 1960s [21–23]. It belongs to Central European forms and is adapted to Polish climatic and soil conditions [24], but it is successfully cultivated in other European countries, as well as in North America, South America and Australia, among others [25–27]. The Henola variety was bred in response to the growing market demand for hemp seeds and oil. It was bred through the positive selection of monoecious plants characterized by the shortest height, well-developed inflorescences and a short vegetation period. In 2017, it was entered into the national Research Centre for Cultivar Testing (RCCT) register [23,27]. It is characterized by a vegetation period shorter by about a month, the technical length of plants almost twice as long, and significantly larger inflorescences than the Białobrzeskie variety (Figure 1) [28]. It is a Polish variety, but, like Białobrzeskie, popular and cultivated around the world [26,27].

Figure 1. Comparison of hemp plants of Białobrzeskie and Henola varieties [Source: own study based on: [29]].

2.2. Collected Data and Methods

When analyzing hemp agricultural technology [29–32], it was concluded that the input data needed to generate neural models should be: soil class, forecrop, number of seeds sown per hectare, weather conditions, degree of qualification and form of harvest. Yield—weight of seeds collected from one hectare of plantation and yield quality—seed germination of a given batch were taken as the output data.

The following data was collected:

- plantation size (ha);
- weight of seeds sown on the plantation (kg);
- soil class—according to the soil quality classification adopted in Poland [33];
- forecrop—a plant grown on the same field in the growing season preceding the hemp cultivation season;
- category—category of seed material sown on a given plantation, according to the Seed Law [34];
- form of harvesting—one- or two-stage harvesting;
- seed moisture (%)—on the basis of data from the ISTA Certificate;
- crop quality—germination in % given on the ISTA Certificate;
- weather conditions—average monthly temperature and monthly rainfall from April to November, based on data provided by the Institute of Meteorology and Water Management—National Research Institute (IM&WM-NRI) on its website [35].

Ultimately, 24 training variables and 336 seed batch cases were included in the dataset. They constituted a training set for ANN models. For data to be entered into STATISTICA 7.1, the file was converted to Comma-Separated Values (CSV) format.

From the training set prepared in this way, 3 ANN models were generated:

1. "Germination and yield 1", with two output variables: yield per hectare and seed germination (%);
2. "Germination 1", with one output variable: seed germination (%);
3. "Yield 1", with one output variable: yield per hectare.

The STATISTICA 7.1 simulator divided the training set into three subsets:

1. Training subset (U) used to teach the network;
2. Validation subset (W), allowing the control of the effects of the learning algorithm during the learning process;
3. Test subset (T)—which allows the assessment of the quality of the generated neural network.

The division into subsets was carried out in the default way for the program, according to the proportion: 2:1:1.

The ANN simulator in the STATISTICA 7.1 package was used for the neural modeling process. The process was carried out in two stages. The former used the Automatic Designer function and the latter used the User Network Designer function.

Using the Automatic Designer function, neural models were generated and analyzed at the next stages. It was assumed that the simulator should test 20 networks of each type and keep the 10 with the best results. The condition of maintaining the network was considered to be a balance between the error and the diversity of the network in order to obtain a wide range of produced models in order to select the optimal topology [36].

The research was carried out using ANN, as the method has been successfully used in the field of agricultural and life sciences for yield prediction and evaluation, and is also excellent for evaluating characteristic variables. PNN (Probabilistic Neural Networks), GRNN (Generalized Regression Neural Networks), RBF (Radial Basis Function Networks) and MLP (Multi-Layer Perceptrons) were tested. Among the networks generated using the Automatic Designer function, the best characteristics were achieved by RBF networks, followed by MLP networks. They are characteristic of non-linear solutions.

After analyzing the models generated using the Automatic Designer function, it was decided to continue the work related to modeling using the User Network Designer function. It was decided to use two networks (RBF and MLP) which achieved the best characteristics in the first stage of research. The best characteristics were achieved by RBF networks, which are shown in results section.

Using this option, 3 RBF models were generated, containing 10 networks each, in which the output variables were again:

1. "Germination and yield 2";
2. "Germination 2";
3. "Yield 2".

The networks were trained with the following algorithms:

1. SS—SubSample;
2. EX—by user (Explicit)—determination of radial deviation;
3. PI—Pseudoinversion.

The error and quality metrics were used to evaluate the models. The error was assumed to be RMSE (Root Mean Square Error), which is represented by the formula:

$$RMSE = \sqrt{\frac{1}{n}\sum_{t=1}^{n}(y_t - y_t^P)^2}$$

Quality is understood as the quality of the network for different subsets obtained during network training. For regression networks, the quotient of standard deviations is given as the network quality.

3. Results

3.1. Qualitative Characteristics and Sensitivity Assessment of the Generated Neural Network Models Created Using the Automatic Designer Function

The 3 best networks were selected for each of the models generated using the *Automatic Designer* function. The number of training cases in the set was 336. The number of training variables was from 23 to 24, depending on the generated model (Table 1).

Table 1. Summary of neural models generated with the Automatic Designer function.

Model	Network	Learning Quality	Validation Quality	Testing Quality	Learning Error	Validation Error	Testing Error	Learning Algorithm
Germination and yield 1	RBF 15:41-3-2:2	0.9626	0.9756	0.9837	0.1359	0.1874	0.1619	KM, KN, PI
Germination 1	RBF 15:41-6-1:1	0.9685	0.9624	0.9252	0.1755	0.1617	0.1742	KM, KN, PI
Yield 1	RBF 17:49-13-1:1	0.7659	1.2494	0.9799	0.0019	0.0035	0.0029	KM, KN, PI

The best results were shown by RBF-type networks. These networks are characterized by one hidden layer with radial neurons. The networks generated using the Automatic Designer function had 3, 6 and 13 neurons in the hidden layer. The networks with 3 and 6 neurons were characterized by low error and high quality. The network with 13 neurons showed signs of overfitting.

In the first model, with two output variables: germination and yield, in the first stage of the research (model: "Germination and yield 1"), for the RBF 15:41-3-2:2 network generated using the Automatic Designer function, the learning quality was 0.9626, validation quality was 0.9756, and test quality was 0.9837. The network learning error for the training set was 0.1359, the validation error was 0.1874, and the test error was 0.1619. KM, KN and PI algorithms were used for the network learning process.

In the model with one output variable, which was germination (model: "Germination 1"), the RBF 15:41-6-1:1 network generated using the Automatic Designer function showed a learning quality of 0.9685, a validation quality of 0.9624, and a test quality of 0.9252. The learning error was 0.1755, the validation error was 0.1617, and the test error was 0.1742. KM, KN and PI algorithms were used for the network learning process.

In the third case, for the network with one output variable: yield (model: "Yield 1"), for the RBF 17:49-15-1:1 model, generated using the Automatic Designer function, the learning quality was 0.7659, the validation quality was 1.2494 and test quality was 0.9799. The learning error was 0.0019, the validation error was 0.0035, and the test error was 0.29. KM, KN and PI algorithms were used for the network learning process.

The "Germination and Yield 1" and "Germination 1" models generated in the first stage of the research, using the Automatic Designer function, were characterized by high quality and low error. The "Yield 1" model showed features of network overfitting, so it was decided to continue the research using the User Network Designer function.

3.2. Qualitative Characteristics and Sensitivity Assessment of the Generated Neural Network Models Created Using the User Network Designer Function

The best 3 networks were selected for each of the models generated using the User Network Designer function. The best characteristics were shown by RBF-type networks. Researchers checked the optimal number of neurons in the hidden layer, gradually increasing it. The best results were obtained with 9 neurons in the hidden layer. With a dozen neurons in the hidden layer, the networks began overfitting.

Networks generated using the User Network Designer function were mostly characterized by higher quality and lower RMSE error than networks generated using the *Automatic Designer* function (Table 2).

Table 2. Comparison of neural models generated with the User Network Designer function.

Model	Network	Learning Quality	Validation Quality	Testing Quality	Learning Error	Validation Error	Testing Error	Learning Algorithm
Germination and yield 2	RBF 22:49-9-2:2	0.9847	0.9934	0.9992	0.1135	0.1096	0.1195	SS, EX, PI
Germination 2	RBF 22:51-9-1:1	0.9841	0.9997	0.9999	0.1867	0.2147	0.2001	SS, EX, PI
Yield 2	RBF 22:45-9-1:1	0.9898	0.9905	0.9790	0.0023	0.0025	0.0020	SS, EX, PI

The RBF 22:49-9-2:2 network, generated in the second stage of the research (model: "Germination and Yield 2") using the User Network Designer function, was characterized by a learning quality 0.9847, a validation quality of 0.9934 and a test quality of 0.9992. The network learning error for the training set was 0.1135, the validation error was 0.1096 and the test error was 0.1195. The SS, EX and PI algorithms were used for the network learning process.

The RBF 22:51-9-1:1 network, generated using the User Network Designer function ("Germination 2" model), had a learning quality of 0.9841, a validation quality of 0.9997 and a test quality of 0.9999. The training, validation and test errors for this network were 0.1867, 0.2147, 0.2001, respectively. The network was trained with SS, EX and PI algorithms.

For RBF 22:45-9-1:1 networks generated using the User Network Designer function (model: "Yield 2"), the learning quality was 0.9898, the validation quality was 0.9905, and the test quality was 0.9790. The errors: learning error 0.0023, validation error 0.0025 and test error 0.0020. The SS, EX and PI algorithms were also used in the learning process of this network.

The "Germination and Yield 1" and "Germination 1" models generated in the first stage of the research, using the Automatic Designer function, were characterized by high quality and low error. The "Yield 1" model showed features of network overfitting. In the second stage of the research, carried out using the User Network Designer function, the models did not show features of network overfitting. In addition, in the "Germination and Yield 2" model, the network quality was higher and the errors were lower than in the "Germination and Yield 1" model. In the "Germination 2" model, the web qualities were higher than in the "Germination 1" model, but the errors were lower in the "Germination 1" model. The "Yield 2" model had a higher learning and validation quality than the "Yield 1" model, and the test quality of both models differed slightly—by 0.01%. The "Yield 2" model was characterized by a higher learning error, but lower validation and testing errors than the "Yield 1" model.

Figure 2 shows screenshots of all 3 generated RBF models (Figure 2).

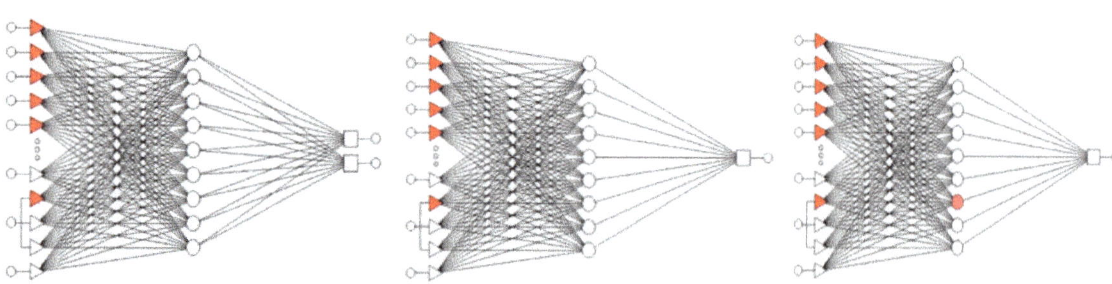

Figure 2. Screenshots of RBF 22:49-9-2:2, RBF 22:51-9-1:1 and RBF 22:45-9-1:1.

3.3. Sensitivity Analysis

An important point in the process of neural modeling is the sensitivity analysis, during which importance ranks are determined for individual variables. This allows the determination of which variables are crucial for the correct learning process and operation of the neural model, and which are of little importance. It is believed that in cases where the error quotient is less than or equal to unity, the removal of the analyzed variable not only has no impact on the operation of the network, but may improve the quality of the generated model. The criterion for the sensitivity analysis was the quotient of the error obtained without the considered variable and the error obtained with the use of all learning variables. On the basis of the quotient, the individual variables were assigned appropriate ranks according to the following rule: the smaller the quotient, the higher the rank. The STATISTICA 7.1 simulator assigns appropriate ranks to individual variables and ranks them, thus supporting the sensitivity analysis process. The lower its rank value, the more important the variable is for the neural modeling process. In the course of the research, a sensitivity analysis of the variables of individual training sets that were involved in the neural modeling process was carried out. Due to this, the information on the level of significance of individual variables was obtained. The results are summarized in Table 3.

Table 3. Sensitivity analysis of the RBF 22:49-9-2:2, RBF 22:45-9-1:1 and RBF 22:51-9-1:1 models.

Network	RBF 22:49-9-2:2		RBF 22:45-9-1:1		RBF 22:51-9-1:1	
Variable	Quotient	Rank	Quotient	Rank	Quotient	Rank
total precipitation _4	1.0622	8	1.0120	6	1.0066	10
total precipitation _5	1.0640	7	1.0120	1	1.0074	2
total precipitation _6	1.0657	2	1.0120	3	1.0074	6
total precipitation _7	1.0657	3	1.0120	4	1.0074	1
total precipitation _8	1.0655	6	1.0112	8	1.0074	5
total precipitation _9	1.0657	1	1.0079	9	1.0074	3
total precipitation _10	1.0657	4	1.0120	2	1.0074	4
total precipitation _11	1.0657	5	1.0120	5	1.0074	7
average monthly temperature _4	1.0221	11	0.9982	18	1.0028	13
average monthly temperature _5	1.0266	10	1.0051	11	1.0039	11
average monthly temperature _6	1.0311	9	1.0079	10	1.0074	8
average monthly temperature _7	1.0207	12	1.0020	12	1.0023	15
average monthly temperature _8	1.0143	15	0.9992	14	1.0016	16
average monthly temperature _9	1.0027	20	0.9979	20	1.0005	21
average monthly temperature _10	1.0057	19	0.9989	15	1.0003	22
average monthly temperature _11	1.0176	14	0.9981	19	1.0014	18
quantity of seeds sown per hectare [kg]	1.0016	21	1.0113	7	1.0071	9
variety	1.0007	22	0.9973	21	1.0006	20
soil class	1.0121	16	0.9967	22	1.0035	12
forecrop	1.0203	13	1.0003	13	1.0027	14
seeds category	1.0078	17	0.9985	16	1.0015	17
harvesting form	1.0062	18	0.9984	17	1.0008	19

On the basis of the performed sensitivity analysis, the ranks of the ANN input variables were determined and they were assigned an appropriate hierarchy. These ranks determine the level of significance of the variables in the context of the quality of operation of the generated neural models.

4. Discussion

Agricultural crops are characterized by frequent non-linearity of processes and phenomena, which makes the relations between them complex and not easy to describe and characterize. Therefore, where traditional statistical methods of describing the studied phenomena fail, the use of artificial intelligence is used [3]. The use of ANN to predict the yield of agricultural plants is more and more often undertaken by researchers from around the world.

Research on the use of ANN in agriculture was conducted, among others, by Medara's team, which worked on data obtained from the Indian Ministry of Agriculture on sugar cane. The studies included data from different regions, which meant differences in the course of weather conditions and sowing dates—similar to the studies presented in this article. Experiments were carried out for 2160 different models. The team was able to successfully model yield with an overall accuracy of 83.49%. The smallest error value achieved was 4.03 [37]. On the other hand, Niedbała and Kozłowski built three independent models for forecasting winter wheat yields. Models were built using ANN with MLP topology based on meteorological data (air temperature and precipitation) and information on applied mineral fertilizers. The lowest error value was 8.85 [38]. The presented research on the prediction of hemp yield quality and quantity is in line with the global trend. RBF-type models generated in the second stage of the research, using the *User Network Designer* function, are characterized by high quality of 97–98%. This quality is higher than the quality of networks generated by researchers conducting research on the yield of, e.g., sugar cane—83.49% [37], and comparable to the quality of the network created by Gandhi's team investigating rice yield—the model of these researchers was characterized by a quality of 97.5% [39].

Research on the use of ANN in forecasting hemp yield was undertaken by Frankowski's team. The results obtained during the experimental plots were used to build a dataset for the ANN. Four input data were adopted: total precipitation, mean temperature, fertilizer and straw yield. Linear, MLP and RBF networks were tested using STATISTICA 7.1. The best results were obtained for linear networks. They were characterized by a quality of 0.910 and a test error of 0.336 [20]. In the research presented in the following article, data from seed orchards were used and the set of input data was significantly expanded. As a result, RBF-type networks were produced with a test quality higher by 0.069 and an error lower by 0.334 than linear networks produced during tests on experimental plots.

A very interesting solution is also the use of AI to predict yields using image analysis. Vijayakumar's team developed three ML-based models for citrus fruit yield prediction based on the use of Unmanned Aerial Vehicle (UAV) imaging and ground imagery. Four ML algorithms were used to generate the models—gradient enhancing regression (GBR), random forest regression (RFR), linear regression (LR) and partial least squares regression (PLSR). The best generated model was characterized by a Mean Absolute Percentage Error (MAPE) of 23.45% [40]. Taking into account the growing popularity of the use of UAV and the opportunities it gives, as well as the often-high variability of, e.g., terrain or soil in one field, the use of this technique supported by ANN seems to be an interesting and legitimate supplement to research on the yielding of agricultural plants, including hemp. They are characterized by the fact that they show differences in development and yield depending on the conditions in the field, which, due to the considerable height of the plants, is practically impossible to determine from the ground level.

As the examples cited show, innovations in the field of technology, including those related to AI methods, are more often and more willingly used in agricultural practice. They are also widely used in crop yield forecasting. Scientists successfully use ANN to forecast the yield of various agricultural crops economically important for a given country or region, and the hemp industry is developing very dynamically around the world.

However, the researchers recognize the limitations of both the use of ANN and those of the current study. For this reason, it is planned to continue the research after extending the dataset with information from the next growing season. Taking into account the variety of available advanced artificial intelligence and machine learning methods, after increasing the dataset, it is also planned to undertake research using other methods, e.g., Deep Learning.

Another problem to solve in yield prediction is the impact of violent weather phenomena, such as hailstorms or exceptionally heavy rains. In the era of progressive climate change, these phenomena are becoming more and more frequent, but at the same time difficult to predict, and their occurrence can significantly mechanically damage or even completely destroy plantations. Therefore, taking up this problem seems to be extremely important and interesting.

5. Conclusions

Based on the conducted research, it was shown that it is possible to create a predictive neural model for assessing the yield of industrial hemp seeds based on a set of characteristics obtained during the agrotechnical process. The information obtained from hemp seed plantations of the Białobrzeskie and Henola varieties was sufficient to build a training set for ANN. The sensitivity analysis carried out showed that in Germination and yield 2 and Germination 2 models, all quotients were higher than unity, and in the Yield 2 model higher or very close to unity, which means that all data from the training set were important for the proper operation of the network.

Modeling using one output variable, which is seed yield, can be used not only in seed orchards, but also in the case of industrial crops, which are gaining more and more importance. Hemp is cultivated on such plantations, e.g., to obtain seeds for food purposes (e.g., for pressing oil or producing dehulled seeds or hemp flour). This branch of the hemp industry is developing very intensively, mainly due to its growing popularity. Therefore, it is extremely important to determine the parameters affecting the yield of seeds.

It is planned to continue the research by extending the dataset with information from the next growing season. This should allow for even better research results and optimization of the set of input data necessary to create a neural model that will be able to forecast the yield of hemp seeds or grains as accurately as possible in the short term.

Author Contributions: Conceptualization, D.S., M.Z. and J.F.; methodology, D.S. and M.Z.; software, D.S. and M.Z.; validation, D.S., M.Z. and J.F.; formal analysis, D.S. and M.Z.; investigation, D.S., M.Z. and J.F.; resources, D.S. and J.F.; data curation, D.S. and M.Z.; writing—original draft preparation, D.S., M.Z. and J.F.; writing—review and editing, D.S., M.Z. and J.F.; visualization, D.S. and M.Z.; supervision, M.Z. and J.F.; project administration, D.S., M.Z. and J.F.; funding acquisition, M.Z. and J.F. All authors have read and agreed to the published version of the manuscript.

Funding: This research received no external funding.

Institutional Review Board Statement: Not applicable.

Informed Consent Statement: Not applicable.

Conflicts of Interest: The authors declare no conflict of interest.

References

1. Kujawa, S.; Niedbała, G. Artificial Neural Networks in Agriculture. *Agriculture* **2021**, *11*, 497. [CrossRef]
2. Przybylak, A.; Bonieki, P.; Zaborowicz, M.; Mo, Z.; Przybył, K. Przykłady wykorzystania modelowanie neuronowego w praktyce rolniczej. *Tech. Rol. Ogrod. Leśna* **2013**, *1*, 21–24.
3. Boniecki, P. *Elementy Modelowania Neuronowego w Rolnictwie*; Wydawnictwo Uniwersytetu Przyrodniczego: Poznań, Poland, 2008.
4. Niedbała, G.; Lenartowicz, T.; Kozłowski, R.J.; Zaborowicz, M. Neural modelling as a prediction method of starch content in potatoes for post-registration and specific agricultural experimentation. *Nauk. Przyr. Technol.* **2015**, *9*, 17. [CrossRef]
5. Francik, S.; Łapczyńska-Kordon, B.; Francik, R.; Wójcik, A. Modeling and Simulation of Biomass Drying Using Artificial Neural Networks. In *Renewable Energy Sources: Engineering, Technology, Innovation; Springer Proceedings in Energy*; Springer: Cham, Switzerland, 2018; pp. 571–581. [CrossRef]
6. Neugebauer, M.; Nalepa, K.; Sołowiej, P. Sieci neuronowe jako narzędzie umożliwiające prognozowanie zapotrzebowania na wodę w uprawach rolnych. *Inżynieria Rol.* **2007**, *2*, 205–210.

7. Frankowski, J.; Zaborowicz, M.; Dach, J.; Czekała, W.; Przybył, J. Biological Waste Management in the Case of a Pandemic Emergency and Other Natural Disasters. Determination of Bioenergy Production from Floricultural Waste and Modeling of Methane Production Using Deep Neural Modeling Methods. *Energies* **2020**, *13*, 3014. [CrossRef]
8. Szwedziak, K.; Polańczyk, E.; Grzywacz, Ż.; Niedbała, G.; Wojtkiewicz, W. Neural Modeling of the Distribution of Protein, Water and Gluten in Wheat Grains during Storage. *Sustainability* **2020**, *12*, 5050. [CrossRef]
9. Szwedziak, K.; Tukiendorf, M. Use of geostatic function to describe wheat grain mass quality. *J. Res. Appl. Agric. Eng.* **2014**, *59*, 126–130.
10. Zaborowicz, M.; Boniecki, P.; Koszela, K.; Przybylak, A.; Przybył, J. Application of neural image analysis in evaluating the quality of greenhouse tomatoes. *Sci. Hortic.* **2017**, *218*, 222–229. [CrossRef]
11. Baraniecki, P.; Latterini, F.; Stefanoni, W.; Frankowski, J.; Wielgusz, K.; Pari, L. Assessment of the Working Performance of an Innovative Prototype to Harvest Hemp Seed in Two Different Conditions of Terrain Slope. *Agronomy* **2022**, *12*, 185. [CrossRef]
12. European Industrial Hemp Association. Available online: www.eiha.org (accessed on 9 March 2023).
13. Hesami, M.; Pepe, M.; Monthony, A.S.; Baiton, A.; Jones, A.M.P. Modeling and optimizing in vitro seed germination of industrial hemp (*Cannabis sativa* L.). *Ind. Crop. Prod.* **2021**, *170*, 113753. [CrossRef]
14. Bose, B.; Priya, J.; Welekar, S.; Gao, Z. Hemp Disease Detection and Classification Using Machine Learning and Deep Learning. In Proceedings of the 2020 IEEE Intl Conf on Parallel & Distributed Processing with Applications, Big Data & Cloud Computing, Sustainable Computing & Communications, Social Computing & Networking, Exeter, UK, 17–19 December 2020; IEEE: Piscataway, NJ, USA, 2020. [CrossRef]
15. Zhu, J.; Yu, T.; Zheng, S.; Niu, C.; Gao, J.; Tang, J. Hemp Disease Detection and Classification Using Machine Learning. In Proceedings of the 2020 International Conferences on Internet of Things (iThings) and IEEE Green Computing and Communications (GreenCom) and IEEE Cyber, Physical and Social Computing (CPSCom) and IEEE Smart Data (SmartData) and IEEE Congress on Cybermatics (Cybermatics), Rhodes, Greece, 2–6 November 2020; IEEE: Piscataway, NJ, USA, 2020. [CrossRef]
16. Ahmad, M.R.; Chen, B.; Dai, J.-G.; Kazmi, S.M.S.; Munir, M.J. Evolutionary artificial intelligence approach for performance prediction of bio-composites. *Constr. Build. Mater.* **2021**, *290*, 123254. [CrossRef]
17. Amaratunga, V.; Wickramasinghe, L.; Perera, A.; Jayasinghe, J.; Rathnayake, U. Artificial Neural Network to Estimate the Paddy Yield Prediction Using Climatic Data. *Math. Probl. Eng.* **2020**, *2020*, 8627824. [CrossRef]
18. Barwicki, J.; Hryniewicz, M.; Grzybek, A. Yield forecasting using artificial intelligence. *Pol. Tech. Rev.* **2020**, *1*, 19–22.
19. Emamgholizadeh, S.; Parsaeian, M.; Baradaran, M. Seed yield prediction of sesame using artificial neural network. *Eur. J. Agron.* **2015**, *68*, 89–96. [CrossRef]
20. Frankowski, J.; Zaborowicz, M.; Sieracka, D.; Łochyńska, M.; Czeszak, W. Prediction of the Hemp Yield Using Artificial Intelligence Methods. *J. Nat. Fibers* **2022**, *19*, 13725–13735. [CrossRef]
21. Vandepitte, K.; Vasile, S.; Vermeire, S.; Vanderhoeven, M.; Van der Borght, W.; Latré, J.; De Raeve, A.; Troch, V. Hemp (*Cannabis sativa* L.) for high-value textile applications: The effective long fiber yield and quality of different hemp varieties, processed using industrial flax equipment. *Ind. Crop. Prod.* **2020**, *158*, 112969. [CrossRef]
22. Mańkowski, J. The Effect of Some Agronomic Factors on the Amount and Quality of Homomorphic Fibre. *Fibres Text. East. Eur.* **2003**, *11*, 20–25.
23. Research Centre for Cultivar Testing. Available online: www.coboru.gov.pl/index_en/ (accessed on 9 March 2023).
24. Grabowska, L.; Jaranowska, B.; Baraniecki, P.; Tymków, J. The Results of Hemp Breeding in Poland. *Natural Fibres* **1998**, *2*, 103–109.
25. Tsaliki, E.; Kalivas, A.; Jankauskiene, Z.; Irakli, M.; Cook, C.; Grigoriadis, I.; Panoras, I.; Vasilakoglou, I.; Dhima, K. Fibre and Seed Productivity of Industrial Hemp (*Cannabis sativa* L.) Varieties under Mediterranean Conditions. *Agronomy* **2021**, *11*, 171. [CrossRef]
26. New Frontier Data. Poland Embraces European Potential for Industrial Hemp. Available online: https://newfrontierdata.com/cannabis-insights/polands-rise-to-a-new-european-hemp-powerhouse/ (accessed on 9 March 2023).
27. Polish Hemp Program. Available online: www.polishhempprogram.com/polish-hemp-program---on-media.html (accessed on 9 March 2023).
28. Burczyk, H.; Oleszak, G. Konopie oleiste (*Cannabis sativa* L. var. olrifera) uprawiane na nasiona do produkcji oleju i biogazu. *Probl. Inżynierii Rol.* **2016**, *94*, 109–116.
29. Burczyk, H.; Frankowski, J. Henola—Pierwsza polska odmiana konopi oleistych. *Zag. Doradz. Rol.* **2018**, *93*, 89–101.
30. Wójtowicz, A.; Strażyński, P.; Mrówczyński, M. (Eds.) *Metodyka Integrowanej Ochrony Konopi dla Doradców*; Instytut Ochrony Roślin—Państwowy Instytut Badawczy: Poznań, Poland, 2018.
31. Grzebisz, W. (Ed.) Rolnictwo cz. VI. Produkcja roślinna. In *Technologie Produkcji Roślinnej*; Hortpress: Warszawa, Poland, 2015; pp. 280–288.
32. Cierpucha, W. (Ed.) *Technologia Uprawy i Przetwórstwa Konopi Włóknistych*; Instytut Włókien Naturalnych i Roślin Zielarskich: Poznań, Poland, 2013; pp. 22–31.
33. Regulation of the Council of Ministers on Soil Classification/Rozporządzenie Rady Ministrów z Dnia 12 Września 2012 r. w Sprawie Gleboznawczej Klasyfikacji Gruntów z Dnia 12 Września 2012 r. (Dz. U. 2012, poz. 1246). Available online: https://isap.sejm.gov.pl/isap.nsf/DocDetails.xsp?id=wdu20120001246 (accessed on 9 March 2023).

34. Seed Law/Ustawa o Nasiennictwie, z dn. 9 Listopada 2012 (Dz. U. 2012 poz. 1512). Available online: https://isap.sejm.gov.pl/isap.nsf/DocDetails.xsp?id=WDU20120001512 (accessed on 9 March 2023).
35. Institute of Meteorology and Water Management—National Research Institute. Available online: www.imgw.pl (accessed on 9 March 2023).
36. Tadeusiewicz, R. Sieci Neuronowe. In *Akademicka Oficyna Wydawnicza*; RM: Warszawa, Poland, 1993.
37. Medar, R.A.; Rajpurohit, V.S.; Ambekar, A.M. Sugarcane Crop Yield Forecasting Model Using Supervised Machine Learning. *Int. J. Intell. Syst. Appl.* **2019**, *11*, 11–20. [CrossRef]
38. Niedbała, G.; Kozłowski, J.R. Application of Artificial Neural Networks for Multi-Criteria Yield Prediction of Winter Wheat. *J. Agric. Sci. Technol.* **2019**, *21*, 51–61.
39. Gandhi, N.; Petkar, O.; Armstrong, L.J. Rice crop yield prediction using artificial neural networks. In Proceedings of the 2016 IEEE Technological Innovations in ICT for Agriculture and Rural Development (TIAR), Chennai, India, 15–16 July 2016; pp. 105–110. [CrossRef]
40. Vijayakumar, V.; Ampatzidis, Y.; Costa, L. Tree-level citrus yield prediction utilizing ground and aerial machine vision and machine learning. *Smart Agric. Technol.* **2023**, *3*, 100077. [CrossRef]

agriculture

MDPI

Article

A Comprehensive Analysis of Machine Learning-Based Assessment and Prediction of Soil Enzyme Activity

Yogesh Shahare [1], Mukund Partap Singh [2], Prabhishek Singh [2], Manoj Diwakar [3], Vijendra Singh [4], Seifedine Kadry [5,6,7,8] and Lukas Sevcik [9,*]

1 Department of Information Technology, Mahatma Gandhi Mission's College of Engineering and Technology (MGMCET), Navi Mumbai 410 209, India
2 School of Computer Science & Engineering Technology, Bennett University, Greater Noida 201310, India
3 Computer Science and Engineering Department, Graphic Era (Deemed to be University), Dehradun 248002, India
4 School of Computer Science, University of Petroleum and Energy Studies, Dehradun 248007, India
5 Department of Applied Data Science, Noroff University College, 4612 Kristiansand, Norway
6 Artificial Intelligence Research Center (AIRC), Ajman University, Ajman 346, United Arab Emirates
7 Department of Electrical and Computer Engineering, Lebanese American University, Byblos 13-5053, Lebanon
8 MEU Research Unit, Middle East University, Amman 11831, Jordan
9 University of Zilina, 010 26 Zilina, Slovakia
* Correspondence: lukas.sevcik@uniza.sk

Abstract: Different soil characteristics in different parts of India affect agriculture growth. Crop growth and crop production are significantly impacted by healthy soil. Soil enzymes mediate almost all biochemical reactions in the soil. Understanding the biological processes of soil carbon and nitrogen cycling requires defining the significance of prospective elements at the play of soil enzymes and evaluating their activities. A combination of Multiple Linear Regression (MLR), Random Forest (RF) models, and Artificial Neural Networks (ANN) was employed in this study to assess soil enzyme activity, including amylase and urease activity, soil physical properties, such as sand, silt, clay, and soil chemical properties, including organic matter (SOM), nitrogen (N), phosphorus (P), soil organic carbon (SOC), pH, and fertility level. Compared to other methods for estimating soil phosphatase, cellulose, and urease activity, the RF model significantly outperforms the MLR model. In addition, due to its ability to manage dynamic and hierarchical relationships between enzyme activities, the RF model outperforms other models in evaluating soil enzyme activity. This study collected 3972 soil samples from 25 villages in the Bhandara district of Maharashtra, India, with chemical, physical, and biological parameters. Overall, 99% accuracy was achieved for cellulase enzyme activity and 94% for N-acetyl-glucosaminidase enzyme activity using the Random Forest model. Crops have been suggested based on the best performance accuracy algorithms and evaluation performance metrics.

Keywords: soil organic matter (SOM); soil enzyme activity (SEA); soil organic carbon (SOC); physical soil features; chemical soil features; machine learning (ML); Artificial Neural Network (ANN)

1. Introduction

Various factors, including agricultural soil, soil management, soil productivity, irrigation, fertilizer, and climate, impact the agriculture sector to produce a good quantity of crops. The primary determinant of an agricultural field is the soil. The capacity of agricultural soil to develop crops depends on the nutrients it contains. Each soil has a variety of physical, chemical, and biological components. Several researchers have been researching agricultural soil to improve soil quality and other factors, but they have not yet achieved suitable outcomes. Artificial intelligence techniques are more useful and innovative. This technique has the best results for improving and growing crops in the agricultural field, which is helpful to the farmers. Soil quality and the amount of farmed land in Maharashtra agriculture have continued to decline due to a lack of expertise and a harsh environment, which significantly impacts economics and crop production [1,2]. Climate change is affecting the agricultural sector, decreasing crop yield, diminishing soil organic carbon (SOC), changing acceptable cropping areas latitudinal, changing growth time, and causing soil degradation [3]. Due to constant degradation and changes in its composition, the state of soil changes over time [4]. Sustaining the productivity of soil for agriculture requires appropriate stability of physical, chemical, and biological elements [5].

Due to a lack of knowledge about agriculture, farmers are unable to identify the deficiency of important nutrients in the soil that are conducive to growing crops. In terms of the farmers' situation, this work has been proposed. This work uses state-of-the-art artificial technology to develop the prediction of a soil fertility and soil enzyme activity model using a soil dataset from Bhandara district, Maharashtra state, which is helpful to farmers in identifying the nutrient deficiency that is present in the soil. This model was developed by using Python programming with Jupiter Notebook.

The assessment of soil minerals is required for compaction characteristics monitoring. Microorganisms make up a large portion of the biological ingredients of soil and contribute more to its strength than physical or chemical constituents. Microorganisms respond quickly to changes in soil structure and become utilized in their surroundings [6]. Farmers are not gaining the appropriate level of crop productivity as an outcome of weather change and biological activity. In this case, soil biochemical analysis, together with soil chemical and physical features, is critical for minimizing and decomposing the nutrient cycle and providing for the crop [7].

Soil enzymes play a vital role in the biogeochemical cycle of "carbon (C), nitrogen (N), and phosphorus (P)" in the soil and can be employed as early indicators of nutrient imbalances caused by climate change [8,9]. Soil enzyme activities include carbon cycle transformations such as C-glucosidase and invertase, as well as general enzyme activity such as dehydrogenase and catalase, and nitrogen cycle transformations such as urease has a direct impact on the nitrogen supply rate in soil, which is commonly employed as a measure of nitrogen deficiency, N-acetyl-glucosaminidase, and protease [10,11]. Soil enzyme activity can help researchers better understand the biological mechanisms of "carbon and nitrogen" transformation and provide guidelines for assessing soil quality in specific areas [12,13].

Several studies have been created and implemented that are connected to estimating soil enzyme activity utilizing various approaches to acquire the results [14]. Based on this research, it is possible to determine the research gaps of soil enzyme activity and how we can increase the biological mechanism associated with specific enzymes in the carbon and nitrogen cycle in the interest of increasing agricultural yield. Multivariate linear regression (MLR) is the most widely used method for estimating soil parameters due to its simple design, quick calculation, and interpretation. On the other hand, MLR is unable to detect nonlinear relationships between responses and environmental variables. As a result, machine learning methods like "Artificial Neural Network (ANN), Support Vector Machine (SVM), Classification and Regression Tree (CART), and Classification Regression Tree (CART)" are increasingly being used in soil property assessment. Machine learning approaches can represent linear and nonlinear correlations between responses and environmental factors and have a simple structure, good fitting ability, and high prediction accuracy [15].

This method offers a novel and more convenient way to assess and estimate soil parameters such as "soil texture, salinity, soil organic carbon, and nitrogen". Random Forest (RF) is a data mining approach developed as an extension of CART [16–18]. The RF model has several advantages over additional statistical modeling techniques, including the capability to represent extremely nonlinear dimensional associations, resistance to "overfitting", relative dependability in the presence of noise data, the creation of an Unbiased Error Rate measure, and the ability to discern the significance of the variables used. As a result, the

RF model has been frequently used to estimate soil properties in multivariate nonlinear data processing. The main contribution of this research paper is as follows:

1. Predict the activity of soil enzymes based on chemical and physical soil parameters.
2. Evaluate the optimal performance model by correlating all factors.
3. Select and determine the optimal artificial approach algorithm for estimating enzyme activity.
4. Compare the performance model-finding algorithm (PSEA-ML and PSEA–ANN).

2. Materials and Methods

The Bhandara district is in the eastern plateau and hills region of the Maharashtra state of India. The study area is located at 20°44′59.99″ N latitude and 79°52′59.99″ E longitude. The elevation of the study area is probably 2000 m, and the average annual temperature and precipitation at the site are 59.6 degrees Celsius and 250 mm [19], respectively. Different types of soils are available in this region, ranging from deep loamy to clay soil mixed with red and black soils. According to the study of soil analogies, different soils are classified as sand, silt, and clay. This study collects various forms of "sand, silt, soil, pH, nitrogen, phosphorus, soil organic matter (SOM), soil organic carbon (SOC), and soil enzyme activity". Soil is used to assess the quality and quantity of each area of land, to determine whether it is balanced or unbalanced, based on the presence of each feature [20].

2.1. Soil Dataset

The soil data used for this research are from different formers with different blocks. There are three types of soil components available in agricultural soil, i.e., physical soil, chemical soil, and organic soil, which are more important for growing crops. Physical soils are identified by properties such as soil texture, soil structure, soil density, and soil temperature. Soil texture is the main property of physical soil. Therefore, soil texture has been considered in this research. The texture of the soil consists of sand, clay, silt, and depth. Many chemical components are available in chemical soil properties; only nitrogen and phosphorus are included here. Similarly, biological soil properties contain many factors, but here only enzymes are included. Some other components like pH value, soil organic carbon (SOC) and soil organic matter (SOM) are included, which is significant to identify the soil enzyme activities. Two types of soil data have been taken in this research; the first type of soil data is for soil fertility, and the second type of soil data are for soil enzyme activity. Soil data are taken from 25 different villages in the Bhandara district of Maharashtra state, India.

Each soil sample presents different chemical, physical, and biological components. The first parameter is Ph value, which is broadly categorized into three categories—neural, acidic, and alkaline—and represented on a scale from 0 to 10. Ph neutral value is around 7, below 7 is acidic, and above 7 Ph value is alkaline. Sand, silt, clay, nitrogen, phosphorus, soil organic matter, and soil organic carbon parameters are represented by their percentage of content available in each soil sample. Based on content analysis of all the parameters identified, the soil fertility level is considered either low, medium, or high. Soil depth represents how much depth is required to remove the soil for soil testing. Soil depth is used for collecting the soil sample for soil testing. The maximum depth of the soil sample is more effective for finding soil enzymes because more soil organisms are available at a greater depth in the soil. A total of 3972 soil sample data were taken with 11 parameters, as shown in Table 1. Soil was collected from each farmer for soil testing of each soil sample, using the farmer's identity to identify which farmer's soil is deficient of nutrients. For verifying and repeatability of soil samples, preprocessing techniques like finding the missing values, repeated values, and converting from categorical values to numerical values were implemented. Based on this assessment and analysis of soil, a soil dataset was generated for developing the proposed research work.

Table 1. Soil sample dataset of Bhandara district, Maharashtra.

S. No.	Ph	Sand (%)	Silt (%)	Clay (%)	N (ppm)	P (ppm)	SOM (ppm)	SOC (ppm)	Fertility Level	Depth (cm/mm)	Soil Enzyme
1	7.70	20.00	43.00	36.00	40.60	12.60	12.60	9.70	Medium	30	Urease
2	6.58	20.00	43.00	36.00	40.60	12.60	12.60	9.70	Medium	30	Urease
3	6.12	20.00	43.00	36.00	40.60	12.60	12.60	9.70	Medium	30	Invertase
4	6.50	20.00	43.00	36.00	40.60	12.60	12.60	9.70	Medium	30	Invertase
5	6.12	20.00	43.00	36.00	40.60	12.60	12.60	9.70	Medium	30	Acid phosphatase
6	6.42	20.00	43.00	36.00	40.60	12.60	12.60	9.70	Medium	30	Acid phosphatase
7	6.24	20.00	43.00	36.00	36.15	8.23	8.23	9.70	Low	30	Urease
8	6.84	20.00	43.00	36.00	0.87	0.87	1.23	17.28	Low	20	Protease
9	6.84	20.00	43.00	36.00	1.20	1.20	1.23	17.28	Medium	20	Protease
10	6.84	33.00	21.00	46.00	1.37	1.37	1.23	17.28	Medium	20	Protease

2.2. Pre-Processing of Soil Dataset

Before applying machine learning algorithms, pre-processing techniques are required to clean the data and convert the non-numeric data into numeric ones; for example, converting the enzyme classification to a numerical form such that 1 indicates the presence of urease and 0 indicates the absence. Similarly, all enzymes need to convert 1 s and 0 s into numerical form. The soil fertility characteristic has three levels—namely low, medium, and high—which also need to be converted into a numerical form, such as low level indicates 0 values, medium level indicates 1 value, and high level indicates 2 values. After completing the pre-processing techniques, 80% of the data is used for the training model and 20% of the data for the testing model.

2.3. Proposed Methodology

This paper proposes a methodology to predict soil enzyme activities using machine learning algorithms (Multiple Linear Regressions (MLR), Random Forest (RF), Extremely Randomized Tree Classifier (ERTC), and Artificial Neural Network (ANN)) by analyzing physical soil characteristics and chemical characteristics. The block diagram of the proposed methodology (PSEA-PC) predicts the soil enzyme activity and crops, as shown in Figure 1.

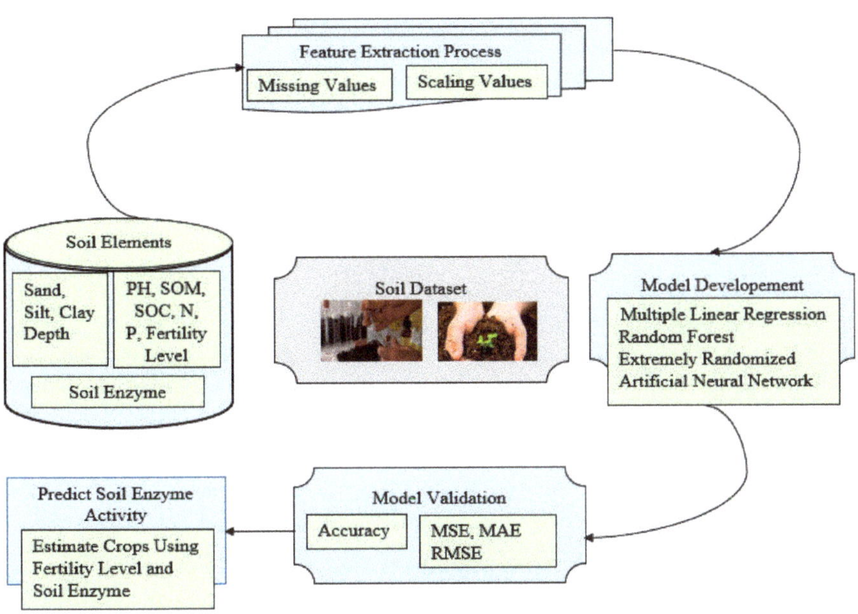

Figure 1. Block diagram of proposed methodology (PSEA-PC).

2.3.1. Development of the ML Model

This section interpreted the different machine learning algorithms with soil enzyme activity.

A. Multiple linear regressions for soil enzyme activity

MLR is a type of supervised machine learning regression technique. Multiple linear regression models are the most suitable technique for predicting soil characteristics [21,22]. Consider a linear regression relationship between numerous independent variables, such as x_1, $x_2, x_3 \ldots .. x_n$, and a dependent variable used (y_{pred}); $\in$ is denoted as the model error term provided in Equation (1):

$$y_{pred} = \sum_{i=1}^{n} b_i x_i + \in = b_0 x_0 + b_1 x_1 + \ldots b_n x_n + \in \quad (1)$$

where Y is the dependent variable or outcome, $x_i(i = 0, 1, 2, 3, \ldots, n)$ are independent variables, c is an intercept, b_i $(i = 0, 1, 2, 3, \ldots, n)$ is the regression coefficient, and ϵ is the residual of regression or error. The cost function (K) is used to find and minimize the error from dependent and independent variables; the best-fit line is provided in Equation (2). The optimized best-fit line is determined using gradient descent utilizing Equation (3), which uses a convergence algorithm for calculating the gradient descent (b_k); detailed discussion is given in Algorithm 1.

$$\mathrm{K}(b_0, b_1) = \frac{1}{2n} \sum_{i=1}^{n} (y\_pred - y)^2 \quad (2)$$

$$\text{Gradient\_descent}\ (b_k) = b_k - \alpha \frac{\partial}{\partial b_k} k\ (b_0, b_1) \quad (3)$$

α implies the learning rate and it could be considered a small range like 0.001, and k implies the feature index number $k = (0, 1, 2, 3, \ldots \ldots \ldots n)$. From Equations (2) and (3), we obtain Equation (4):

$$b_k = b_k - \frac{\alpha}{n} \sum_{i=1}^{n} (\ y\_pred - y) \quad (4)$$

Algorithm 1 PSEA–MLR-I (Predict the soil enzyme activity using Multiple Linear Regression)

Target: Optimal combination of response variables and enzyme activity in the soil
Input: N = (PH, Sand, Silt, Clay, N, P, SOM, SOC, Fertility level, and Depth)
Output: K = (Predict soil enzyme activity)
1: Initialization of all N and K soil data parameters
2: Pre-processing of the soil dataset with N and K parameters
3: Randomly select 80% soil dataset for training and 20% soil dataset for testing purposes
4: Apply MLR-supervised ML algorithms on a given data set
5: Compute the Accuracy, MSE, RMSE, and MAE of the model
6: Predict soil enzyme activity
7: End

B. Random Forest for soil enzyme activity

The Random Forest model is a multivariate technique that was created to improve the efficiency and accuracy of Classification and Regression Trees (CART). This model combines numerous Classification and Regression Tree algorithms and random variable selection and bagging to make each Classification and Regression Tree more fulfilled. Simultaneously, random feature extraction and bagging techniques cause every factor in the Random Forest to have a smaller correlation [23,24]. Calculate information gain (IG) using the entropy method of all splitting feature data given in Equation (5) and Algorithm 2.

$$\text{IG (IDV, DV) = Entropy (IDV) − Entropy (IDV, DV)} \quad (5)$$

For Binary Tree Gini, the importance of two child nodes is provided in Equation (6).

$$mi_{j=}We_jC_j - We_{left(j)}\ C_{left(j)} - We_{right(j)}\ C_{right(j)} \quad (6)$$

The importance of each feature of the decision tree can be calculated in Equation (7)

$$Fe_{i=}\frac{\sum_{i=1}^{n} node\ j\ splits\ on\ feature\ i\ mi_j}{\sum_{k\ \in all\ nodes}\ mi_j} \quad (7)$$

IDV implies an independent variable (x_i features), DV is a dependent variable (y_i features), mi_j implies the importance of node j, $We_{(j)}$ considers the weight number of samples reaching node j, $We_{left(j)}$ represents a left child node that is split on node j, $C_{(j)}$ implies an impurity value of node j, and Fe_i is the importance of features i; see i Equations (8) and (9).

$$Narm_{i=}\frac{Fe_i}{\sum_{k\ \in all\ nodes}\ Fe_i} \quad (8)$$

$$RFfe_i = \frac{\sum_{k\ \in all\ nodes}\ Narm_i}{\text{IDV}} \quad (9)$$

Algorithm 2 PSEA–RF-II (Predict soil enzyme activity using Random Forest algorithms)

Target: Optimal combination of response variables and enzyme activity in the soil
Input: N = (PH, Sand, Silt, Clay, N, P, SOM, SOC, Fertility level, and Depth)
Output: K = (Predict soil enzyme activity)
1: Initialization of all N and K soil data parameters
2: Pre-processing of the soil dataset with N and K parameters
3: Randomly select 80% soil dataset for training and 20% soil dataset for testing purposes
4: Apply RF-supervised ML algorithms on a given data set
5: Compute the Accuracy, MSE, RMSE, and MAE of the Model
6: Predict soil enzyme activity
7: End

This algorithm has been used for predicting soil enzyme activity using Random Forests. First, initialize the soil parameters including the chemical, physical, and biological components. Then, use the pre-processing techniques for cleaning the data, converting the categorical to numerical, and determining the missing and null data. Next, select 80% of the data for training and 20% for testing and use the Random Forest model to identify the best accuracy using evaluation metrics like MSE, RMSE, and MAE.

C. Extremely Randomized Trees Classifiers for soil enzyme activity

Extremely Randomized Trees Classifiers are a form of ensemble classification algorithm that outputs a classification result by combining the outcomes of several de-correlated decision trees aggregated in a "forest." (Algorithm 3). It is conceptually identical to a Random Forest Classifier, apart from how the decision trees in the forest are constructed. Create an additional tree classifier based on each decision tree's original dataset [25]. Using mathematical notation, randomly choose n features from a collection of all features offered by each tree for splitting the data and obtaining the best feature of all trees (Gini Index). This decision tree creates a multi-correlated feature from various random samples. First, we need to calculate the entropy of each feature based on mathematical Equation (10):

$$\text{Entropy(s)} = \sum_{i=1}^{m} -p_i log_2\ (p_i) \quad (10)$$

where Entropy(s) is a random sample of each feature of the tree, m is the number of unique classification labels, p_i is the proportion of each row with the target label i.

Algorithm 3 PSEA–ERT-III (Predict soil enzyme activity using extremely randomized tree classifiers)

Target: Optimal combination of response variables and enzyme activity in the soil
Input: N = (PH, Sand, Silt, Clay, N, P, SOM, SOC, Fertility level, and Depth)
Output: K = (Predict soil enzyme activity)
1: Initialization of all N and K soil parameters
2: Pre-processing of the soil dataset with N and K parameters
3: Randomly select 80% soil dataset for training and 20% soil dataset for testing purposes
4: Apply ERT-supervised ML algorithms on a given data set
5: Compute the Accuracy, MSE, RMSE, and MAE of the Model
6: Predict soil enzyme activity
7: End

2.3.2. Artificial Neural Network for Soil Enzyme Activity

An Artificial Neural Network (ANN) is a system that divides artificial neurons into three layers (input, hidden, and output). During the training of the ANN approach when used for regression analysis, the basic parameters of artificial neurons, such as weight, threshold, and activation functions, were tuned [26–28]. Given input soil properties, this ANN technique predicts soil enzyme activity (Algorithm 4). The ANN model with "relu" and "sigmoid" activation functions was utilized to add a hidden layer. The soil enzyme activities convert the cycle of carbon (C), nitrogen (N), and phosphorus (P) with C-glucosidase, invertase, dehydrogenase, catalase, urease, N-acetyl-glucosaminidase, protease, which are predicted using 80% of the training dataset and 20% of the testing dataset. Figure 2 shows the structure of an Artificial Neural Network Model. An Artificial Neural Network is divided into two techniques for passing data known as forwarding propagation; the forward propagation considered a perceptron is provided in Equations (11)–(15).

$$\sum = (x_1 * we_1) + (x_2 * we_2) + \ldots\ldots.. + (x_n * we_n) \tag{11}$$

$$x\ we_i = (x_1 * we_1) + (x_2 * we_2) + \ldots\ldots.. + (x_n * we_n) \tag{12}$$

$$\sum = x\ we_i \tag{13}$$

$$Z = x\ we_i + B \tag{14}$$

$$y_{pred} = \sigma(z) = \frac{1}{1 + e^{-z}} \tag{15}$$

where σ indicates an activation function of the neural model, we_i is the identified weight feature value, y_{pred} represents a predicted value, and B implies a bias of the neural model. For implementing the Artificial Neural Network, various parameters are required like the input layer, hidden layer, and output layer. Equation (11) shows the addition of all the input features with weight to calculate the average of features. Equation (12) is determined to simplify all the features with feature weight values. Equation (13) is determined to optimize the feature along with weight values in a single process. Equation (14) produces the outcomes by adding the bias (B) with all features and weight values. Equation (15) shows the prediction of the result based on all features using the sigmoid function.

Algorithm 4 PSEA–ANN-IV (Predict soil enzyme activity using ANN)

Target: Optimal combination of response variables and enzyme activity in the soil
Input: N = (Sand, Silt, Clay, PH, SOM, SOC, Available Nitrogen, Available Phosphorus, Depth)
Output: K = (Predict soil enzyme activity)
1: Initialization of all N and K parameters
2: If (Sn ! = Sc)
3: Then pre-process and scale the data
4: Otherwise, go to step 5
5: Choose and select x and y variable
6: Split 80% data for training and 20% data for testing
7: Add first and second hidden layer (activation function = relu)
8: Add output hidden layer (activation function = sigmoid)
9: Compile and validate data
10: Select the epoch and calculate the accuracy
11: Predict soil enzyme activity
12: End
(Where Sn = scale features from soil dataset, Sc = scale soil enzyme features of the dataset)

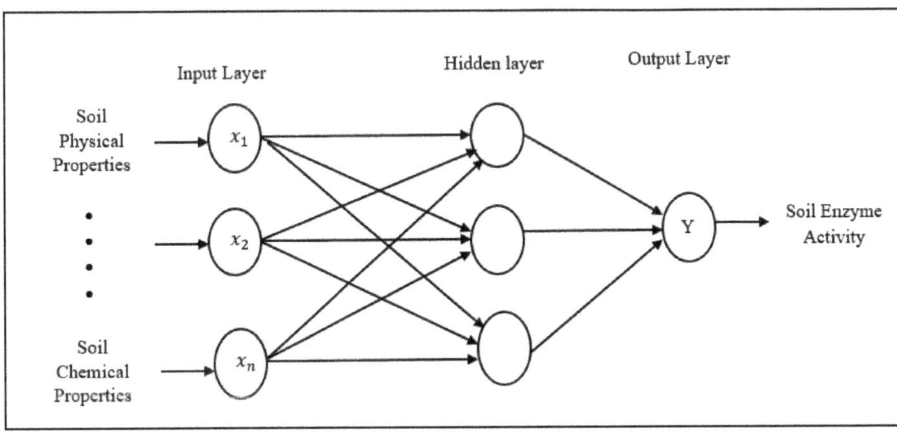

Figure 2. Structure of ANN Model.

3. Result and Discussion

The descriptive statistics of soil enzyme activity with soil properties datasets are shown in Table 2. The descriptive statistics summary includes the following parameters: Minimum (Min), Maximum (Max), Standard Mean Value (Mean), and Standard Deviation (SD). Python with jupyter notebook was used to implement the descriptive statistical analysis, machine learning model, and Artificial Neural Network on Windows 10. In this summary, calculate the descriptive statistics of PH, Sand, Silt, Clay, Available Nitrogen, Available Phosphorus, SOM, SOC, Depth, Soil Fertility Level, and Soil Enzyme Activity. The PH standard deviation value was smaller than the mean (SD Mean), and the soil enzyme activity was SD > Mean. The mean value of available nitrogen and depth soil parameters was extremely covariate (Mean > 100%) compared to other parameters, while the standard deviation of available nitrogen and available phosphorus was highly covariate (SD > 100%) compared to certain other factors [29]. Figure 3 illustrates the very positive and strongly negative correlations of a soil dataset in matrix format and according to this correlation, the correlation between sand depth and clay soil qualities is substantially negatively correlated.

Table 2. Statistics summary of soil physical, chemical features, and soil enzyme activity.

Soil Properties	Mean	Min	Max	SD
PH	6.3314	3.8000	8.7900	0.7280
Sand	38.1617	8.0000	90.0000	19.3332
Silt	31.7412	4.0000	67.0000	14.5515
Clay	32.1268	11.0000	47.0000	10.5310
Available Nitrogen	19.1986	0.6900	317.0000	51.6195
Available Phosphorus	9.9923	0.6900	146.5000	23.0569
SOM	5.1652	0.6900	127.2000	11.1173
SOC	7.6628	0.6900	127.2000	14.1052
Depth	19.2412	10.0000	30.0000	5.4071
Soil Fertility Level (Low, Medium, and High)	1.2921	0.0000	2.0000	0.8235
Soil Enzyme	9.3529	0.0000	19.0000	18.0000

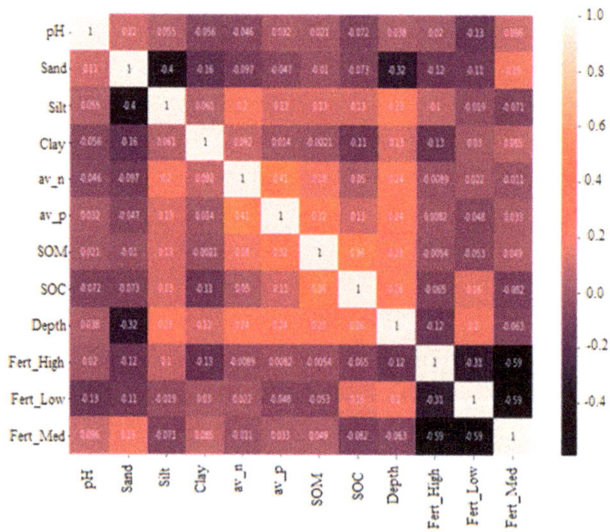

Figure 3. Heatmap of the positive and negative correlation of all soil features.

Model Validation

In this study, the dataset collected approximately 3972 soil sample datasets, of which we used the training dataset of randomly selected records, accounting for approximately 80% of the total records, to develop Multiple Linear Regression, Random Forest, extra tree classifier, and Artificial Neural Network models, and the testing dataset included the remaining 20% of the records to verify the model's estimation accuracy for soil enzyme activities [30]. The performance of the ML and ANN models was assessed using the coefficient of determination (R2), mean absolute error (MAE), root mean square error (RMSE), classification report, and confusion matrix in Table 3. The following are the evaluation performance indices calculated to validate the models provided in Equations (16)–(18).

$$R^2 = 1 - \frac{\sum_{i=1}^{n} \left(pre_i - \overline{obs_i}\right)^2}{\sum_{i=1}^{n} \left(obs_i - \overline{obs_i}\right)^2} \quad (16)$$

$$MAE = \frac{1}{n} \sum_{i=1}^{n} | \ pre_i - obs_i \ | \quad (17)$$

$$RMSE = \sqrt{\frac{1}{n}\sum_{1=1}^{n}(obs_i \;- pre_i\;)^2} \tag{18}$$

where pre_i and obs_i are the predicted and observed values, respectively, from a random sample of i, $\overline{obs_i}$ is the mean value observation, and n is the total number of sample records.

Table 3. Confusion matrix with classification metrics report.

		Predicted Soil Enzyme Activity Data		
		Positive	**Negative**	
Actual Soil Enzyme Activity Data	Positive	True Positive (TP)	False Negative (FN)	Sensitivity $\frac{TP}{(TP+FN)}$
	Negative	False Positive (FP)	True Negative (TN)	Specificity $\frac{TN}{(TN+FP)}$
Classification Metrics		Precision $\frac{TP}{(TP+FP)}$	Negative Predictive Value $\frac{TN}{(TN+FN)}$	Accuracy $\frac{TP+TN}{(TP+TN+FP+FN)}$

These tables have identified the confusion matrix with classification reports. There are four parameters: true positive, true negative, false positive, and false negative. Based on these parameters, the actual observation and predicted observation has been measured. This study identifies the actual soil enzyme activity and predicts the soil enzyme activity based on the confusion matrix.

This correlation aims to identify important features that may be used to implement and estimate soil enzyme activity using numerical data from the training and testing datasets. Figure 4 depicts the dataset's summary of soil enzyme activity. In this study, 3972 soil samples were collected from the Bhandara district of Maharashtra to predict the activity of each enzyme factor related to the carbon, nitrogen, and phosphorus nutrient cycle, including urease, acid phosphatase, invertase, alkaline phosphatase, phosphatase, protease, cellulose, N-acetyl-glycosaminidases, and C-glucosidase [31,32]. This study used a large amount of urease and cellulose soil enzyme activity for prediction, with 80% of the dataset being used for training and 20% being used for testing to determine the best soil enzyme activity solution.

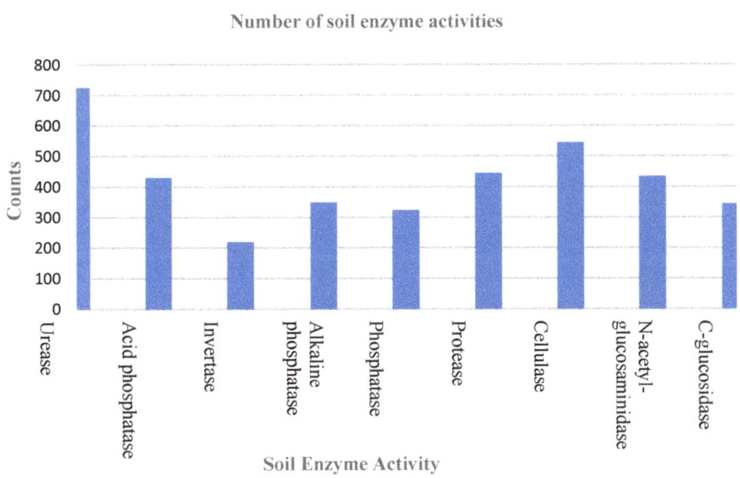

Figure 4. Summary of soil enzyme activities in the dataset.

Figure 5 depicts the graphical representations of soil samples, including the soil's physical, chemical, and enzyme activity. This is used to forecast soil enzyme activity based on each soil component's amount, determining what proportion of soil components are available in the soil, as well as soil enzyme activity and fertility level. Based on the summary of all values of soil factors, levels of urease, acid phosphatase, invertase, alkaline phosphatase, phosphatase, protease, cellulose, N-acetyl-glucosaminidase, and C-glucosidase are predicted.

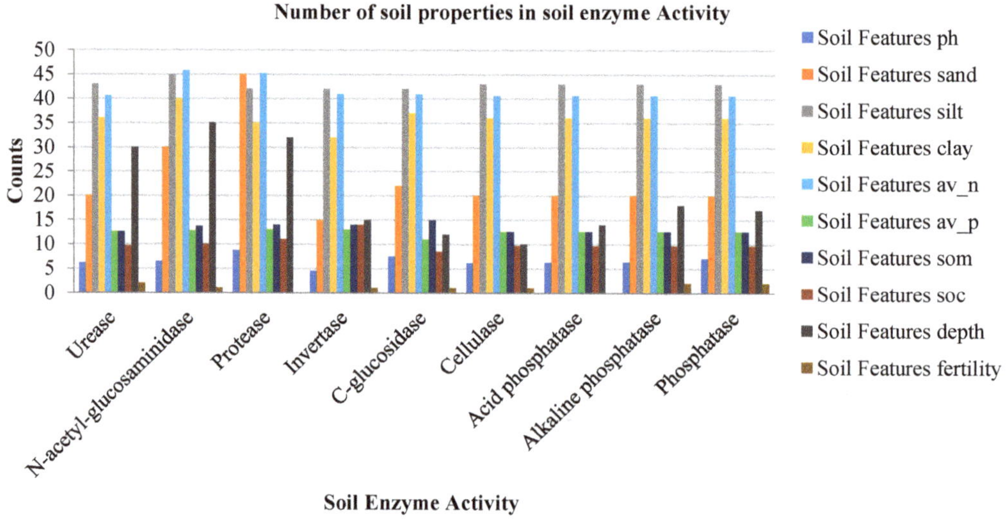

Figure 5. Summary of soil enzyme activities with soil properties.

Figure 6 shows a graphical depiction of the machine learning model classification, which includes Random Forest, additional tree classifier, and regression—including Multiple Linear Regression and Artificial Neural Network—to predict soil enzyme activity. In comparison to other classifications, urease soil enzyme activity found the RF and additional tree models had good accuracy. The RF model was found to be superior in terms of acid phosphatase activity, invertase, alkaline phosphatase, phosphatase, protease, and N-acetyl-glucosaminidase. The MLR model was best suited for cellulose, and the ANN model seemed good for C-glucosidase. According to the implementation results, the Random Forest model outperformed other models in terms of identifying soil quality and enhancing agricultural productivity for a specific location. Using Multiple Linear Regression with training and testing datasets, we were able to estimate the cellulose soil enzyme activity with high accuracy.

Classification and regression approaches were utilized in this work to determine the most effective approach for estimating soil enzyme activity. For predicting the activity of urease, acid phosphatase, invertase, alkaline phosphatase, phosphatase, protease, cellulose, N-acetyl-glucosaminidase, and C-glucosidase soil enzymes, RF, MLR, Extra Tree, and ANN models were used in classification, while RF and MLR were used in regression.

Multiple Linear Regression and Random Forest models were employed in this investigation to determine which soil enzyme activity had the best performance, including MSE, MAE, and RMSE characteristics. Figure 7 shows how MLR's evaluation performance metrics are represented. MSE's urease and N-acetyl-glucosaminidase soil enzyme activity was found to be good, meaning there was minimal error compared to others (0.0549 and 0.0449). For MAE, a slight error of phosphatase and C-glucosidase (0.111567) was discovered rather than the activity of other enzymes. C-glucosidase (0.1342) had a small error in RMSE compared to other enzymes. Random Forest regression techniques are the best appropriate model for predicting soil enzyme activity. In Table 4, we estimate the soil enzyme activity by analyzing the MSE, RMSE, and MAE of the Multiple Linear Regression approach.

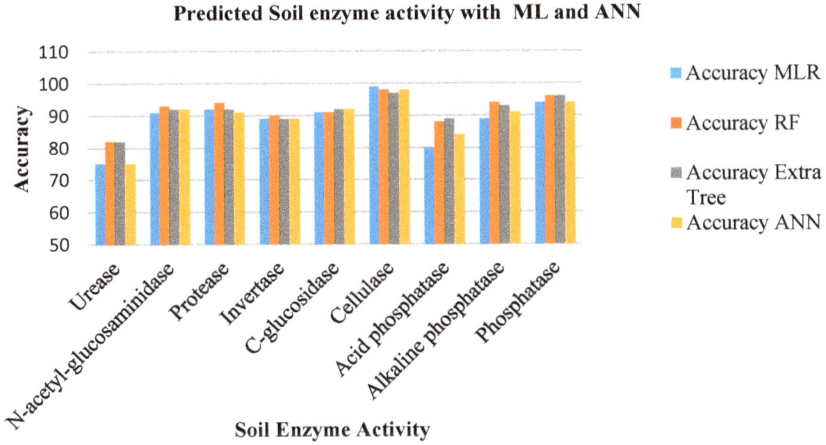

Figure 6. Predicted soil enzyme activity with ML and ANN.

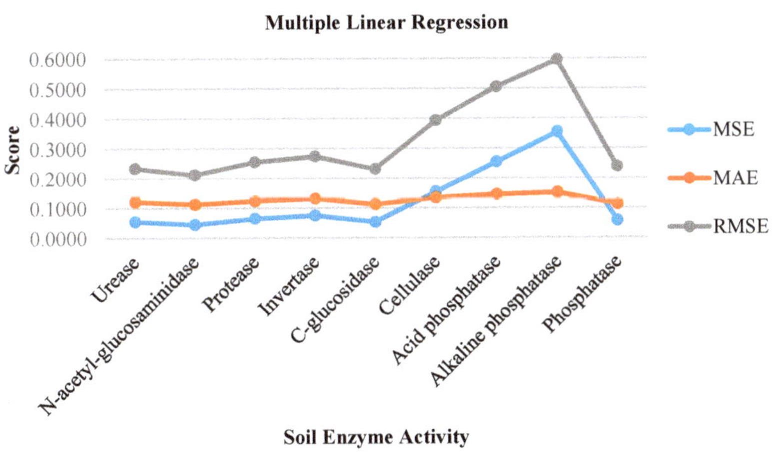

Figure 7. Evaluation performance metrics of MLR.

Figure 8 shows that the MAE, RMSE, and MSE parameters were used to construct the evaluation performance metrics. Alkaline phosphatase soil enzyme activity was found to be an excellent dependent variable for MSE prediction outcomes, with a lower error (0.0146) than other variables. C-glucosidase (0.0162) soil enzyme activity was discovered to have a lower error than the others in MAE performance criteria. Alkaline phosphatase soil enzyme activity had a lower error (0.1209) than the others in terms of RMSE performance metrics.

Table 4. Compare the MSE, MAE, and RMSE of MLR for PSEA-ML.

Soil Enzyme	MSE	MAE	RMSE
		Multiple Linear Regression	
Urease	0.0549	0.1216	0.2343
N-acetyl-glucosaminidase	0.0449	0.1126	0.2119
Protease	0.0649	0.1232	0.2547
Invertase	0.0749	0.1316	0.2737
C-glucosidase	0.0535	0.1116	0.2313
Cellulase	0.1549	0.1342	0.3936
Acid phosphatase	0.2549	0.1452	0.5049
Alkaline phosphatase	0.3549	0.1516	0.5957
Phosphatase	0.0569	0.1116	0.2385

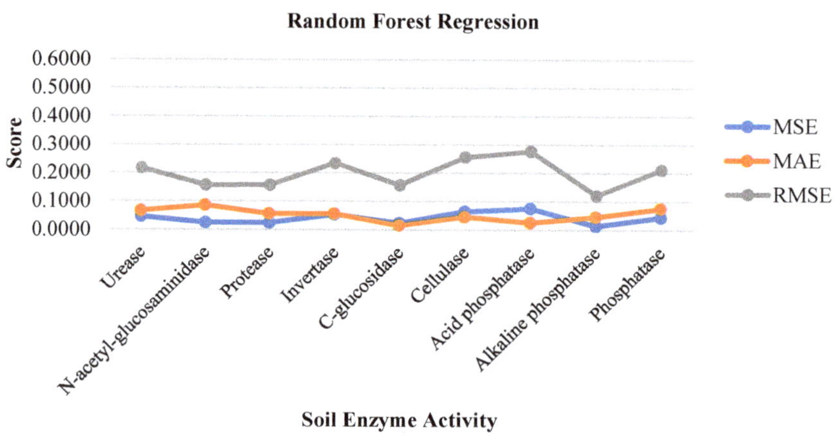

Figure 8. Evaluation performance metric of Random Forest.

In Table 5, the soil enzyme activity is estimated by analyzing MSE, RMSE, and MAE of the Random Forest regression approach. The result demonstrated the identification of the best optimal outcome of soil enzyme activity.

Table 5. Compare the MSE, MAE, and RMSE of the Random Forest approach for PSEA-ML.

Soil Enzyme	MSE	MAE	RMSE
		RF	
Urease	0.0466	0.0662	0.2159
N-acetyl-glucosaminidase	0.0245	0.0862	0.1564
Protease	0.0246	0.0569	0.1569
Invertase	0.0545	0.0566	0.2334
C-glucosidase	0.0246	0.0162	0.1569
Cellulase	0.0655	0.0462	0.2559
Acid phosphatase	0.0765	0.0262	0.2765
Alkaline phosphatase	0.0146	0.0462	0.1209
Phosphatase	0.0446	0.0762	0.2112

In Table 6, the measured MAE, RMSE, and MSE parameters were used to construct the evaluation performance metrics. Cellulase soil enzyme activity was found to be an excellent dependent variable for MSE prediction outcomes, with a lower error (0.0259) than other variables. Cellulase soil enzyme activity was discovered to have a lesser error (0.0752) than the others in the MAE performance criteria. Cellulase and N-acetyl-glucosaminidase soil enzyme activity had less error (0.1609 and 0.2493, respectively) than the others in terms of RMSE performance metrics.

Table 6. Compare the MSE, MAE, and RMSE of the extra tree regressor approach for PSEA-ML.

Soil Enzyme	MSE	MAE	RMSE
		Extra Tree Regressor	
Urease	0.1761	0.7610	0.4196
N-acetyl-glucosaminidase	0.0622	0.2746	0.2493
Protease	0.0777	0.0777	0.2787
Invertase	0.1036	0.1036	0.3219
C-glucosidase	0.0829	0.0829	0.2879
Cellulase	0.0259	0.0752	0.1609
Acid phosphatase	0.0907	0.1473	0.3012
Alkaline phosphatase	0.0618	0.0999	0.2486
Phosphatase	0.0500	0.0814	0.2236

The Artificial Neural Network model implemented using Python programming with Keras and TensorFlow library in Python was built using three layers: input layer, hidden layer, and output layer. Programming used dense layers for the fully connected neural network. The dense layer considered an input layer with 10 units (input features) with the 'relu' activation function, dense_1 represented the hidden layer with the 'relu' activation function, and the output layer used 1 unit (output features) with the 'sigmoid' function. For compiling, the ANN model used an 'adam' optimizer for reducing the error/loss with 'binary_crossentropy' loss. This model executed 32 batch sizes and 100 epochs for calculating the loss score, training, and validating accuracy. This model achieved 99% accuracy in cellulase enzyme activity. Figure 9 shows the ANN model summary and Figure 10 shows the epoch generation with ANN model loss and validation accuracy.

```
Model: "sequential"
_________________________________________________________________
Layer (type)                 Output Shape              Param #
=================================================================
dense (Dense)                (None, 10)                130
_________________________________________________________________
dense_1 (Dense)              (None, 10)                110
_________________________________________________________________
dense_2 (Dense)              (None, 1)                 11
=================================================================
Total params: 251
Trainable params: 251
```

Figure 9. Artificial Neural Network Model Summary.

Table 7 measures the MAE, RMSE, and MSE parameters used to construct the evaluation performance metrics. Protease soil enzyme activity was found to be an excellent dependent variable for MSE prediction outcomes, with a lower error (0.1384) than other variables. C-glucosidase soil enzyme activity was discovered to have a lower error (0.1443) than the others in the MAE performance criteria. C-glucosidase and protease soil enzyme activity had a lower error (0.3799) than the others in terms of RMSE performance metrics.

```
Epoch 1/100
25/25 [==============================] - 3s 4ms/step - loss: 0.6193 - accuracy: 0.7355
Epoch 2/100
25/25 [==============================] - 0s 7ms/step - loss: 0.4433 - accuracy: 0.9342
Epoch 3/100
25/25 [==============================] - 0s 6ms/step - loss: 0.3224 - accuracy: 0.9884
Epoch 4/100
25/25 [==============================] - 0s 4ms/step - loss: 0.2390 - accuracy: 0.9923
Epoch 5/100
25/25 [==============================] - 0s 8ms/step - loss: 0.1805 - accuracy: 0.9923
Epoch 6/100
25/25 [==============================] - 0s 4ms/step - loss: 0.1382 - accuracy: 0.9923
Epoch 7/100
25/25 [==============================] - 0s 4ms/step - loss: 0.1083 - accuracy: 0.9923
Epoch 8/100
25/25 [==============================] - 0s 6ms/step - loss: 0.0865 - accuracy: 0.9923
Epoch 9/100
25/25 [==============================] - 0s 4ms/step - loss: 0.0711 - accuracy: 0.9923
Epoch 10/100
```

Figure 10. ANN model of epoch generation with loss and validation accuracy.

Table 7. Compare the MSE, MAE, and RMSE of the Artificial Neural Network approach for PSEA-ANN.

Soil Enzyme	MSE	MAE ANN	RMSE
Urease	0.1818	0.3464	0.4263
N-acetyl-glucosaminidase	0.2353	0.4732	0.4850
Protease	0.1383	0.3421	0.3718
Invertase	0.3328	0.5687	0.5768
C-glucosidase	0.1443	0.1443	0.3798
Cellulase	0.3826	0.3826	0.6185
Acid phosphatase	0.1913	0.4138	0.4373
Alkaline phosphatase	0.3402	0.5735	0.5832
Phosphatase	0.4263	0.6417	0.6529

Table 8 lists the specific crops depending on all soil enzyme activity and soil fertility levels (low, medium, and high). Because some soil components are particularly low in low fertility levels, a balanced number of factors, as well as enzyme activity, is required. The soil enzyme activity—such as urease, invertase, C-glucosidase, and acid phosphatase, which are associated with low fertility levels—is estimated. Based on this prediction, cucumber, maize, peanut pepper, soybean, and sugarcane were identified as crops that are useful for harvesting and increasing crop productivity [33–35]. Based on the analysis of soil fertility level and soil enzyme activity using machine learning algorithms, specific crops are suggested for increasing crop productivity. Each crop is suggested for a different soil fertility level—low, medium, and high—for each soil enzyme activity. For example, potato crop requires a high fertility level with protease and phosphatase soil enzyme activity.

According to this prediction, the activity of soil enzymes such as N-acetyl-glucosaminidase, cellulase, and alkaline phosphatase—which correspond to a medium fertility level—was estimated. The activity of the N-acetyl-glucosaminidase soil enzyme indicated crops such as chickpea cotton, rice wheat, peanut, and soybean. Cucumbers, maize, peanut, pepper, soybean, and sugarcane are likely have cellulase and alkaline phosphatase soil enzymes. The activity of soil enzymes such as protease and phosphatase—which are associated with high fertility levels—is estimated, and specific crops are chosen based on this prediction. Potato, cotton, sugarcane, maize, soybean, and pear were indicated by protease, and phosphatase soil enzyme activity indicated the crops potato, cotton, sugarcane, maize, soybean, and pear.

Table 8. List of specific crops based on soil fertility level and soil enzyme activity.

Fertility Level	Soil Enzyme Activity	Crops					
Low	Urease	Cucumber	Maize	Peanut	Pepper	Soybean	Sugarcane
Medium	N-acetyl-glucosaminidase	Chickpea	Cotton	Rice	Wheat	Peanut	Soybean
High	Protease	Potato	Cotton	Sugarcane	Maize	Soybean	Pear
Low	Invertase	Cucumber	Maize	Peanut	Pepper	Soybean	Sugarcane
Low	C-glucosidase	Cucumber	Maize	Peanut	Pepper	Soybean	Sugarcane
Medium	Cellulase	Cucumber	Maize	Peanut	Pepper	Soybean	Sugarcane
Low	Acid phosphatase	Cucumber	Maize	Peanut	Pepper	Soybean	Sugarcane
Medium	Alkaline phosphatase	Cucumber	Maize	Peanut	Pepper	Soybean	Sugarcane
High	Phosphatase	Potato	Cotton	Sugarcane	Maize	Soybean	Pear

4. Conclusions

This research was performed to evaluate soil enzyme activities, which included nine target features of soil enzymes such as urease, acid phosphatase, invertase, alkaline phosphatase, phosphatase, protease, cellulose, N-acetyl-glucosaminidase, and C-glucosidase, as well as chemical factors such as PH, SOC, SOM, available nitrogen, and available phosphorus; the physical factors were sand, silt, clay, and depth of soil for soil testing. Machine learning models such as MLR, RF, and extra tree classification techniques were compared with the ANN model for estimating soil enzyme activity. The best model was determined using a classification report, confusion matrix, and evaluation performance regression techniques such as MSE, MAE, and RMSE. According to the experimental results, the Random Forest model seems to be the most suitable model for determining the optimal soil enzyme activities as compared to other classification models. MAE, RMSE, and MSE were used to obtain good results in the MLR and RF regression techniques. Specific crops were recommended based on soil fertility levels, which are divided into three categories: low, medium, and high. Each soil level revealed a varied soil enzyme activity with a given crop, which is extremely beneficial to farmers in terms of enhancing crop output and determining soil quality.

Future work will include collecting additional soil enzyme activity classification of soil samples from various regions, estimating activities using various artificial methodologies, and recommending certain crops with fertilizer doses.

Author Contributions: Conceptualization, Y.S. and M.P.S.; methodology, P.S. and M.D.; software, V.S.; validation, S.K.; formal analysis, L.S. All authors have read and agreed to the published version of the manuscript.

Funding: This research received no external funding.

Institutional Review Board Statement: Not applicable.

Data Availability Statement: Not applicable.

Conflicts of Interest: The authors declare no conflict of interest.

References

1. Luis Moreno, J.; Bastida, F.; Díaz-López, M.; Li, Y.; Zhou, Y.; López-Mondéjar, R.; Benavente-Ferraces, I.; Rojas, R.; Rey, A.; García-Gil, J.C.; et al. Response of soil chemical properties, enzyme activities and microbial communities to biochar application and climate change in a Mediterranean agroecosystem. *Geoderma* **2022**, *407*, 115536. [CrossRef]
2. Tajik, S.; Ayoubi, S.; Lorenz, N. Soil microbial communities affected by vegetation, topography and soil properties in a forest ecosystem. *Appl. Soil Ecol.* **2020**, *149*, 103514. [CrossRef]
3. Xie, X.; Wu, T.; Zhu, M.; Jiang, G.; Xu, Y.; Wang, X.; Pu, L. Comparison of random forest and multiple linear regression models for estimation of soil extracellular enzyme activities in agricultural reclaimed coastal saline land. *Ecol. Indic.* **2021**, *120*, 106925. [CrossRef]
4. Pukalchik, M.A.; Katrutsa, A.M.; Shadrin, D.; Terekhova, V.A.; Oseledets, I.V. Machine learning methods for estimation the indicators of phosphogypsum influence in soil. *J. Soils Sediments* **2019**, *19*, 2265–2276. [CrossRef]
5. Zhang, L.; Chen, X.; Xu, Y.; Jin, M.; Ye, X.; Gao, H.; Chu, W.; Mao, J.; Thompson, M.L. Soil labile organic carbon fractions and soil enzyme activities after 10 years of continuous fertilization and wheat residue incorporation. *Sci. Rep.* **2020**, *10*, 11318. [CrossRef]

6. Jha, S.K.; Ahmad, Z. Soil microbial dynamics prediction using machine learning regression methods. *Comput. Electr. Agric.* **2018**, *147*, 158–165. [CrossRef]
7. Archana, K.; Saranya, K.G. Crop Yield Prediction, Forecasting and Fertilizer Recommendation using Voting Based Ensemble Classifier. *SSRG Int. J. Comput. Sci. Eng.* **2020**, *7*, 5–8.
8. Aponte, H.; Meli, P.; Butler, B.; Paolini, J.; Matus, F.; Merino, C.; Cornejo, P.; Kuzyakov, Y. Meta-analysis of heavy metal effects on soil enzyme activities. *Sci. Total Environ.* **2020**, *737*, 139744. [CrossRef]
9. dos Santos Teixeira, A.F.; Silva, S.H.G.; de Carvalho, T.S.; Silva, A.O.; Guimarães, A.A.; de Souza, F.M. Soil physicochemical properties and terrain information predict soil enzymes activity in phytophysiognomies of the Quadrilátero Ferrífero region in Brazil. *Catena* **2021**, *199*, 105083. [CrossRef]
10. Curtright, A.J.; Tiemann, L.K. Meta-analysis dataset of soil extracellular enzyme activities in intercropping systems. *Data Brief* **2021**, *38*, 107284. [CrossRef]
11. Jian, S.; Li, J.; Chen, J.; Wang, G.; Mayes, M.A.; Dzantor, K.E.; Hui, D.; Luo, Y. Soil extracellular enzyme activities, soil carbon and nitrogen storage under nitrogen fertilization: A meta-analysis. *Soil Biol. Biochem.* **2016**, *101*, 32–43. [CrossRef]
12. Piotrowska-Długosz, A.; Kobierski, M.; Długosz, J. Enzymatic activity and physicochemical properties of soil profiles of luvisols. *Materials* **2021**, *14*, 6364. [CrossRef] [PubMed]
13. Tajik, S.; Ayoubi, S.; Nourbakhsh, F. Prediction of soil enzymes activity by digital terrain analysis: Comparing artificial neural network and multiple linear regression models. *Environ. Eng. Sci.* **2012**, *29*, 798–806. [CrossRef]
14. Zhang, H.; Wu, P.; Yin, A.; Yang, X.; Zhang, M.; Gao, C. Prediction of soil organic carbon in an intensively managed reclamation zone of eastern China: A comparison of multiple linear regressions and the random forest model. *Sci. Total Environ.* **2017**, *592*, 704–713. [CrossRef]
15. Karthigadevi, K. Random Forest Classification Algorithm for Agricultural Data Analysis in Tirunelveli District. *J. Xi'an Univ. Archit. Technol.* **2020**, *12*, 418–432.
16. Hemageetha, N.; Nagalakshmi, N. Classification Techniques in Analysis of Salem District Soil condition for Cultivation of Sunflower. *Int. J. Comput. Sci. Eng.* **2018**, *6*, 642–646. [CrossRef]
17. Rajamanickam, J.; Savitha, D.M. Predictive model construction for prediction of soil fertility using decision tree machine learning algorithm. *Kongunadu Res. J.* **2021**, *8*, 30–35.
18. Zhao, D.; Arshad, M.; Li, N.; Triantafilis, J. Predicting soil physical and chemical properties using Vis-NIR in Australian cotton areas. *Catena* **2021**, *196*, 104938. [CrossRef]
19. Kashiwar, S.R.; Kundu, D.M.C.; Dongarwar, U.R. Soil fertility appraisal of Bhandara block of Maharashtra using geospatial techniques. *Int. J. Chem. Stud.* **2020**, *8*, 2570–2576. [CrossRef]
20. Geng, X.; Zhu, C.; Zhang, J.; Xiong, Z. Prediction of Soil Fertility Change Trend Using a Stochastic Petri Net. *J. Sign. Process. Syst.* **2021**, *93*, 285–297. [CrossRef]
21. Pinheiro, H.S.K.; de Carvalho, W., Jr.; da Silva Chages, C.; dos Anjos, L.H.C.; Owens, P.R. Prediction of topsoil texture through regression trees and multiple linear regressions. *Rev. Bras. Cienc. Solo* **2018**, *42*, e0170167. [CrossRef]
22. Zhang, H.; Yin, S.; Chen, Y.; Shao, S.; Wu, J.; Fan, M.; Chen, F.; Gao, C. Machine learning-based source identification and spatial prediction of heavy metals in soil in a rapid urbanization area, eastern China. *J. Clean. Prod.* **2020**, *273*, 122858. [CrossRef]
23. Deng, X.; Ma, W.; Ren, Z.; Zhang, M.; Grieneisen, M.L.; Chen, X.; Fei, X.; Qin, F.; Zhan, Y.; Lv, X. Spatial and temporal trends of soil total nitrogen and C/N ratio for croplands of East China. *Geoderma* **2020**, *361*, 114035. [CrossRef]
24. Xu, Y.; Wang, X.; Bai, J.; Wang, D.; Wang, W.; Guan, Y. Estimating the spatial distribution of soil total nitrogen and available potassium in coastal wetland soils in the Yellow River Delta by incorporating multi-source data. *Ecol. Indic.* **2020**, *111*, 106002. [CrossRef]
25. Sirsat, M.S.; Cernadas, E.; Fernández-Delgado, M.; Khan, R. Classification of agricultural soil parameters in India. *Comput. Electr. Agric.* **2017**, *135*, 269–279. [CrossRef]
26. Kalkhajeh, Y.K.; Arshad, R.R.; Amerikhah, H.; Sami, M. Comparison of multiple linear regressions and artificial intelli-gence-based modeling techniques for prediction the soil cation exchange capacity of Aridisols and Entisols in a semi-arid region. *Aust. J. Agric. Eng.* **2012**, *3*, 39–46.
27. Van Klompenburg, T.; Kassahun, A.; Catal, C. Crop yield prediction using machine learning: A systematic literature review. *Comput. Electr. Agric.* **2020**, *177*, 105709. [CrossRef]
28. Were, K.; Bui, D.T.; Dick, Ø.B.; Singh, B.R. A comparative assessment of support vector regression, artificial neural networks, and random forests for predicting and mapping soil organic carbon stocks across an Afromontane landscape. *Ecol. Indic.* **2015**, *52*, 394–403. [CrossRef]
29. Swapna, B.; Manivannan, S.; Nandhinidevi, R. Prediction of Soil Reaction (Ph) and Soil Nutrients Using Multivariate Statistics Techniques for Agricultural Crop and Soil Management. *Int. J. Adv. Sci. Technol.* **2020**, *29*, 1900–1912.
30. Munawar, A.A.; Yunus, Y.; Devianti; Satriyo, P. Calibration models database of near-infrared spectroscopy to predict agricul-tural soil fertility properties. *Data Brief* **2020**, *30*, 105469. [CrossRef]
31. Kivlin, S.N.; Treseder, K.K. Soil extracellular enzyme activities correspond with abiotic factors more than fungal community composition. *Biogeochemistry* **2014**, *117*, 23–37. [CrossRef]
32. Weintraub, S.R.; Wieder, W.R.; Cleveland, C.C.; Townsend, A.R. Organic matter inputs shift soil enzyme activity and allocation patterns in a wet tropical forest. *Biogeochemistry* **2013**, *114*, 313–326. [CrossRef]

33. Panchamurthi, S. Soil Analysis and Prediction of Suitable Crop for Agriculture using Machine Learning. *Int. J. Res. Appl. Sci. Eng. Technol.* **2019**, *7*, 2328–2335. [CrossRef]
34. Priya, P.; Muthaiah, U.; Balamurugan, M. Predicting yield of the crop using machine learning algorithms. *Int. J. Eng. Sci. Res. Technol.* **2018**, *7*, 1–7.
35. Sukhadia, K.; Chaudhari, M.B. A Survey on Rice Crop Yield Prediction in India Using Improved Classification Technique. Int. *J. Sci. Res. Comput. Sci. Eng. Inf. Technol.* **2019**, *5*, 501–507.

agriculture

MDPI

Article

A Lightweight Pest Detection Model for Drones Based on Transformer and Super-Resolution Sampling Techniques

Yuzhe Bai [1,†], Fengjun Hou [1,†], Xinyuan Fan [1], Weifan Lin [1], Jinghan Lu [1], Junyu Zhou [1], Dongchen Fan [2] and Lin Li [1,*]

[1] China Agricultural University, Beijing 100083, China; byz0871@cau.edu.cn (Y.B.); fxy07dr@cau.edu.cn (X.F.); lujinghan2019@cau.edu.cn (J.L.); cau_zhoujy@163.com (J.Z.)

[2] School of Computer Science and Engineering, Beihang University, Beijing 100191, China; 213352411@buaa.edu.cn

* Correspondence: lilinlsl0726@cau.edu.cn

† These authors contributed equally to this work.

Abstract: With the widespread application of drone technology, the demand for pest detection and identification from low-resolution and noisy images captured with drones has been steadily increasing. In this study, a lightweight pest identification model based on Transformer and super-resolution sampling techniques is introduced, aiming to enhance identification accuracy under challenging conditions. The Transformer model was found to effectively capture spatial dependencies in images, while the super-resolution sampling technique was employed to restore image details for subsequent identification processes. The experimental results demonstrated that this approach exhibited significant advantages across various pest image datasets, achieving Precision, Recall, mAP, and FPS scores of 0.97, 0.95, 0.95, and 57, respectively. Especially in the presence of low resolution and noise, this method was capable of performing pest identification with high accuracy. Furthermore, an adaptive optimizer was incorporated to enhance model convergence and performance. Overall, this study offers an efficient and accurate method for pest detection and identification in practical applications, holding significant practical value.

Keywords: smart agriculture; pest detection; Transformer; super resolution

1. Introduction

With the continuous advancement of agricultural technology, drones have been progressively adopted as efficient automation tools in various agricultural operations [1], including sowing, fertilization, and monitoring. In particular, for crop health monitoring, drones have demonstrated immense potential and value. Pests, as one of the primary threats in agricultural production, pose serious risks to crop health. While pesticides can address some pest issues [2], timely and effective pest detection remains paramount for pest prevention and control.

Traditional pest detection methods often rely on manual inspections [3] and solar tracking [4]. Not only these methods exhibit low efficiency, but also their accuracy is constrained by human experience and the intensity of manual labor, leading to potential oversights. Furthermore, the frequency and scope of manual inspections are limited, preventing extensive, high-frequency pest monitoring, especially given the small size of pests [5]. This limitation can result in missing optimal opportunities for prevention and control during the initial stages of pest outbreaks. While pheromone-based pest detection methods exist [6], they are specific to particular pests [7], offering limited versatility.

The rapid advancement of computer vision technology in recent years has introduced new avenues for smart agriculture [8–11]. Through image recognition and deep learning techniques, high-efficiency and accurate identification of pests can be achieved [12]. In this realm, researchers from various countries have embarked on several investigations. Liang,

Quanjia, developed a rice pest recognition model based on an improved YOLOv7 algorithm. By employing the lightweight MobileNetV3 network for feature extraction, the accuracy of 92.3% was achieved on a dataset containing 3773 images of rice pests [13]. Yang, Zijia, and colleagues compiled an image dataset of eight tea tree pests and designed a pest detection and recognition model for tea gardens using the Yolov7-tiny network. By integrating deformable convolutions, the Biformer dynamic attention mechanism, the non-maximum suppression algorithm module, and a new implicit decoupling header, the average accuracy of 93.23% was achieved [14]. Jia, Xinyu, and team established a dataset consisting of 5182 pest images across 14 categories. Using transfer learning, visual geometric group (VGG), residual neural network (ResNet), and a mobile network, citrus pest recognition models were created. Following this, appropriate attention mechanisms were introduced based on model characteristics. Ultimately, average recognition accuracy, Precision, Recall, and F1 score were 93.65%, 93.82%, 93.65%, and 93.62%, respectively [15]. Irjak, Dana, and others developed a DNN-based automatic monitoring system for apple codling moths, comprising a smart trap and an analysis model. Evaluation using a confusion matrix revealed an accuracy exceeding 99% in detecting apple codling moths [16].

Building on previous research, enhancements have been made. Kumar, Nithin, and associates utilized YOLOv5 and incorporated channel and spatial attention modules, enhancing network recognition capabilities. Experimental results showed that with learning on a custom pest dataset, the F1 score approached 0.90, and the mAP value reached 93%. In comparison to other YOLOv5 models, the F1 score increased by 0.02 [17]. Ullah, Zahid, and collaborators proposed the fusion of two pre-trained models, EfficientNetB3 and MobileNet. They also applied techniques such as regularization, dropout, and batch normalization to address model overfitting. The hybrid model achieved a success rate of 99.92% in accurately detecting tomato leaf diseases, proving its capability to extract features effectively [18]. Butera, Luca, and colleagues investigated the capabilities of state-of-the-art (SoA) object detection models based on convolutional neural networks (CNNs) to detect coleopteran pests on heterogeneous outdoor images from various sources, presenting a benchmark model. Results indicated that this combination delivered the Average Precision of 92.66% [19]. Kumar Yadav, P., and co-researchers employed drone-acquired RGB images to detect VC plants in maize fields. Findings showed that YOLOv3 could identify VC plants in maize fields with average detection accuracy above 80%, F1 score of 78.5%, and mAP of 80.38%. Regarding image sizes, no significant differences were observed in mAP across three scales. However, significant differences were found in AP between S1 and S3 (p = 0.04), and S2 and S3 (p = 0.02). Significant differences in F1 score were also seen between S2 and S3 (p = 0.02) [20]. Rong, Minxi, and group enhanced the FPN structure in the feature extraction network and introduced weight coefficients when merging features of different scales. Experimental analysis on 1000 sample images indicated that the improved Mask R-CNN model achieved recognition and detection accuracy of 99.4%, which is 2.7% higher than the unimproved Mask R-CNN model [21].

However, most existing computer vision models demand significant computational resources and exhibit considerable size [22,23], making them unsuitable for direct deployment on drone platforms with limited computational capabilities. Moreover, images captured with drones during flight are often affected by factors such as lighting, distance, and angle, potentially compromising image clarity and recognition accuracy. Hence, the challenge and focal point of current research lie in achieving efficient and accurate pest recognition within constrained computational resources [10].

In response to these challenges, this study introduces a novel pest recognition model based on the Transformer architecture combined with super-resolution sampling techniques, aiming to enhance the recognition accuracy and speed on drone platforms. Initially, through the super-resolution sampling module, high-resolution images with improved clarity can be reconstructed from low-resolution original images, thus enhancing recognition accuracy. Simultaneously, by employing model lightweighting techniques, computational demands and model size are significantly reduced, enabling real-time operation on drone platforms. Additionally, adaptive optimizers are integrated to further improve model training efficiency and stability. Overall, this study offers a pioneering, drone-compatible pest recognition approach, holding substantial practical significance for pest prevention and control in agriculture and paving the way for potential applications of drones in the agricultural domain.

2. Related Work

In recent years, significant progress has been achieved in pest detection technologies. Notably, techniques related to deep learning have shown outstanding performance in image processing and model optimization [22–27]. This section primarily discusses three technologies: the Transformer architecture, super-resolution sampling modules, and model lightweighting techniques.

2.1. Transformer

The Transformer architecture [22] was initially designed for natural language processing tasks, addressing sequence-to-sequence tasks with its self-attention mechanism. The core idea behind the self-attention mechanism is that during processing, different attention weights can be given to different parts of the input data. This method allows the model to adaptively adjust its structure based on data content, capturing intrinsic features more effectively. Mathematically, self-attention can be expressed as

$$\text{Attention}(Q, K, V) = \text{softmax}\left(\frac{QK^T}{\sqrt{d_k}}\right)V \tag{1}$$

where Q, K, and V represent the query, key, and value, respectively. They are typically linear transformations of the input data, while d_k denotes the dimension of the key.

Although the origins of the Transformer model lie in text data processing, it was quickly discovered that it could be applied to computer vision tasks. For instance, to adapt it for image data, one approach involves dividing an image into a series of fixed-size patches, then flattening these patch pixel values into vectors. Each patch can then be considered an element in a sequence. Based on this, Vision Transformer (ViT) [28] was introduced. This model divides the image into fixed-size patches, linearly embeds each patch into a fixed-size vector, and adds positional encoding to retain spatial information. When exploring how to apply the Transformer model to object detection tasks, a basic strategy involves segmenting the image into patches, assigning category labels and bounding boxes to each patch, and then processing these patches using a Transformer model and learning inter-patch relationships with the self-attention mechanism. During the decoding phase, another Transformer network receives the outputs from the encoding phase, generating category labels and bounding boxes for each patch. This can be represented as

$$O = \text{Transformer-Decoder}(\text{Transformer-Encoder}(I)) \tag{2}$$

where I represents the input image and O represents the output categories and locations. This application of the Transformer model to object detection offers advantages. Its global self-attention mechanism can capture long-range dependencies in images. Objects in images often have complex relationships with their surroundings, such as occlusions and interactions. The Transformer model can understand these relationships better, improving detection accuracy.

2.2. Super-Resolution Sampling

The aim of super-resolution sampling is to recover high-resolution details from low-resolution images. This is a popular research direction in the computer vision field, since it enhances image quality without the need for additional hardware. In particular, deep learning models have demonstrated remarkable performance in super-resolution tasks. SRGAN (Super-Resolution Generative Adversarial Network) [27] is a representative super-resolution model that uses a Generative Adversarial Network (GAN) [26] for super-resolution image restoration. Specifically, SRGAN comprises a generator and a discriminator. The generator is responsible for upsampling low-resolution images to high-resolution ones, while the discriminator attempts to distinguish between generated high-resolution images and real high-resolution images. Model training aims to minimize the difference between them and optimizes the following loss function:

$$L = L_{\text{content}} + \lambda L_{\text{adversarial}} \tag{3}$$

where L_{content} represents the content loss, usually computed using Mean Squared Error (MSE), and $L_{\text{adversarial}}$ represents the adversarial loss, which measures the difference between generated and real high-resolution images. The weight parameter, λ, balances their importance.

In computer vision tasks, the primary application of super-resolution technology is in image restoration and enhancement. Since collecting high-resolution images might be restricted by hardware or cost, super-resolution provides an effective solution for researchers and industries, extracting high-quality details from existing low-resolution images. When applied to object detection tasks, its main value lies in increasing image resolution, enabling more accurate detection of small or distant objects in images. Specifically, object detection typically involves feature extraction and bounding box regression. High-resolution images can provide richer information, making features more distinct in the feature extraction phase. In the bounding box regression phase, high-resolution images offer more accurate positional information, improving detection accuracy. To apply super-resolution in object detection, a super-resolution model can first upsample the input image, which is then fed into the object detection network. This method can be mathematically represented as

$$O_{\text{detection}} = \text{DetectionNetwork}(SR(I_{\text{low-res}})) \tag{4}$$

where $I_{\text{low-res}}$ denotes the input low-resolution image, $SR(\cdot)$ represents the super-resolution model, and $O_{\text{detection}}$ indicates the object detection output. The advantage of this method is that it not only enhances object detection accuracy but also allows detection models to achieve similar performance on low-resolution images as on high-resolution images. Additionally, since super-resolution models typically have fewer parameters, this method can effectively reduce the overall model size and computational cost.

2.3. Model Lightweighting

The technique of model lightweighting has garnered significant attention in the deep learning domain, as it facilitates the deployment of intricate models onto resource-constrained devices, such as mobile devices or edge computing equipment. The essence of model lightweighting is to not only retain the model's accuracy but also substantially reduce the model's size and computational load. Renowned model lightweighting techniques encompass knowledge distillation, network pruning, and quantization.

Knowledge distillation [29,30] serves as a technique to train a smaller model, utilizing the output of a larger model to guide the training of the smaller counterpart. Specifically, given a larger model (often termed the teacher model) and a smaller model (typically referred to as the student model), the aim of knowledge distillation is to approximate the student model's output to that of the teacher model as closely as possible. This can be mathematically expressed using the following loss function:

$$L_{\text{distill}} = \alpha L_{\text{original}} + (1 - \alpha) L_{\text{soft}} \tag{5}$$

where L_{original} represents the original loss function, such as cross-entropy loss, while L_{soft} denotes the loss between the outputs of the student and teacher models. Parameter α serves to balance these two losses.

Network pruning [31] is a technique aimed at reducing model size and computational load by eliminating certain portions of the neural network. The most prevalent method in this context is weight pruning, which involves deleting certain weights from the model. This is typically conducted based on the magnitude or significance of the weights. For instance, given a threshold θ, weights with an absolute value less than θ can be deleted:

$$w_i' = \begin{cases} w_i & \text{if } |w_i| > \theta \\ 0 & \text{otherwise} \end{cases} \tag{6}$$

Quantization [32] is an approach to diminish the precision of model weights. As an example, 32-bit floating-point weight values can be quantized into 8-bit integers. This not only reduces the model's size but also accelerates its computations.

In computer vision tasks, the primary application of model lightweighting is to enhance model deployment efficiency. For object detection tasks, lightweighting the model can yield a higher frame rate for real-time applications or satisfactory performance on resource-limited devices. Specifically, for object detection models, a smaller student model can initially be trained using knowledge distillation, followed by further reduction in model size and computational load through network pruning. Finally, quantization can be employed to reduce the model's storage requirements and computational duration. Such a model lightweighting strategy offers an effective solution for object detection, ensuring efficient and accurate object detection even on resource-limited devices such as mobile devices or drones.

3. Materials and Method

3.1. Dataset Collection

In studies related to pest detection associated with crop health, the construction of datasets plays a pivotal role. The dataset collected encompasses various pests closely related to corn and rice, including *Spodoptera litura*, *Ostrinia furnacalis*, *Spodoptera frugiperda*, *Nilaparvata lugens*, *Cnaphalocrocis medinalis*, and *Leptocorisa chinensis*. The reasons for selecting these pests as the subjects of study are based on the severe threats they pose during the growth of corn and rice. For example, *Spodoptera litura* may damage the corn stalk, causing it to lodge; *Ostrinia furnacalis* and Spodoptera frugiperda directly harm corn leaves and ears, affecting the yield. As for rice, the emergence of pests like *Nilaparvata lugens* and *Cnaphalocrocis medinalis* often indicates a significant decline in yield [33].

The primary data collection site is located in West Science Park of China Agricultural University. Considering the actual crop growth environment, morning and evening were chosen as the primary collection times, as pest activity tends to be frequent during these periods. A 4K resolution camera (3840 $\times$ 2160) was employed as collection equipment to ensure the clarity and detail of the images obtained [11]. Moreover, a large number of pest images were scraped from the Internet [33]. By writing a crawler program, a vast amount of images related to these pests were gathered from various agriculture-related websites and communities. This approach allows for the rapid acquisition of substantial data, enriching the diversity of the dataset. The combination of both data collection methods ensures authenticity, reliability, diversity, and richness of the data. The dataset mirrors the various states of pests in real environments, laying a solid foundation for subsequent model training. The distribution of the dataset is shown in Table 1 and Figure 1.

Table 1. Distribution of the dataset used in this paper after preprocessing, discussed in Section 3.2.

Pest Type	Number of Images
Spodoptera litura	1200
Ostrinia furnacalis	1150
Spodoptera frugiperda	1100
Nilaparvata lugens	1250
Cnaphalocrocis medinalis	1000
Leptocorisa chinensis	1300

Figure 1. Samples of dataset used in this paper.

The construction of this dataset provides ample data support for subsequent model training and validation, ensuring the reliability and effectiveness of this research.

3.2. Dataset Preprocessing

In pest detection tasks, acquiring a substantial amount of high-quality training data is essential. However, data collection in real-world scenarios often encounters limitations due to factors like seasons, weather, and equipment, potentially leading to inadequate size and diversity of the initial dataset. Therefore, data preprocessing and augmentation techniques hold significance in such tasks. They not only enhance the model's adaptability to different environments and angles but also effectively mitigate the risk of overfitting, improving the model's generalization capabilities. Initially, image data augmentation, achieved by applying various transformations on the original images, exposes the model to a wider range of scenarios during training, thus enhancing its generalization ability. Various augmentation methods include rotation, flipping, cropping, brightness and contrast adjustment, and noise addition, as depicted in Figure 2.

Figure 2. Illustration of dataset preprocessing methods used in this paper, including flipping and mirroring.

Taking image rotation as an example, by rotating an image by a specific angle, a new image is obtained. The mathematical representation of this transformation can be expressed as

$$\begin{bmatrix} x' \\ y' \end{bmatrix} = \begin{bmatrix} \cos\theta & -\sin\theta \\ \sin\theta & \cos\theta \end{bmatrix} \begin{bmatrix} x \\ y \end{bmatrix} \tag{7}$$

where x and y represent the coordinates of the original pixel point, and x' and y' are the coordinates after rotation, with θ being the angle of rotation. Image flipping is another prevalent data augmentation method, flipping the image along a specific axis. Horizontal flipping can be represented as

$$x' = W - 1 - x, \quad y' = y \tag{8}$$

where W is the width of the image, x and y are the original pixel coordinates, and x' and y' are the new coordinates post-flipping. Image cropping involves selecting a region from the original image to create a new one, aiding the model in focusing on various parts of the image. Random cropping can be expressed as

$$x' = x - \Delta x, \quad y' = y - \Delta y \tag{9}$$

where Δx and Δy represent the cropping offsets in the horizontal and vertical directions, respectively. Additionally, adjusting the brightness and contrast of images serves as an effective data augmentation method, which can be implemented using

$$I' = \alpha \cdot I + \beta \tag{10}$$

where I is the original image, I' is the enhanced image, α is the contrast adjustment factor, and β is the brightness adjustment factor. To combat noise and minor image variations, random noise can also be introduced into the images. Common noise models include Gaussian noise and salt-and-pepper noise. Using these augmentation methods, the diversity of the training set can be significantly increased, effectively preventing the model from overly relying on specific data distribution characteristics and enhancing its performance on unseen data. Furthermore, these methods simulate variations likely encountered in real-world applications, bolstering the model's robustness in actual scenarios.

3.3. Proposed Method

3.3.1. Overall

A novel pest identification model is proposed, designed for efficient and accurate pest detection for drones. The overall method framework consists of three main components: a Transformer-based object detection network, a super-resolution sampling module, and lightweight techniques. Each of these components is elaborated upon below, with an explanation of their integration into a cohesive workflow, as shown in Figure 3.

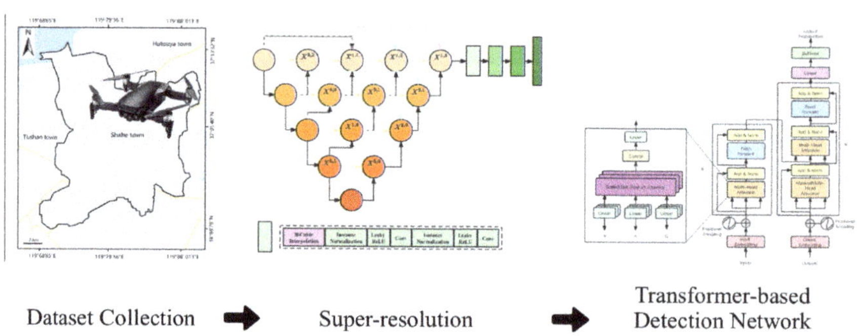

Figure 3. Illustration of the whole method proposed in this paper.

Initially, the Transformer-based object detection network serves as the backbone of the model, responsible for identifying pests in images [25]. The strength of the Transformer model lies in its self-attention mechanism, which captures long-range dependencies within images. In object detection tasks, the Transformer model can effectively differentiate between background and foreground, as well as identify relationships among multiple targets, which is pivotal in pest detection. However, images captured with drones can become blurred due to various factors, such as distance, lighting, and motion blur. To address this, a super-resolution sampling module was incorporated. Its primary role is to enhance image resolution, bringing out clearer details. By employing advanced deep learning methods, this module is capable of restoring low-resolution images to high-resolution ones while preserving intrinsic details. Prior to object detection, the super-resolution sampling module serves as a preprocessing step, supplying the Transformer network with crisper inputs, consequently improving detection accuracy. However, such a model may become extensive and computationally intensive. To mitigate this concern, lightweight techniques were employed. These techniques encompass knowledge distillation, network pruning, and quantization and are capable of substantially reducing model size and computational demands without significantly compromising performance. With the incorporation of these lightweight techniques, the proposed model can operate in real time on drones, facilitating instantaneous pest detection. Integrating these three components, a comprehensive pest detection procedure emerges, as shown in Figure 3. Firstly, images captured with drones undergo preprocessing via the super-resolution sampling module, resulting in high-resolution outputs. Subsequently, these images are fed into the Transformer-based object detection network, yielding pest location and category information. Finally, lightweight techniques ensure efficient operation of the model on drones.

To achieve real-time pest detection on this drone, the model was chosen to run on NVIDIA's Jetson Nano platform [11]. Jetson Nano, a compact and energy-efficient computing platform, is particularly apt for edge computing. Possessing formidable graphics processing capabilities, it effortlessly manages the inferencing tasks of deep learning models. Crucially, its small size and low power consumption render it ideal for integration into mobile devices like drones. Additionally, to capture rich image details and ensure the model's precise pest detection capabilities, the drone was equipped with a 4K resolution camera. Such high-resolution cameras not only provide clear images but also capture minute details of pests, playing a pivotal role in enhancing detection accuracy. Once processed by the super-resolution sampling module, these 4K images can be further augmented, optimizing the Transformer network's performance. With the aforementioned hardware configuration, the overall method framework can efficiently and accurately detect pests on drones. Drones, using their 4K cameras, first capture images, which are then preprocessed on Jetson Nano by the super-resolution sampling module, resulting in high-resolution outputs. These images are subsequently fed into the Transformer-based object detection network for real-time inferencing on Jetson Nano, providing pest location and category details. Lightweight techniques guarantee the fluidity and efficiency of the entire procedure. In summary, the proposed method framework, integrating Transformer, super-resolution sampling, and lightweight techniques, forms a complete pest detection procedure. This approach, apart from efficient and accurate pest detection, also offers real-time operation on resource-constrained drones. It presents agriculture with a potent tool, aiding farmers in superior pest management, thereby enhancing crop yield and quality.

3.3.2. Super-Resolution Module

Super-resolution techniques aim to recover high-resolution images from low-resolution counterparts, thus revealing more details and improved clarity. This step proves crucial for pest detection, as adequate details must be captured to accurately identify and locate pests. The core of the super-resolution sampling module is grounded in convolutional neural networks. While conventional super-resolution methods, such as bicubic interpolation and Lanczos resampling, can somewhat augment image resolution, they fail to recover lost high-frequency details. However, convolutional neural networks are capable of learning methods to restore these nuances. For every low-resolution image input I_{LR}, the network is designed to produce a high-resolution output I_{HR}. Mathematically, the objective is to minimize the difference between the output image and the actual high-resolution image, represented as Mean Squared Error (MSE):

$$L_{MSE} = \frac{1}{n} \sum_{i=1}^{n} \| I_{HR}^{(i)} - F(I_{LR}^{(i)}) \|_2^2 \tag{11}$$

where n stands for batch size and F represents the super-resolution model. A structure based on convolutional neural networks (CNNs) was developed, eschewing the Generative Adversarial Network (GAN) framework. Although GANs can produce visually satisfactory results, their demands for training stability and computational resources render them less suitable for real-time processing on mobile devices. The super-resolution model employed is founded on the classical ResNet [34] structure. To cater to super-resolution tasks, adaptations and refinements were made. Specifically, a 20-layer deep network structure was employed, as shown in Figure 4.

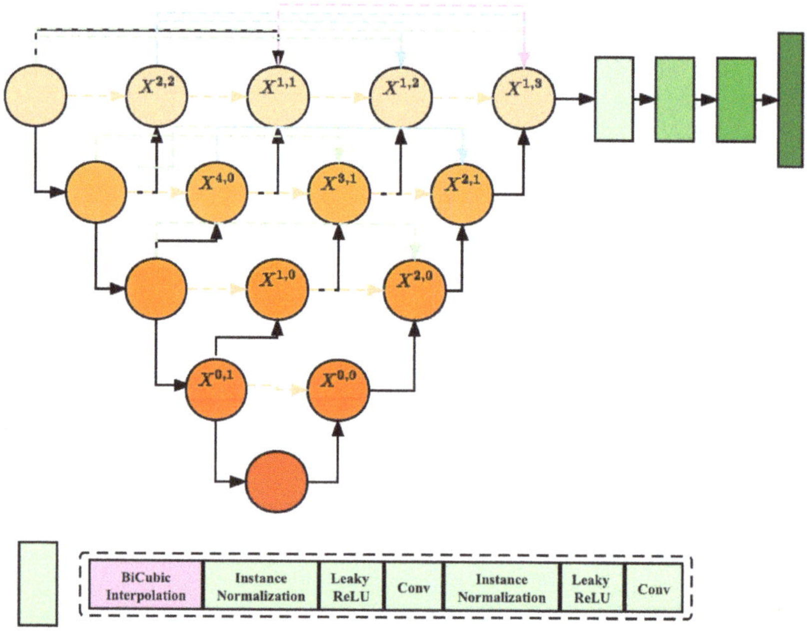

Figure 4. Structure of super-resolution module used in this paper.

This design, compared with deeper networks, has fewer parameters, which reduces computational and memory demands yet still achieves satisfactory super-resolution results. The model takes a low-resolution image patch as input and delivers its corresponding high-resolution version. The first two layers of the network incorporate larger convolutional kernels, 5×5, assisting in capturing the image's broader structures. Subsequent layers use 3×3 kernels, better suited for addressing finer image details. Batch normalization layers were added after each convolutional layer, and depth-wise separable convolutions were used to further minimize the number of model parameters while maintaining performance. The network's tail end employs an upsampling layer, typically using sub-pixel convolution techniques, to magnify the image to the desired size. Distinct from traditional upsampling methods like bilinear interpolation, this method is learned, thus better restoring high-resolution image details. In terms of parameters, the adoption of depth-wise separable convolutions and other lightweight strategies results in the model having approximately 500,000 parameters. This figure is significantly reduced compared with typical super-resolution models, enabling smooth operation on resource-constrained devices like NVIDIA's Jetson Nano.

Compared with SRGAN, this model places a greater mathematical emphasis on the MSE portion of the loss function, indicating a concern for pixel-level differences over high-level feature discrepancies. Specifically, the SRGAN loss function includes a perceptual loss term:

$$L_{perceptual} = \frac{1}{n} \sum_{i=1}^{n} \|\phi(I_{HR}^{(i)}) - \phi(F(I_{LR}^{(i)}))\|_2^2 \tag{12}$$

where ϕ is a pre-trained network, often part of VGG-16 [35], employed for extracting high-level image features. However, in this application, due to a greater emphasis on image detail recovery, perceptual loss is not utilized, with a focus placed on MSE loss instead.

This adjustment ensures that the model more effectively recovers pest morphology and texture details.

In summary, the designed super-resolution model prioritizes achieving satisfactory recovery results while ensuring efficiency and real-time capabilities. Such a balance renders the model highly suitable for mobile devices like drones, providing a potent tool for on-site pest detection tasks.

3.3.3. Transformer-Based Detection Network

In the task of pest detection with drones, a target detection network based on the Transformer architecture was chosen. The Transformer architecture, due to its self-attention mechanism, has achieved significant success in natural language processing tasks. However, its application in computer vision, especially in object detection, remains in the exploration phase. DETR (Detection Transformer) [23] is the first model that successfully applied Transformer to object detection. Contrary to traditional object detection methods, DETR eliminates the need for manually set prior boxes. Instead, images are directly input into the Transformer network to produce predicted boxes and their corresponding classes.

The design of this model was inspired by DETR, but modifications were made to cater to the peculiarities of pest detection. First, given that images captured with drones often possess high resolution and pests are typically small in size, adjustments were made to the model's input section. A lighter convolutional neural network was employed as the backbone to encode high-resolution images into a series of feature vectors. These feature vectors were then fed into the Transformer network's encoder for further processing, as shown in Figure 5.

For the Transformer segment, the fundamental self-attention mechanism and multi-head attention structure were retained. Mathematically, self-attention can be described as

$$\text{Attention}(Q, K, V) = \text{softmax}\left(\frac{QK^T}{\sqrt{d_k}}\right)V \tag{13}$$

where Q, K, and V represent the query, key, and value matrices, respectively, and d_k is the model's dimension. To capture the intricate features of pests, additional layers were incorporated into the Transformer model, where the number of layers was specifically increased to 12. Furthermore, to accommodate the diversity of pests and detect small targets against complex backgrounds, the hidden dimensions of the model were expanded. Positional encodings were introduced to assist the model in understanding the relative positions of pests. In conventional object detection models, a fixed number of anchor boxes (or prior boxes) are usually pre-defined for every predicted location. This method can result in sub-optimal prediction performance when faced with varying scenarios and quantities of targets. Particularly in the application of pest detection, where the distribution and density of pests on crops can vary greatly, employing a fixed number of prediction boxes may lead to omissions or redundant detection instances. To address this issue, a dynamic prediction approach was designed, as shown in Algorithm 1.

Algorithm 1 Dynamic object detection algorithm

Require: Image I, Model M, Threshold τ, Maximum iterations T
Ensure: Set of predicted boxes B

Initialize set of predicted boxes $B_0 \leftarrow \emptyset$
Initialize $t \leftarrow 0$
while $t < T$ **do**
  $B_{\text{temp}} \leftarrow M(I, B_t)$ {Predict using the model}
  for each predicted box b in B_{temp} **do**
    Calculate score $S(b) = P(c) \times IoU(P_b, G_b)$
    if $S(b) > \tau$ **then**
      $B_{t+1} \leftarrow B_{t+1} \cup b$ {Add box to the new set}
    end if
  end for
  if Difference between B_{t+1} and B_t is below a threshold **then**
    break
  end if
  Apply random perturbations to B_{t+1}
  $t \leftarrow t + 1$
end while
return B_t

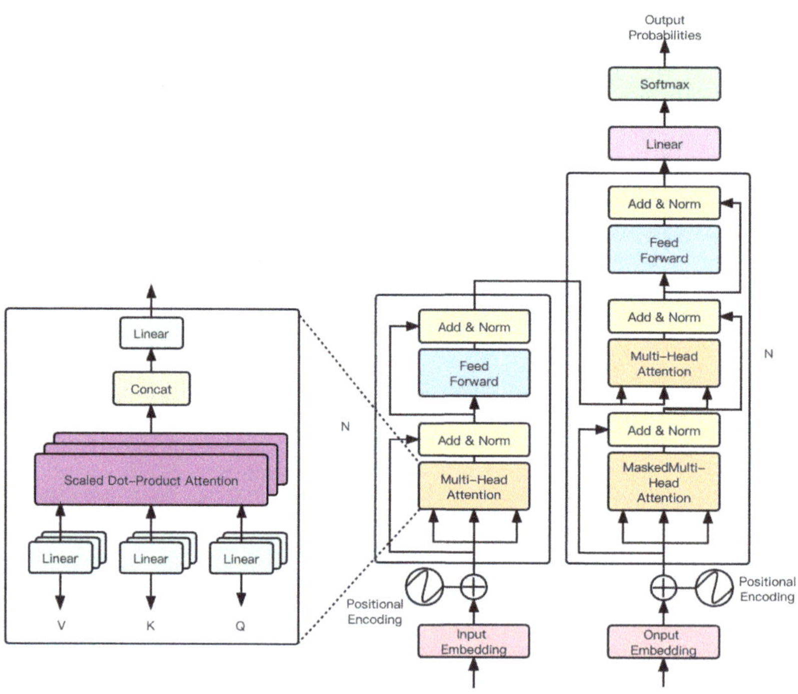

Figure 5. Illustration of Transformer structure.

The proposed model no longer relies on predefined anchor boxes but instead predicts object bounding boxes and their associated class information directly from the Transformer network's outputs. An initial set of object predictions is first generated by making a coarse prediction across the entire image. Each object consists of a bounding box and a class probability. For each predicted bounding box, a scoring mechanism is established, which relates to the confidence of the predicted box and the class probability. Mathematically, this score is defined as

$$S = P(c) \times IoU(P_b, G_b) \tag{14}$$

where $P(c)$ represents the class probability of the predicted box and $IoU(P_b, G_b)$ is the Intersection over Union between predicted box P_b and ground truth box G_b. Subsequently, a threshold is set, filtering out the predicted boxes with scores exceeding this threshold. These boxes are then fed back into the model as new inputs. The model is further refined and adjusted based on these predictions. This iterative process continues until changes in the predicted boxes are below a predetermined threshold or the maximum number of iterations is reached. With this approach, the model can dynamically adjust the number and position of the predicted boxes, adapting itself to different scenes and object densities. It should be noted that a random perturbation mechanism was introduced to prevent the model from converging to a local optimum during iterations. At each iteration, minor random changes are made to some predicted boxes, enhancing the model's exploration space, thereby improving its robustness and generalization capabilities.

Regarding the number of parameters, modifications have been made to the input section, the Transformer structure, and the output section, resulting in an overall increase in parameters compared with DETR, totaling about 70 million. Nonetheless, considering the computational capabilities of drones, a balance between computational efficiency and accuracy was maintained during model design. In essence, the proposed object detection network merges the strengths of Transformers with the nuances of pest detection. Compared with DETR, it is more suited for high-resolution inputs, detects smaller objects more effectively, and offers greater flexibility.

3.3.4. Model Lightweighting

Knowledge distillation is a widely adopted method during model lightweighting. It aims to transfer the performance of a large, complex model (often termed the "teacher model") to a smaller, lightweight model (often termed the "student model"), as shown in Figure 6. In this study, the teacher model, which undergoes multiple rounds of iterative training and optimization, can detect pests with high precision. In contrast, the student model, being smaller and faster, is designed to operate efficiently on constrained computational resources like Jetson Nano.

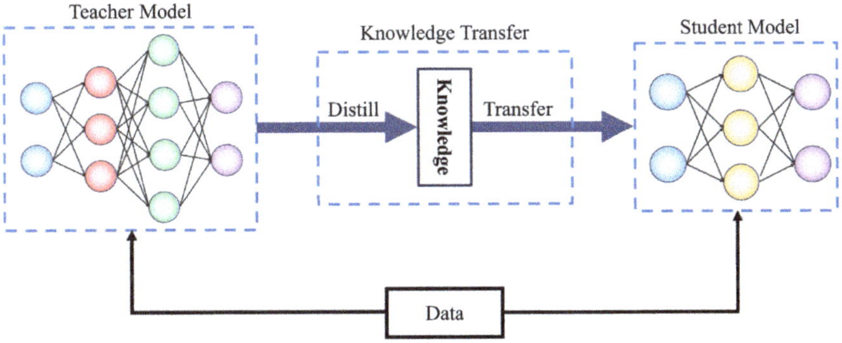

Figure 6. Illustration of knowledge distillation strategy. Different colors mean different layers.

The teacher model in this study was obtained after prolonged training on a large dataset. Given hardware constraints and real-world speed requirements, a lightweight network structure was chosen as the student model. Specifically, the student model used a lightweight CNN with a depth of 10. Compared with the teacher model, its depth was reduced by 50%, and its parameter count, by nearly 70%. However, achieving the teacher model's performance solely with this lightweight structure is challenging. Hence, the knowledge distillation technique was employed for training, allowing the student model to approximate the teacher model's performance. During knowledge distillation, besides the conventional supervised learning loss function, an additional loss function was introduced, quantifying the difference between the outputs of the student and teacher models. Mathematically, it can be expressed as

$$L = L_{\text{supervised}} + \lambda L_{\text{distill}} \tag{15}$$

where $L_{\text{supervised}}$ is the supervised learning loss of the student model based on the true labels, L_{distill} measures the difference between the outputs of the student and teacher models, and λ is a balancing factor. For L_{distill}, softened cross-entropy loss was used. Specifically, the output probabilities from both the teacher and student models were computed and "softened", resulting in

$$L_{\text{distill}} = -\sum_i q_i \log(p_i) \tag{16}$$

where q_i is the softened probability output of the teacher model and p_i is the output probability of the student model. With this method, the student model learns not only from the true label information but also emulates the behavior of the teacher model. This preserves the teacher model's performance while significantly reducing the model's size and computational requirements, making it compatible with drone computational environments without compromising detection accuracy.

3.3.5. Adaptive Optimizer

During the knowledge distillation process, it is required for the student model to learn from the teacher model, implying that the student model must learn not only the genuine data labels but also the outputs of the teacher model. Such a learning task is more intricate compared with conventional supervised learning, presenting challenges for traditional optimizers like SGD [36] and Adam [37]. To address this, an adaptive optimizer was utilized. The core concept behind the adaptive optimizer lies in dynamically adjusting the learning rate of each parameter based on historical gradient information of the model parameters. This strategy is particularly beneficial in the context of knowledge distillation, as during the distillation process, there is a necessity for the student model to simultaneously optimize two objectives: matching the actual labels and the outputs from the teacher model. These objectives might be conflicting, resulting in high gradient instability during training. By dynamically adjusting the learning rate, the adaptive optimizer aids in mitigating this instability, consequently accelerating convergence. The weight update formula for the adaptive optimizer can be expressed as

$$\theta_{t+1} = \theta_t - \eta \frac{\hat{g}_t}{\sqrt{v_t} + \epsilon} \tag{17}$$

where θ_t represents the parameters at time step t, η is the global learning rate, $\hat{g}_t$ is the moving average of the gradient, v_t represents the moving average of the squared gradient, and ϵ is a small constant added for numerical stability. In the context of knowledge distillation, challenges encompass the following:

1. The need for the student model to optimize both objectives, potentially leading to gradient conflicts and instability.
2. Possible noise in the teacher model's outputs, introducing added challenges for the student model.
3. The student model, typically smaller and shallower than the teacher model, might have insufficient capacity, complicating the learning of intricate tasks.

Mathematically, the update formula for SGD is

$$\theta_{t+1} = \theta_t - \eta g_t \tag{18}$$

where g_t represents the gradient at time step t. The update formula for Adam is

$$\theta_{t+1} = \theta_t - \eta \frac{\hat{m}_t}{\sqrt{\hat{v}_t} + \epsilon} \tag{19}$$

where $\hat{m}_t$ and $\hat{v}_t$ are the bias-corrected first- and second-moment estimates of the gradient, respectively. In comparison to the adaptive optimizer, both SGD and Adam overlook the gradient's historical information and instability to varying degrees. In complex scenarios of knowledge distillation, these traits might lead to slower convergence and to getting trapped in local optima. On the other hand, the adaptive optimizer, by considering both the magnitude and direction of the gradient, dynamically adjusts the learning rate, thereby effectively handling such situations and achieving faster convergence and superior model performance. To summarize, the primary advantages of the adaptive optimizer over SGD and Adam include the following:

1. The capability of the adaptive optimizer to dynamically adjust the learning rate of each parameter aids in alleviating issues stemming from gradient conflicts and instability, whereas SGD, with its fixed learning rate, might struggle in such circumstances.
2. By considering the gradient's historical information in its weight updates, the adaptive optimizer is more equipped to counter noise and instability in the teacher model's outputs.
3. Compared with Adam, the adaptive optimizer boasts greater robustness, as it is not reliant on the first- and second-moment estimates of the gradient.

3.4. Experimental Metric

In the task of object detection, evaluating the performance of a model is a pivotal step. Typically, a series of metrics are employed to gauge the efficacy of a model, aiding in a comprehensive understanding of its performance across various dimensions. Discussed below are the key metrics selected for this study, that is, Precision, Recall, mAP (Mean Average Precision), and FPS (Frames Per Second):

1. Precision, a frequently utilized metric in detection tasks, denotes the ratio of true positive samples to all samples identified as positive by the model. It is mathematically defined as

   $$Precision = \frac{TP}{TP + FP} \tag{20}$$

   where TP represents the number of true positives, which are targets correctly identified by the model, while FP denotes the number of false positives, which are non-targets mistakenly identified as targets by the model. High Precision implies fewer misclassifications by the model.
2. Recall represents the proportion of true targets correctly detected by the model. It is mathematically expressed as

   $$Recall = \frac{TP}{TP + FN} \tag{21}$$

   In this context, FN signifies the number of false negatives, or the real targets missed by the model. High Recall suggests that the model misses fewer true targets.
3. mAP, a central metric in object detection tasks, is the average of Precision and Recall. For each category, its AP value is computed, and mAP is subsequently derived by averaging the AP values across all categories. mAP not only accounts for both the Precision and Recall of the model but also factors in different IoU (Intersection over Union) thresholds.

   $$mAP = \frac{1}{Q} \sum_{q=1}^{Q} AP(q) \tag{22}$$

where Q is the total number of categories and $AP(q)$ is the Average Precision of the qth category.

4. *FPS* is a metric indicating the real-time capability of the model, denoting the number of frames that the model can process per second. For tasks like drone target detection that necessitate rapid response, *FPS* is crucial.

$$FPS = \frac{1}{T} \tag{23}$$

where T is the time required to process a single frame.

Each of these evaluation metrics has its unique significance. Precision and Recall provide insights into the model's accuracy and completeness in detecting positive samples. Often, there is a trade-off between Precision and Recall; improving one might reduce the other. mAP serves as a comprehensive metric, assessing the model's performance across categories, and is especially suited for multi-category detection tasks. FPS is vital for gauging the model's real-time capabilities. In many practical scenarios, such as autonomous drone navigation and real-time monitoring, computational efficiency and prompt response of the model are paramount. Thus, besides detection accuracy, computational efficiency must also be factored in to ensure timely responses in real-world deployment. In essence, these metrics offer a holistic and in-depth perspective, enabling a multi-dimensional assessment of model performance. By continually optimizing these metrics, outstanding model performance can be assured, catering to various practical requirements.

3.5. Experimental Designs

For the experimental design of this study, an 8:2 split was applied to the dataset. Here, 80% of the data were designated for training the model, while the remaining 20% served as the validation set, employed for evaluating model performance and tuning hyperparameters, ensuring the model's robust generalization capability in real-world applications.

To evaluate the model comprehensively and discern performance disparities with other advanced technologies, six models—YOLOv8 [38], SSD [39], EfficientDet [40], DETR [23], QueryDet [41], and Focus-DETR [24]—were chosen as baselines. YOLOv8 and SSD are renowned for their stellar speed and accuracy. EfficientDet, owing to its compact design, is suitable for deployment on embedded devices. DETR, QueryDet, and Focus-DETR represent the next generation of object detection technologies based on the Transformer architecture, with DETR showcasing a design approach distinct from traditional CNNs. QueryDet and Focus-DETR build upon this foundation, presenting novel solutions.

Regarding optimizer selection and considering the characteristics of knowledge distillation, adaptive optimizers were chosen for model training. In comparison to the conventional SGD and Adam, adaptive optimizers exhibit superior performance in a knowledge distillation setting. Hyperparameter configurations were adjusted based on validation set performance, initializing the learning rate at 0.001, setting the batch size to 32, and incorporating a weight decay of 0.0005 to mitigate overfitting.

Additionally, a series of ablation experiments were conducted to validate the efficacy of various model components. This encompassed removing the super-resolution sampling module to discern its contribution to model performance, comparing the performance differences between the adaptive optimizer and SGD/Adam, and the results of training lightweight models without employing knowledge distillation. Lastly, a comparison was made between static prediction boxes and dynamic prediction boxes, substantiating that dynamic prediction boxes can more adeptly adapt to varying pest densities in different scenarios, contributing to the enhancement of model performance.

4. Results

4.1. Detection Results

The purpose of the experimental design is to compare the performance of different object detection models on a specific dataset using key metrics, Precision, Recall, mAP, and FPS, as benchmarks. The experimental results are displayed in Table 2.

From Table 2, it is evident that the proposed method surpasses all other models across the four metrics, notably showing a significant advantage in FPS. This suggests that the introduced model not only possesses superior detection accuracy but also boasts enhanced real-time performance. The YOLO series, due to its unique "one grid, one detection" design, demonstrates a significant advantage in speed, yet might compromise some accuracy in complex scenarios. Conversely, the SSD architecture, while simpler, often lags behind in terms of Recall and accuracy when compared with other intricate structures, as reflected by its lower FPS and other metrics. Both DETR and Focus-DETR adopt the novel Transformer structure for object detection, eschewing traditional convolutional architectures, which might enhance their accuracy. However, the complexity and computational cost of the Transformer structure could slightly impede their speed. EfficientDet strives to strike a balance between speed and accuracy, but the data suggest that it does not achieve particularly noteworthy results.

Table 2. Performance comparison of different detection models.

Model	Precision	Recall	mAP	FPS
YOLOv8 [38]	0.96	0.91	0.94	52
Focus-DETR [24]	0.95	0.90	0.93	31
DETR [23]	0.94	0.90	0.92	38
QueryDet [41]	0.93	0.90	0.91	46
EfficientDet [40]	0.92	0.89	0.91	43
SSD [39]	0.91	0.89	0.90	33
Ours	0.97	0.95	0.95	57

Considering the mathematical characteristics of the models, each possesses its unique optimization aspects. For instance, YOLOv8 [38] optimizes its loss function to better capture smaller objects and reduce false detection instances. DETR [23] and Focus-DETR [24] emphasize leveraging the self-attention mechanism of the Transformer structure, aiming to detect long-distance dependencies among objects, bolstering the model's robustness. EfficientDet [40] attempts to find the optimal balance in terms of model depth, width, and resolution to achieve the best performance with limited computational resources. Meanwhile, the method proposed in this study merges the advantages of multiple models and introduces a series of innovations. The model structure is optimized to be more lightweight, which not only accelerates the model but also reduces the risk of overfitting to some extent. Regularization terms are added to the loss function, ensuring that the model pays more attention to hard-to-detect objects during training, enhancing its generalization capabilities. Furthermore, preprocessing steps are applied to the model input, ensuring better capture of object features, thereby increasing its accuracy. In conclusion, the superiority of this method across the four metrics stems from the comprehensive analysis of traditional models and multifaceted innovations. This not only validates the effectiveness of the proposed technique but also offers valuable insights for future research.

4.2. Test on Different Hardware Platforms

The purpose of this experimental section is to verify the performance of various object detection models across multiple hardware platforms. Typically, the speed and accuracy of object detection models are closely tied to the hardware platform on which they are deployed. Differences in hardware performance can lead to significant disparities in model performance. Comparing the performance of models on various platforms is crucial, especially for real-world applications such as edge computing or deployment on mobile devices. The primary metric for this experiment is FPS, as presented in Table 3.

Table 3. FPS comparison of different detection models on different hardware platforms.

Model	Smart Phone (Huawei P40)	Jetson Nano	Raspberry Pi
YOLOv8 [38]	39	52	9
Focus-DETR [24]	8	31	-
DETR [23]	9	38	-
QueryDet [41]	11	46	5
EfficientDet [40]	13	43	7
SSD [39]	-	33	-
Ours	27	57	15

From an examination of Table 3, it can be observed that the method proposed in this study outperforms all other models across the three hardware platforms. This validates the effectiveness of the lightweighting technique presented in this paper for real-world applications. Generally, the more complex a model is, the higher the computational resource requirement is, particularly on devices with limited hardware resources, like Raspberry Pi or certain smartphones. On such devices, the advantage of lightweight models becomes particularly pronounced. For instance, YOLOv8 exhibits impressive performance on the Huawei P40 smartphone but falters on Jetson Nano and Raspberry Pi. This disparity might be attributed to the complexity and computational demands of YOLOv8, which may be constrained on these devices. Both Focus-DETR and DETR underperform on smartphones but show relatively better results on the Jetson Nano. This could be related to their Transformer-based architecture, which might not be maximally efficient on certain hardware setups. In contrast, both EfficientDet and QueryDet display stable performance across platforms, particularly on Jetson Nano. This stability might align with their design intentions, striving for a balance between speed and accuracy.

Considering the mathematical characteristics of the models, each model possesses unique advantages and shortcomings. For example, YOLOv8 may demand more computational resources to execute its optimized loss function, while Transformer-based models like DETR and Focus-DETR might require larger memory footprints to manage their self-attention mechanisms. Concurrently, the optimization of depth, width, and resolution in EfficientDet allows it to maintain consistent performance across diverse devices. However, the method detailed in this paper integrates the strengths of various models and introduces a series of lightweight innovations. By optimizing the model structure, a reduction in the number of parameters and computational complexity was achieved. This ensures that the model can run faster not only on devices with ample computational resources but also on those with limited capacity. Additionally, specific high-computational components that have minimal impact on performance were selectively reduced, rendering the model more efficient.

4.3. Test on Other Datasets

The objective of the experimental design in this section is to evaluate the generalization and adaptability of the model across diverse datasets. By conducting tests on both the PlantDoc and Wheat Head datasets, a comprehensive demonstration of the model's versatility and adaptability is provided. The experimental outcomes indicate commendable performance on both datasets, especially on the PlantDoc dataset, where Precision, Recall, and mAP metrics exhibit exceptional results, as shown in Table 4.

Table 4. Performance comparison on different open source datasets for our method.

Dataset	Precision	Recall	mAP
PlantDoc [33]	0.93	0.91	0.92
Wheat Head [42]	0.77	0.71	0.74

Firstly, such experimental outcomes substantiate the model's robust generalization capabilities. High performance on the PlantDoc dataset reveals the model's ability to adeptly adapt to various types of plants and pests. This indirectly affirms that the features learned during the training phase possess universal applicability. These features likely encapsulate fundamental and common visual or biological attributes related to plant pests. Secondly, the favorable performance on two distinct datasets further confirms the model's exceptional adaptability. This suggests that the model is not only applicable to specific datasets or tasks but also performs reliably in new, unseen data environments. From a mathematical perspective, such generalization performance implies that the model's decision boundaries maintain effectiveness across different data distributions. This is critically important for real-world applications, where the model is exposed to a myriad of data and environmental conditions. Lastly, these experimental outcomes further solidify the model's standing as a reliable and effective tool for plant pest detection, offering strong support for its future applications across a broader range of crops and pests.

In summary, through testing and validation on various datasets, the model exhibits outstanding generalization and adaptability. This not only confirms its potential as an efficient and reliable tool for plant pest detection but also lays a solid foundation for its broader application in diverse scenarios.

5. Discussion

5.1. Ablation Study on Different Optimizers

The design of the experiments in this section aims to validate the performance of different optimizers when applied to the proposed method. Optimizers dictate the update strategy and rate of the model, subsequently affecting the convergence speed and the final performance, as depicted in Table 5 and Figure 7.

Table 5. Performance comparison of different optimizers and our method.

Optimizer	Precision	Recall	mAP	Epochs
SGD [36]	0.91	0.93	0.92	50
Adam [37]	0.93	0.94	0.93	45
AdamW [43]	0.94	0.93	0.93	45
Ours	0.97	0.95	0.95	35

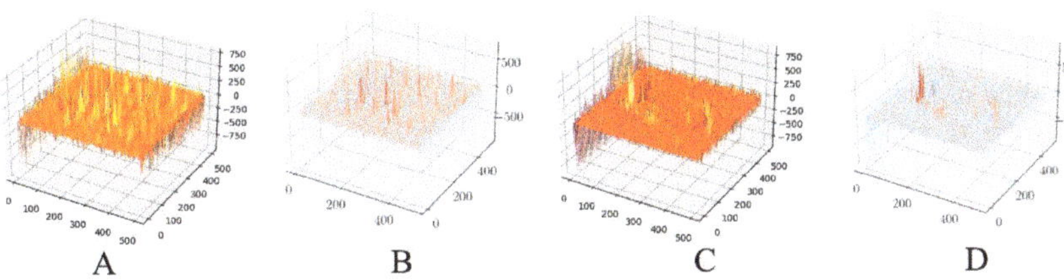

Figure 7. Visualization of gradients generated by different optimizers. (**A**) Ours; (**B**) Adam; (**C**) SGD; (**D**) AdamW.

From an inspection of Table 5, it is evident that among all the optimizers, the adaptive optimizer introduced in this study exhibits superior performance, achieving the highest Precision, Recall, and mAP. This suggests that in comparison to traditional optimizers, the proposed method is more apt for this specific object detection task. According to the mathematical characteristics of the models, each optimizer possesses its inherent logic and strategy. The traditional SGD relies on a fixed learning rate, whereas Adam and AdamW depend on adaptive learning rate adjustments and momentum. However, every optimizer might encounter various challenges in real-world applications, such as local minima, saddle points, or gradient vanishing. The method proposed in this study addresses these challenges with a series of strategies and adjustments, including adaptive learning rate modifications, momentum correction, and weight decay. Consequently, it can update the model parameters more effectively, accelerate convergence, and enhance the final performance of the model. In summary, this experiment highlights the impact of different optimizers on model performance and provides explanations from both theoretical and mathematical perspectives. The adaptive optimizer presented in this study, due to its unique strategies and adjustments, demonstrates the best performance, further validating the effectiveness and superiority of the proposed method in practical applications.

5.2. Ablation Study on Super-Resolution Module

This section was designed to validate the performance of various super-resolution strategies with the proposed method, especially considering low-resolution or compressed images. The results are presented in Table 6.

Table 6. Performance comparison of different super-resolution strategies and our method.

Optimizer	Precision	Recall	mAP
None	0.90	0.88	0.89
SRGAN [27]	0.94	0.92	0.93
Super-resolution module	0.97	0.95	0.95

Upon examination of Table 6, it is evident that the model's performance is the most compromised when no super-resolution strategy is employed. This underscores the importance of high resolution in object detection. SRGAN, a super-resolution approach based on Generative Adversarial Networks, has previously demonstrated effectiveness across numerous tasks. In this experiment, SRGAN indeed enhanced the Precision, Recall, and mAP of the model. Nonetheless, the super-resolution module proposed in this study outperformed all other strategies, suggesting deeper optimization tailored for this specific object detection task.

From a mathematical perspective, SRGAN leverages Generative Adversarial Networks to amplify image details with the primary intent of making the super-resolved image perceptually closer to the genuine high-resolution counterpart. However, the adversarial nature of GANs might introduce certain unrealistic details, potentially compromising the accuracy of object detection. In contrast, the super-resolution module presented in this study, while addressing perceptual image quality, places a heightened emphasis on the restoration of authentic details. This is possibly achieved using more intricate feature extraction and the fusion of multi-scale information, ensuring that the elevation in resolution does not come at the cost of genuine object detail fidelity. Such findings further affirm that in practical applications, employing an appropriate super-resolution strategy is pivotal for enhancing object detection performance on low-resolution or compressed images. The super-resolution module introduced in this study, with its unique design and optimization, successfully addresses this challenge.

5.3. Ablation Study on Lightweighting Methods

The primary objective of the experimental design in this chapter is to investigate the impact of lightweighting techniques on model performance, with a specific focus on the trade-off between speed (FPS (Frames Per Second)) and model metrics (Precision, Recall, mAP). This is particularly important for practical applications where lightweight models are often more suitable for resource-constrained environments such as embedded or mobile devices.

As observed from Table 7, the model without lightweighting demonstrates the highest Precision, Recall, and mAP but performs relatively poorly in terms of FPS, reaching only 33 FPS. This indicates that while the model exhibits high performance, the computational complexity is also increased, resulting in slower processing speed. However, in real-world applications, especially those requiring rapid response, FPS is an important metric that cannot be ignored. When knowledge distillation is employed as a lightweighting method, the model experiences an increase in FPS to 57, while the drop in Precision, Recall, and mAP is relatively minor. This suggests that knowledge distillation effectively enhances the model's processing speed while maintaining high performance. Knowledge distillation works by extracting knowledge from a larger, high-performing model (teacher model) to train a smaller, faster model (student model), enabling the student model to maintain high performance levels while reducing computational load. Quantization, another lightweighting technique, achieves the FPS value of 52 but experiences a more significant decline in Precision and mAP. Quantization reduces the bit width of model weights, thereby decreasing the model size and computational complexity. This usually comes at the cost of some performance sacrifice but significantly improves the processing speed. As shown by the experimental results, quantization elevates FPS while having a more substantial impact on model performance. When both knowledge distillation and quantization are combined (All), the model reaches the highest FPS, 73, but there is a decline in Precision, Recall, and mAP. This represents a typical trade-off scenario, where the model achieves significant improvements in processing speed at the expense of some performance loss.

Table 7. Performance comparison of different super-resolution strategies and our method.

Lightweighting Method	Precision	Recall	mAP	FPS
None	0.97	0.96	0.97	33
Knowledge distillation	0.97	0.95	0.95	57
Quantization	0.93	0.95	0.94	52
All	0.91	0.92	0.91	73

From a mathematical and algorithmic perspective, lightweighting usually involves pruning and quantizing model structures and parameters, which alter the model's mathematical properties and decision boundaries. Therefore, different lightweighting methods have varying degrees of impact on model performance. For instance, knowledge distillation often involves techniques such as soft labels and temperature scaling, which can somewhat maintain the complexity of the model's decision boundary, thus retaining higher performance levels during the lightweighting process. In contrast, quantization is a more "rigid" method of pruning and could significantly alter the model's decision boundaries, leading to performance degradation.

In summary, this experiment comprehensively explores the influence of different lightweighting methods on model performance and processing speed. The results not only reveal the trade-offs between performance and speed for various lightweighting strategies but also provide valuable insights for selecting appropriate lightweighting methods in practical applications. These findings facilitate the broader deployment of models in resource-constrained environments, especially in scenarios requiring fast and efficient processing.

5.4. Limitations and Future Works

Despite the superior performance demonstrated in previous sections, certain limitations of the proposed method are recognized. Firstly, even though the super-resolution module can effectively recover true details, its performance might be compromised on images with specific low resolution or high noise levels. Super-resolution techniques always grapple with the trade-off between accuracy and perceptual quality, and under extreme conditions, they might not consistently achieve optimal restoration results. Moreover, while the proposed adaptive optimizer exhibited commendable convergence speed and performance, its superiority might be challenged on certain intricate datasets or model architectures. Real-world data often exhibit considerable diversity and complexity, which could potentially affect the stability and effectiveness of the optimizer. Additionally, this research primarily focused on object detection tasks. However, the applicability and efficacy of the method on other tasks, such as image segmentation, facial recognition, or action detection, remain to be validated.

By addressing these limitations, clear directions for future research emerge. On one hand, further exploration into the super-resolution module is warranted, especially regarding how to better balance accuracy and perceptual quality for images under extreme conditions, ensuring both detailed and authentic image restoration. For the adaptive optimizer, future efforts could concentrate on enhancing its stability and performance on a broader and more complex array of datasets. Given the current research limitations, there is potential for applying the proposed method to other computer vision tasks to ascertain its universality. Furthermore, integration with other advanced techniques, such as neural network architecture search or knowledge distillation, might further boost the effectiveness of the method. Lastly, considering computational resources and efficiency, future endeavors could investigate how to reduce the computational load and model parameters while maintaining or even elevating performance. Such advancements would not only cater to the needs of mobile devices or edge computing but also promote the practicality and ubiquity of the method.

6. Conclusions

With the widespread application of drone technology in agriculture, ecology, and other fields, there has been a growing demand for pest detection and identification. In particular, lightweight pest identification models suitable for deployment on drones hold significant application value. They can efficiently perform pest detection in real time or nearly in real time, providing a timely decision-making basis for agricultural pest control. However, images captured with drones often suffer from challenges like low resolution, compression, and noise. Ensuring accurate and swift pest identification under these adverse conditions has been a longstanding technical challenge.

To address the aforementioned problems, a lightweight pest identification model based on Transformer and super-resolution sampling techniques is proposed in this study. Initially, the Transformer model, a powerful sequence-to-sequence model, was identified to be especially apt for capturing various spatial dependencies in images, thereby enhancing the accuracy of identification. Meanwhile, the super-resolution sampling technique focuses on addressing issues of low resolution and noisy images, restoring image details and furnishing subsequent identification processes with clearer and more accurate image data. Comparisons were made between the proposed method and other traditional methods in experiments. The results indicated that on various pest image datasets, this approach demonstrated significant advantages in terms of Precision, Recall, mAP, and FPS, achieving scores of 0.97, 0.95, 0.95, and 57, respectively. Especially for images affected by low resolution and noise, the super-resolution module was found capable of effectively restoring true details, while Transformer ensured high-accuracy pest identification even under such circumstances. Additionally, an in-depth exploration was conducted on model optimization in this study, leading to the proposal of an adaptive optimizer. It displayed commendable convergence and performance on intricate datasets and model structures.

Considering the complexity and diversity of real-world data for future research directions, further optimization of the super-resolution module could be conducted to handle even more extreme conditions. Also, in light of computational resources and efficiency, further lightweighting and optimization of the model could be explored. In summary, the lightweight pest identification model introduced in this study, based on Transformer and super-resolution sampling techniques, not only addresses the challenges of low resolution and noisy images but also offers a high-accuracy and efficient method for pest identification, holding significant value and implications for practical applications in pest detection and identification.

Author Contributions: Conceptualization, Y.B., D.F. and L.L.; Methodology, Y.B., X.F. and D.F.; Software, F.H., X.F. and W.L.; Validation, X.F.; Formal analysis, W.L. and J.L.; Investigation, J.L.; Data curation, Y.B., F.H. and J.Z.; Writing—original draft, Y.B., Fengjun Hou, X.F., W.L., J.L., J.Z., D.F. and L.L.; Writing—review & editing, W.L. and L.L.; Visualization, J.L. and J.Z.; Supervision, L.L.; Project administration, J.Z., D.F. and L.L.; Funding acquisition, L.L. All authors have read and agreed to the published version of the manuscript.

Funding: This research received no external funding.

Data Availability Statement: Not applicable.

Conflicts of Interest: The authors declare no conflict of interest.

References

1. Kumar, Y.P.; Alex, T.J.; Hardin, R.; Searcy, S.W.; Braga-Neto, U.; Popescu, S.C.; Martin, D.E.; Rodriguez, R.; Meza, K.; Enciso, J. Detecting volunteer cotton plants in a corn field with deep learning on UAV remote-sensing imagery. *Comput. Electron. Agric.* **2023**, *204* , 107551. [CrossRef]
2. Liu, K.; Qi, Z.; Tan, L.; Yang, C.; Hu, C. Mixed Use of Chemical Pesticides and Biopesticides among Rice;Crayfish Integrated System Farmers in China: A Multivariate Probit Approach. *Agriculture* **2023**, *13*, 1590. [CrossRef]
3. Group, O.O.P.M. Bayer AG's MagicTrap Rapidly Detects Pest Infestations and Provides Optimum Protection for the Canola Crop. *Outlooks Pest Manag.* **2022**, 33.
4. Kanwal, T.; Rehman, S.U.; Ali, T.; Mahmood, K.; Villar, S.G.; Lopez, L.A.D.; Ashraf, I. An Intelligent Dual-Axis Solar Tracking System for Remote Weather Monitoring in the Agricultural Field. *Agriculture* **2023**, *13*, 1600. [CrossRef]
5. Ye, Y.; Huang, Q.; Rong, Y.; Yu, X.; Liang, W.; Chen, Y.; Xiong, S. Field detection of small pests through stochastic gradient descent with genetic algorithm. *Comput. Electron. Agric.* **2023**, *206*, 107694. [CrossRef]
6. Shakirzyanova, G.; Nabiev, A.; Kholbekov, O.; Abdukakharov, V. Pheromone Monitoring in the Granaries of Uzbekistan. *Agric. Sci.* **2023**, *14*, 499–508. [CrossRef]
7. Zapponi, L.; Nieri, R.; Zaffaroni-Caorsi, V.; Pugno, N.M.; Mazzoni, V. Vibrational calling signals improve the efficacy of pheromone traps to capture the brown marmorated stink bug. *J. Pest Sci.* **2023**, *96*, 587–597. [CrossRef]
8. Zhang, Y.; Wa, S.; Liu, Y.; Zhou, X.; Sun, P.; Ma, Q. High-accuracy detection of maize leaf diseases CNN based on multi-pathway activation function module. *Remote Sens.* **2021**, *13*, 4218. [CrossRef]
9. Zhang, Y.; Li, M.; Ma, X.; Wu, X.; Wang, Y. High-Precision Wheat Head Detection Model Based on One-Stage Network and GAN Model. *Front. Plant Sci.* **2022**, *13*, 787852. [CrossRef]
10. Zhang, Y.; Wa, S.; Zhang, L.; Lv, C. Automatic plant disease detection based on tranvolution detection network with GAN modules using leaf images. *Front. Plant Sci.* **2022**, *13*, 875693. [CrossRef]
11. Zhang, Y.; Wang, H.; Xu, R.; Yang, X.; Wang, Y.; Liu, Y. High-Precision Seedling Detection Model Based on Multi-Activation Layer and Depth-Separable Convolution Using Images Acquired by Drones. *Drones* **2022**, *6*, 152. [CrossRef]
12. Wang, J.; Wang, P.; Tian, H.; Tansey, K.; Liu, J.; Quan, W. A deep learning framework combining CNN and GRU for improving wheat yield estimates using time series remotely sensed multi-variables. *Comput. Electron. Agric.* **2023**, *206*, 107705. [CrossRef]
13. Jia, L.; Wang, T.; Chen, Y.; Zang, Y.; Li, X.; Shi, H.; Gao, L. MobileNet-CA-YOLO: An Improved YOLOv7 Based on the MobileNetV3 and Attention Mechanism for Rice Pests and Diseases Detection. *Agriculture* **2023**, *13*, 1285. [CrossRef]
14. Yang, Z.; Feng, H.; Ruan, Y.; Weng, X. Tea Tree Pest Detection Algorithm Based on Improved Yolov7-Tiny. *Agriculture* **2023**, *13*, 1031. [CrossRef]
15. Jia, X.; Jiang, X.; Li, Z.; Mu, J.; Wang, Y.; Niu, Y. Application of Deep Learning in Image Recognition of Citrus Pests. *Agriculture* **2023**, *13*, 1023. [CrossRef]
16. Čirjak, D.; Aleksi, I.; Lemic, D.; Pajač Živković, I. EfficientDet-4 Deep Neural Network-Based Remote Monitoring of Codling Moth Population for Early Damage Detection in Apple Orchard. *Agriculture* **2023**, *13*, 961. [CrossRef]
17. Kumar, N.; Nagarathna; Flammini, F. YOLO-Based Light-Weight Deep Learning Models for Insect Detection System with Field Adaption. *Agriculture* **2023**, *13*, 741. [CrossRef]
18. Ullah, Z.; Alsubaie, N.; Jamjoom, M.; Alajmani, S.H.; Saleem, F. EffiMob-Net: A Deep Learning-Based Hybrid Model for Detection and Identification of Tomato Diseases Using Leaf Images. *Agriculture* **2023**, *13*, 737. [CrossRef]
19. Butera, L.; Ferrante, A.; Jermini, M.; Prevostini, M.; Alippi, C. Precise Agriculture: Effective Deep Learning Strategies to Detect Pest Insects. *IEEE CAA J. Autom. Sin.* **2022**, *9*, 246–258. [CrossRef]
20. Pest Detect pre-launch for silverleaf whitefly. *Aust. Cottongrower* **2023**, *11*, 159 .
21. Rong, M.; Wang, Z.; Ban, B.; Guo, X. Pest Identification and Counting of Yellow Plate in Field Based on Improved Mask R-CNN. *Discret. Dyn. Nat. Soc.* **2022**, *2022*, 1913577. [CrossRef]

22. Vaswani, A.; Shazeer, N.; Parmar, N.; Uszkoreit, J.; Jones, L.; Gomez, A.N.; Kaiser, Ł.; Polosukhin, I. Attention is all you need. *Adv. Neural Inf. Process. Syst.* **2017**, *30, 271.*
23. Carion, N.; Massa, F.; Synnaeve, G.; Usunier, N.; Kirillov, A.; Zagoruyko, S. End-to-end object detection with transformers. In *European Conference on Computer Vision*; Springer: Berlin/Heidelberg, Germany, 2020; pp. 213–229.
24. Zheng, D.; Dong, W.; Hu, H.; Chen, X.; Wang, Y. Less is More: Focus Attention for Efficient DETR. *arXiv* **2023**, arXiv:2307.12612.
25. Zhang, Y.; Liu, X.; Wa, S.; Chen, S.; Ma, Q. GANsformer: A detection network for aerial images with high performance combining convolutional network and transformer. *Remote Sens.* **2022**, *14*, 923. [CrossRef]
26. Goodfellow, I.; Pouget-Abadie, J.; Mirza, M.; Xu, B.; Warde-Farley, D.; Ozair, S.; Courville, A.; Bengio, Y. Generative adversarial nets. *Adv. Neural Inf. Process. Syst.* **2014**, *27, 351.*
27. Ledig, C.; Theis, L.; Huszár, F.; Caballero, J.; Cunningham, A.; Acosta, A.; Aitken, A.; Tejani, A.; Totz, J.; Wang, Z.; et al. Photo-realistic single image super-resolution using a generative adversarial network. In Proceedings of the IEEE Conference on Computer Vision and Pattern Recognition, Honolulu, HI, USA, 21–26 July 2017; pp. 4681–4690.
28. Dosovitskiy, A.; Beyer, L.; Kolesnikov, A.; Weissenborn, D.; Zhai, X.; Unterthiner, T.; Dehghani, M.; Minderer, M.; Heigold, G.; Gelly, S.; et al. An image is worth 16x16 words: Transformers for image recognition at scale. *arXiv* **2020**, arXiv:2010.11929.
29. Hinton, G.; Vinyals, O.; Dean, J. Distilling the knowledge in a neural network. *arXiv* **2015**, arXiv:1503.02531.
30. Lin, X.; Wa, S.; Zhang, Y.; Ma, Q. A dilated segmentation network with the morphological correction method in farming area image Series. *Remote Sens.* **2022**, *14*, 1771. [CrossRef]
31. Zhang, Y.; Liu, X.; Wa, S.; Liu, Y.; Kang, J.; Lv, C. GenU-Net++: An Automatic Intracranial Brain Tumors Segmentation Algorithm on 3D Image Series with High Performance. *Symmetry* **2021**, *13*, 2395. [CrossRef]
32. Zhang, Y.; He, S.; Wa, S.; Zong, Z.; Lin, J.; Fan, D.; Fu, J.; Lv, C. Symmetry GAN Detection Network: An Automatic One-Stage High-Accuracy Detection Network for Various Types of Lesions on CT Images. *Symmetry* **2022**, *14*, 234. [CrossRef]
33. Singh, D.; Jain, N.; Jain, P.; Kayal, P.; Kumawat, S.; Batra, N. PlantDoc: A Dataset for Visual Plant Disease Detection. In Proceedings of the 7th ACM IKDD CoDS and 25th COMAD, New York, NY, USA, 5–7 January 2020; CoDS COMAD 2020; pp. 249–253. [CrossRef]
34. He, K.; Zhang, X.; Ren, S.; Sun, J. Deep residual learning for image recognition. In Proceedings of the IEEE Conference on Computer Vision and Pattern Recognition, Las Vegas, NV, USA, 27–30 June 2016; pp. 770–778.
35. Simonyan, K.; Zisserman, A. Very deep convolutional networks for large-scale image recognition. In Proceedings of the International Conference on Learning Representations (ICLR), Banff, AB, Canada, 14–16 April 2014.
36. Ruder, S. An overview of gradient descent optimization algorithms. *arXiv* **2016**, arXiv:1609.04747.
37. Kingma, D.P.; Ba, J. Adam: A method for stochastic optimization. *arXiv* **2014**, arXiv:1412.6980.
38. Terven, J.; Cordova-Esparza, D. A comprehensive review of YOLO: From YOLOv1 to YOLOv8 and beyond. *arXiv* **2023**, arXiv:2304.00501.
39. Liu, W.; Anguelov, D.; Erhan, D.; Szegedy, C.; Reed, S.; Fu, C.Y.; Berg, A.C. Ssd: Single shot multibox detector. In Proceedings of the Computer Vision–ECCV 2016: 14th European Conference, Amsterdam, The Netherlands, 11–14 October 2016; Proceedings, Part I 14; Springer: Berlin/Heidelberg, Germany, 2016; pp. 21–37.
40. Tan, M.; Pang, R.; Le, Q.V. Efficientdet: Scalable and efficient object detection. In Proceedings of the IEEE/CVF Conference on Computer Vision and Pattern Recognition, Seattle, WA, USA, 13–19 June 2020; pp. 10781–10790.
41. Yang, C.; Huang, Z.; Wang, N. QueryDet: Cascaded sparse query for accelerating high-resolution small object detection. In Proceedings of the IEEE/CVF Conference on Computer Vision and Pattern Recognition, New Orleans, LA, USA, 18–24 June 2022; pp. 13668–13677.
42. Kaggle. Global Wheat Detection. 2020. Available online: https://www.kaggle.com/datasets/vbookshelf/global-wheat-head-dataset-2021 (accessed on 10 September 2023).
43. Loshchilov, I.; Hutter, F. Decoupled weight decay regularization. *arXiv* **2017**, arXiv:1711.05101.

agriculture

MDPI

Article

Combining Neural Architecture Search with Knowledge Graphs in Transformer: Advancing Chili Disease Detection

Boyu Xie [1,†], Qi Su [1,†], Beilun Tang [1], Yan Li [1], Zhengwu Yang [1], Jiaoyang Wang [1], Chenxi Wang [1], Jingxian Lin [2] and Lin Li [1,*]

[1] China Agricultural University, Beijing 100083, China; xieboyu@cau.edu.cn (B.X.); suqi2021@cau.edu.cn (Q.S.); tangbl2023@163.com (B.T.); libyan_cau@163.com (Y.L.); yemyoung@cau.edu.cn (Z.Y.); wjy20180926@163.com (J.W.); wangchenxi@cau.edu.cn (C.W.)

[2] School of Computer Science and Engineering, Beihang University, Beijing 100191, China; linjingxian2018@163.com

[*] Correspondence: lilinlsl0726@cau.edu.cn

[†] These authors contributed equally to this work.

Abstract: With the advancement in modern agricultural technologies, ensuring crop health and enhancing yield have become paramount. This study aims to address potential shortcomings in the existing chili disease detection methods, particularly the absence of optimized model architecture and in-depth domain knowledge integration. By introducing a neural architecture search (NAS) and knowledge graphs, an attempt is made to bridge this gap, targeting enhanced detection accuracy and robustness. A disease detection model based on the Transformer and knowledge graphs is proposed. Upon evaluating various object detection models on edge computing platforms, it was observed that the dynamic head module surpassed the performance of the multi-head attention mechanism during data processing. The experimental results further indicated that when integrating all the data augmentation methods, the model achieved an optimal mean average precision (mAP) of 0.94. Additionally, the dynamic head module exhibited superior accuracy and recall compared to the traditional multi-head attention mechanism. In conclusion, this research offers a novel perspective and methodology for chili disease detection, with aspirations that the findings will contribute to the further advancement of modern agriculture.

Keywords: chili disease identification; knowledge graphs; Transformers; neural architecture search; focal loss

1. Introduction

Peppers, as one of the widely cultivated crops globally, not only possess significant economic value but also serve as indispensable ingredients in numerous traditional dishes [1]. However, during the growth process, peppers are vulnerable to various diseases, which can profoundly impact their yield and quality, leading to substantial economic losses for farmers and the entire agricultural supply chain [2].

Conventional crop disease detection primarily relies on agricultural experts' expertise and manual observation [3]. Such methods are time-consuming and inefficient, falling short of meeting the demands for large-scale and real-time disease detection. With the rapid advancements in information technology and computer vision [4–6], newer techniques exhibit strengths in efficiency, accuracy, and scalability, significantly enhancing the accuracy and efficiency of disease detection [7].

Zeng et al. combined convolutional neural networks and transfer learning to detect plant diseases by inspecting plant leaves, achieving an impressive accuracy rate of 99.5% [8]. Li et al. developed an MTC-YOLOv5n model for cucumber disease detection based on YOLOv5, incorporating coordinate attention (CA) and Transformer to reduce distractions and enhance model precision, further lightweighting the model for mobile deployment [9]. Abbas, Amreen, and colleagues utilized conditional generative adversarial networks (C-GAN) to generate synthetic images of tomato plant leaves, subsequently training a DenseNet model to classify ten types of tomato diseases, achieving an accuracy rate of 97.11% [10]. Sun et al. proposed a real-time lightweight model for apple disease detection, MEAN-SSD, detecting five common apple diseases with an mAP of 83.12% and a speed of 12.53 FPS [11]. To address plant-disease identification in complex field scenarios, Wang et al. introduced a dual-stream hierarchical bilinear pooling model, primarily enhancing network layer information interaction capabilities for fine-grained recognition [12].

Knowledge graph technology, as an emerging method of data organization and representation, offers an intuitive and structured visualization of complex data relations [13]. In agriculture, knowledge graphs can consolidate information and knowledge related to crop growth, diseases, fertilization, and irrigation, offering decision-making support to farmers and agricultural experts and assisting in better crop management and disease prevention [14]. Zhou et al. created a knowledge graph for specific diseases of tomatoes and cucumbers. By integrating image modality, text modality, and knowledge graphs, an ITK-Net crop disease identification model was established, achieving 99.63% accuracy [15]. Zhu et al. addressed fruit-pest problems by first constructing a lychee knowledge graph, then using a VGG-16 model for disease and pest recognition, achieving a 94.9% accuracy rate [16]. Guan et al. constructed an agricultural knowledge graph, then used a CNN-DNN-BiLSTM network for fruit tree pest detection, comparing their results with the VGG network and BiLSTM network, showcasing the superiority of their model over traditional deep learning models [17].

Combining the knowledge graph technology with computer vision for pepper disease detection not only facilitates rapid and accurate disease identification but also offers targeted recommendations and methods for disease treatment and management [18]. For instance, using related information from the knowledge graph, specific fertilization, irrigation, and disease treatment recommendations can be provided to farmers, aiming to prevent and control diseases proactively. Furthermore, by merging computer vision and knowledge graph techniques, predictions on the occurrence, development, and spread trends of diseases can be made, granting more scientific and precise decision-making support for agricultural production and management [19].

Based on the aforementioned discussions, the primary objective of this study is to investigate the roles and impacts of a neural architecture search and knowledge graphs in chili disease detection tasks on model performance. By comparing with baseline models, this research seeks to ascertain whether these two mechanisms can enhance the model's efficacy, thereby introducing a high-precision and rapid method for chili disease detection. A chili disease identification system based on a neural architecture search and knowledge graphs was constructed, leveraging the strengths of both to elevate the efficiency and accuracy of disease detection. The main innovations and contributions are as follows:

1. A neural architecture search is applied to pepper disease image detection for the first time, automatically optimizing the model structure to achieve heightened detection accuracy.
2. A wealth of knowledge about pepper diseases is consolidated using knowledge graphs, enriching the background information and treatment recommendations for the identification results.
3. A novel method of combining Transformer in object detection is introduced and further optimized through the neural architecture search.
4. To capture subtle features in pepper disease images, a dynamic head structure is designed, and an advanced focal loss function is introduced.
5. Comprehensive experimental verification demonstrates the system's superior performance across various hardware platforms.

This research holds practical value for pepper cultivators and provides new research insights and technical references for the disease detection of other crops. Through this study, the aspiration is to propel agricultural disease detection into a more intelligent and accurate new era.

2. Related Work

2.1. Application of Neural Architecture Search in Deep Learning

The core idea of the neural architecture search (NAS) lies in the automated search for the optimal structure of deep learning models. Given the vast model space encompassing thousands of possible combinations, the goal of NAS is to identify the most performant model structure among these combinations [20].

For convolutional neural networks (CNN), which have been extensively applied to image processing tasks with remarkable success [21–23], traditional CNN models such as VGG [24] and ResNet [25] have their structures manually crafted based on researchers' insights. However, as the tasks become increasingly complex, the manual design of network architecture has become more challenging. This is where the potential of NAS is realized. NAS endeavors to explore different combinations of convolutional kernel sizes, layer counts, and connection strategies to automatically discover the most fitting CNN architecture for specific tasks [20]. The fundamental optimization problem for NAS can be expressed as [20]

$$\arg\min_{\alpha} \mathcal{L}(f(w^*(\alpha), \alpha); \mathcal{D}_{val}), \tag{1}$$

where α denotes the network structure parameters, $w^*(\alpha)$ represents the weights given the network structure parameters α, $\mathcal{L}$ is the loss function, and $\mathcal{D}_{val}$ stands for the validation set.

On the other hand, due to its self-attention mechanism, the Transformer model has shown superior performance on sequence data and has been widely adopted for natural language processing tasks [26,27]. Similar to CNNs, the structure of Transformer models can also be optimized using NAS. In NAS for Transformers, common alterations include the number of attention heads, model depth, and feed-forward neural network dimensions. For instance, through NAS, a more compact Transformer model can be discovered that maintains a performance close to the original model while significantly reducing computational requirements. The optimization problem can also be represented as in Equation (1).

In summary, NAS provides an effective method for automatically optimizing deep learning models such as CNNs and Transformers. By facilitating automated search processes, not only can NAS identify high-performing model architectures, but it can also save significant time and effort for researchers. With the further development of NAS techniques, it is anticipated that more high-performance, computationally efficient deep learning models will emerge.

2.2. Application of Knowledge Graphs in Agricultural Tasks

Knowledge graphs, as structured knowledge organization methods, have been increasingly recognized in agricultural tasks [28]. The knowledge ecosystem in agriculture is intricate, encompassing soil types, climatic conditions, crop varieties, and pest species. The strategic combination of this information determines the ultimate result of agricultural production. Knowledge graphs can structure and visualize this data, providing potent decision support for agricultural production and research.

Consider the core agricultural task of disease prediction. Traditional methods [3] largely rely on empirical knowledge, whereas knowledge graphs integrate multi-faceted data, such as historical records, soil testing outcomes, and weather forecasts, offering a more precise model for disease prediction.

2.2.1. Data Annotation Process

The data related to diseases are first gathered from various sources, potentially including reports from agricultural departments, research papers from experts, and field experiment data. These datasets are then subjected to preprocessing tasks, such as data cleaning and format conversion. Subsequently, with the aid of expert knowledge and semi-automated tools, these datasets are annotated to establish relationships (e.g., causality or correlation) between various factors, such as soil type or climatic conditions and diseases.

2.2.2. Model Input

When constructing knowledge graph models, inputs mainly comprise numerical or categorical information of various factors, such as soil type (sandy, clay, or loamy) and climatic conditions (temperature, humidity, rainfall, etc.). Moreover, historical records of disease occurrences, such as the incidence rate or disease type from the previous quarter, can also be integrated.

2.2.3. Model Output

The model output predominantly pertains to predictions related to disease occurrences, which include the likelihood of the disease manifesting, potential disease types, and probabilities associated with each type. These outputs can offer farmers targeted preventive and treatment recommendations.

Mathematically, the construction of a knowledge graph can be perceived as a graph model, where nodes represent various factors or diseases, and edges signify their relationships. For disease prediction, a probabilistic model, such as a Bayesian network, can be devised to depict the probabilistic relationships between various factors and diseases. Specifically, given the observed values of factors x, the probability of disease occurrence can be expressed as

$$P(y|x) = \frac{P(x|y)P(y)}{P(x)}, \quad (2)$$

where y represents the event of disease occurrence, $P(x|y)$ denotes the probability of observing factor x given the disease occurrence, $P(y)$ is the prior probability of disease occurrence, and $P(x)$ is the marginal probability of factor x.

To conclude, knowledge graphs have found broad applications in agricultural tasks [29–31], especially in disease prediction. By organizing and integrating diverse information in a structured manner, knowledge graphs not only enhance the accuracy of disease prediction but also deliver robust decision support for agricultural production and research.

3. Materials

3.1. Data Entry for Knowledge Graphs

Knowledge graphs have demonstrated their immense value in various tasks within the current AI research, especially in the identification of chili pepper diseases [29], where they can provide rich semantic background knowledge. Detailed below is the methodology employed to construct and utilize the knowledge graph to aid in disease detection from image datasets.

3.1.1. Knowledge Graph Construction

Initially, the core entities of the knowledge graph were determined, including "Disease", "Pathogen", "Affected Part", and "Treatment Method". These entities are vital factors in disease identification and treatment. Each entity possesses associated attributes, such as the "Name", "Incubation Period", and "Typical Symptoms" of a disease. Subsequently, relationships between these entities were established. For instance, a "Pathogen" might "Cause" a certain "Disease", and a "Disease" might "Affect" a certain "Part", as illustrated in Figure 1.

$$\mathcal{G} = \{\mathcal{E}, \mathcal{R}, \mathcal{A}\} \quad (3)$$

where $\mathcal{E}$ represents the set of entities, $\mathcal{R}$ denotes the set of relationships, and $\mathcal{A}$ stands for the set of attributes.

Figure 1. Illustration of the knowledge graphs generated in this paper. Leaf blight: This can be caused by various pathogens in different crops, but in chili peppers, it might be due to Alternaria solani or Phytophthora capsici. Black spot: Black spot is typically associated with roses and is caused by the fungus Diplocarpon rosae. In chili peppers, a disease with similar symptoms might be caused by a different pathogen, so it is important to accurately diagnose the disease. Brown spot: This could refer to bacterial leaf spot caused by Xanthomonas campestris pv. vesicatoria in chili peppers. Black mold: This is usually referring to the sooty mold that grows on the honeydew produced by insects. Early blight: This is typically caused by the fungus Alternaria solani.

3.1.2. Knowledge Graph Application

The knowledge graph provides not only detailed information about chili pepper diseases but also equips the model with semantic background knowledge. When a suspected disease region is detected in an image by the model, this region is associated with the disease entity in the knowledge graph to gather more information about that disease. For instance, upon detecting a disease, information such as its typical symptoms, potential pathogens, affected parts, and recommended treatment methods can be retrieved from the knowledge graph. To realize this functionality, a mapping function M was defined that takes the model's output and associates it with the knowledge graph [28]:

$$\mathcal{I} = M(\mathcal{O}, \mathcal{G}) \tag{4}$$

where $\mathcal{O}$ is the model's output, $\mathcal{G}$ is the knowledge graph, and $\mathcal{I}$ is the information associated with the knowledge graph.

3.1.3. Adapting Image Datasets

During model training on the image dataset, both image annotations and information from the knowledge graph were utilized as auxiliary inputs. Specifically, for each image, the disease entities and attributes related to them were retrieved and input into the model along with the image. To facilitate this functionality, an input function I was defined, which takes the image data and its related knowledge graph information, generating the model's input:

$$x' = I(x, \mathcal{I}) \tag{5}$$

where x is the image data, $\mathcal{I}$ is the information related to the knowledge graph, and x' is the model's input. In conclusion, the knowledge graph plays a pivotal role in chili pepper disease identification. It enriches the model with semantic background knowledge and amplifies the model's inference capabilities.

3.2. Image Dataset Collection and Annotation

For the training of the chili pepper disease identification model, a substantial amount of annotated image data were necessary. Initially, a plethora of chili images were gathered from multiple online agricultural databases, as shown in Table 1. These images covered different growth stages, lighting conditions, and shooting angles, ensuring data diversity, as showcased in Figure 2.

Table 1. Distribution of the images in the dataset used in this paper.

Kind	Number of Images before Augmentation	Number of Images after Augmentation
Black Mold	331	852
Brown Spot	219	719
Black Spot	486	1046
Leaf Blight	173	683
Early Blight	320	905

Figure 2. Samples of chili pepper disease images dataset in this paper. (Leaf blight, black spot, brown spot, black mold, early blight.)

Following data collection, a team comprising agricultural experts and data annotators was assembled. Using annotation tools, they annotated each image for the location and category of diseases. Each disease region was represented with a bounding box, accompanied by a specific disease name, as illustrated in Figure 3.

Figure 3. Illustration of the annotation screenshot by LabelImg [32] application.

Mathematically, this representation can be expressed as

$$D = \{(x_i, y_i)\}_{i=1}^{N} \tag{6}$$

where x_i is the i-th image, y_i denotes its corresponding disease annotation, inclusive of bounding boxes and category labels, and N is the total number of images.

3.3. Dataset Augmentation

Data augmentation is a common technique in deep learning, allowing for increased data diversity without actually expanding the dataset, thereby enhancing the model's generalization capabilities. Given the characteristics of agricultural images, such as variations in lighting and obstructions, it was decided to employ data augmentation to bolster the model's robustness against these factors. Several augmentation techniques were applied, including random cropping, rotation, scaling, brightness, and contrast adjustments, as depicted in Figure 4.

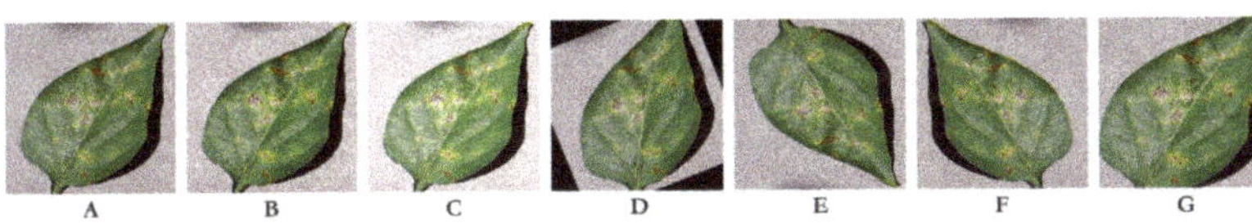

Figure 4. Demo of different augmentation methods used in this paper. (**A**) is the original image; (**B**) is contract augmentation; (**C**) is brightness augmentation; (**D**) is rotation augmentation; (**E**) is flipping vertically augmentation; (**F**) is flipping horizontally augmentation; (**G**) is cropping augmentation.

Mathematically, given an image x, an augmentation function $\mathcal{T}$ was defined that takes image x and produces the augmented image x':

$$x' = \mathcal{T}(x) \tag{7}$$

Through data augmentation, a vast number of images slightly different from the original yet retaining the same semantic essence can be generated. This not only amplifies the volume of training data but also aids the model in learning more robust features, thereby enhancing its performance in real-world scenarios. In summary, through the construction of the knowledge graph and the collection, annotation, and augmentation of datasets, a rich and diverse training dataset was provided for the task of chili pepper disease identification. This laid a solid foundation for the training and evaluation of the model, ensuring commendable results in practical applications.

4. Proposed Method

4.1. Overall

A comprehensive framework that integrates both the Transformer and knowledge graph models is proposed in this study, aiming for efficient and accurate detection of chili pepper diseases. To fully exploit both the image data of chili pepper diseases and the related knowledge information, a two-stage model, has been designed, as depicted in Figure 5.

Initially, image features of chili pepper diseases are extracted using the Transformer model after a CNN module, as shown in Figure 5. Subsequently, by incorporating the knowledge graph model, related knowledge information is integrated to provide a more comprehensive and precise decision support for disease identification. The input to the model is twofold: the first being the image data of chilies, encompassing images of both healthy and diseased peppers; the second pertains to knowledge information related to diseases, which might encompass aspects such as disease types, pathogens, influencing factors, and mechanisms of disease onset. The model output is the identification result of the chili pepper diseases, covering disease type, the likelihood of occurrence, and related knowledge information. These outputs can offer farmers targeted prevention and treatment recommendations.

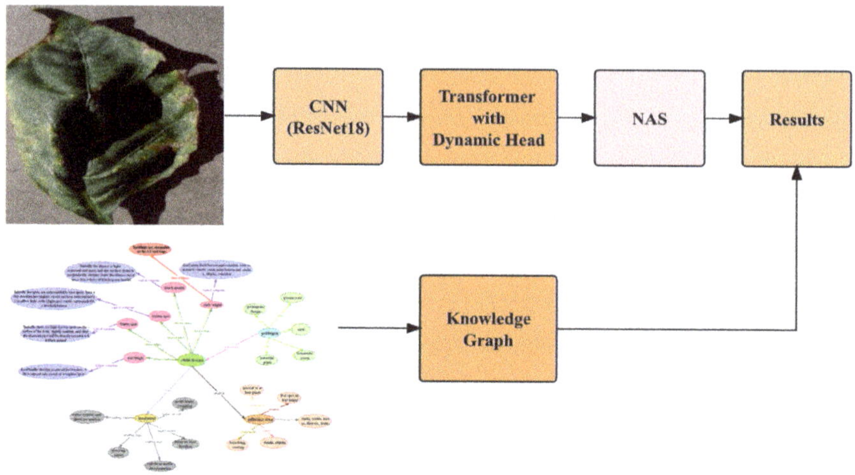

Figure 5. Illustration of the whole method proposed in this paper.

4.2. NAS for Performance Optimization

To further enhance the performance of the model, neural architecture search (NAS) technology is applied for the structural search of the entire model, as shown in Figure 6. Specifically, a search space is first defined, encompassing various potential structural configurations of the Transformer model, such as its depth, number of attention heads, and dimensions of the feed-forward neural network. Then, using NAS, the model structure best suited for the chili pepper disease detection task is autonomously sought. Furthermore, to integrate the knowledge graph model, structural configurations related to knowledge information, such as the embedding methods for knowledge nodes and the association methods between knowledge and features, are also incorporated into the search space. Through the autonomous search with NAS, not only can the most performant model structure be found, but the optimal way to integrate knowledge information with features can also be determined.

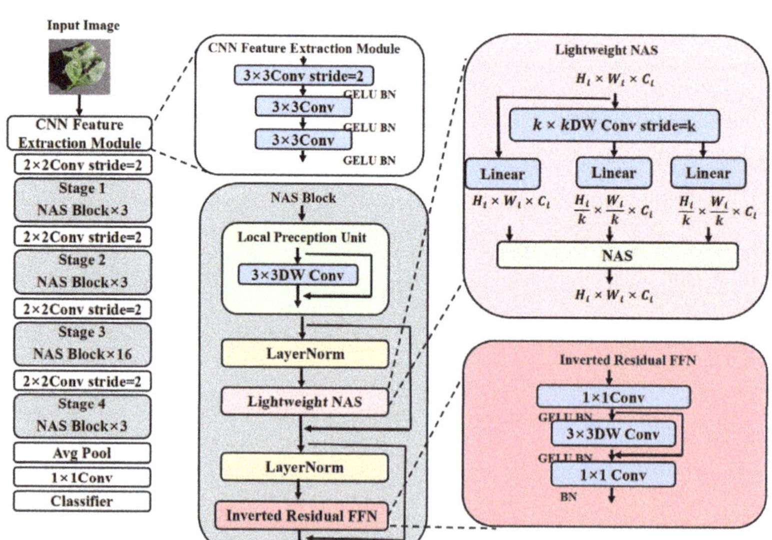

Figure 6. Illustration of the NAS module used in this paper. GELU [33] means Gaussian error linear units. The NAS block details are shown in the gray bounding box.

In summary, the method proposed in this study is a comprehensive framework that integrates both the Transformer and knowledge graph models. Through its two-phase model design, it harnesses both the image data of chili pepper diseases and the related knowledge information, providing comprehensive and precise decision support for disease identification. Moreover, by employing NAS technology, the model's performance is further optimized, adapting it more closely to the characteristics and demands of the chili pepper disease detection task.

4.3. Integration of Knowledge Graph

The value of knowledge graphs in various AI applications is gradually gaining recognition among researchers [30,31]. In the task of chili pepper disease identification, the knowledge graph can equip the model with rich prior knowledge and background information, aiding the model in better understanding and identifying diseases. The input for this module is the raw features from the object detection model and the entity and relationship information related to chilies in the knowledge graph, as shown in Figure 1. The output is the enhanced features after integrating with the knowledge graph, as shown in Figure 1.

Entities and relationships related to chilies are first extracted from the knowledge graph to build a disease-attribute subgraph. Then, a graph neural network (GNN) [29] is used to encode this subgraph, obtaining the embedding representation for each disease. Mathematically, this process can be represented as

$$h_v^{(l+1)} = \sigma\left(\sum_{u \in \mathcal{N}(v)} W^{(l)} h_u^{(l)}\right), \tag{8}$$

where $h_v^{(l)}$ denotes the embedding of node v at layer l, $\mathcal{N}(v)$ represents the set of neighbors of node v, $W^{(l)}$ is the weight matrix at layer l, and σ is an activation function. Subsequently, the obtained disease embeddings are fused with the raw features from the object detection model through a fully connected layer, mathematically expressed as

$$f' = \text{ReLU}(W_f f + b_f + h), \tag{9}$$

where f is the raw feature, h is the disease embedding, and W_f and b_f are the weight matrix and bias, respectively. The design of this fusion module is driven by the intent to leverage prior knowledge and background information from the knowledge graph to enhance the model's comprehension capability. Conventional object detection models only learn features from images and lack a deep understanding of the reasons behind and impacts of the diseases. The knowledge graph, on the other hand, can provide the model with this invaluable information, aiding the model in better distinguishing between various diseases, thereby enhancing identification accuracy. It not only equips the model with rich information from the knowledge graph, enhancing its understanding of the diseases, but also introduces a novel, more potent feature representation method, allowing the model to learn features not just from images but also from the knowledge graph. Finally, integrating the knowledge graph provides a more stable and robust feature representation, ensuring the model's robust performance even in the face of noisy or incomplete data.

4.4. Transformer in Focus Detection Task

The recent computer vision research has extensively focused on the Transformer model due to its unique self-attention mechanism. Initially, Transformers were designed for handling natural language, aiming to capture long-range dependencies in texts. However, it was discovered by researchers that its self-attention property is also highly suited for image processing tasks, especially in scenarios that necessitate capturing long-distance relationships between different parts of an image. In the chili pepper disease identification task presented in this study, morphological features of diseases can appear anywhere in the image, and there might exist correlations or structural dependencies between these locations. For instance, the onset of a disease on one side of the chili could imply the emergence of disease symptoms on the opposite side. Traditional CNN models, focusing primarily on local features, might miss such global, long-distance dependencies. This is where the Transformer's uniqueness in object detection, especially in chili pepper disease identification, comes into play.

In the design of this study, a CNN, ResNet18 [25], is initially utilized to extract basic image features, which are then fed as inputs to the Transformer module [26], as depicted in Figure 7.

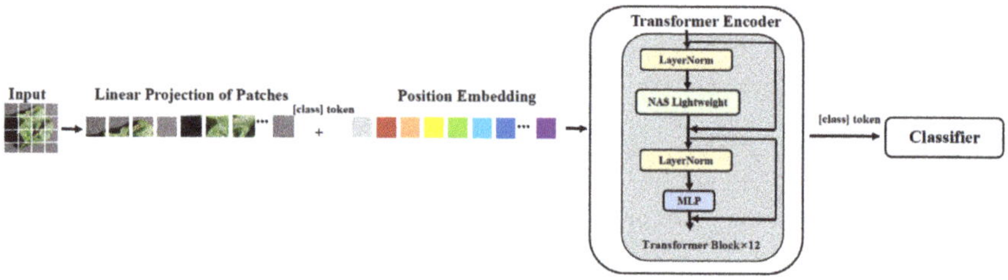

Figure 7. Illustration of the Transformer architecture used in our model.

The Transformer module [26] converts the feature map into a sequential format, with each pixel point acting as an element in the sequence. These elements undergo processing via the self-attention mechanism, resulting in new feature representations. These new features not only amalgamate local information but also integrate global, long-range information. Specifically, the self-attention mechanism in the Transformer can be mathematically represented as

$$\text{Attention}(Q, K, V) = \text{softmax}\left(\frac{QK^T}{\sqrt{d_k}}\right)V, \tag{10}$$

where Q, K, and V are the query, key, and value derived from the input features via linear transformation, respectively, and d_k denotes the dimension of the key. Adopting this design, which combines the local feature extraction capabilities of CNN and the long-distance relationship capturing abilities of the Transformer, allows for a more accurate identification of chili disease features. Furthermore, applying NAS on this structure can further optimize the model, automatically searching for the network configuration most suited for chili disease identification, thereby enhancing the accuracy of identification.

4.4.1. Dynamic Head for Tiny Focus Feature

In object detection, especially in detecting tiny focus features, the conventional multi-head self-attention [26] used in Transformer models might encounter certain limitations. To more precisely capture these small yet pivotal features, a design called "Dynamic Head" is proposed, intended to replace the original multi-head attention mechanism, as shown in Figure 8.

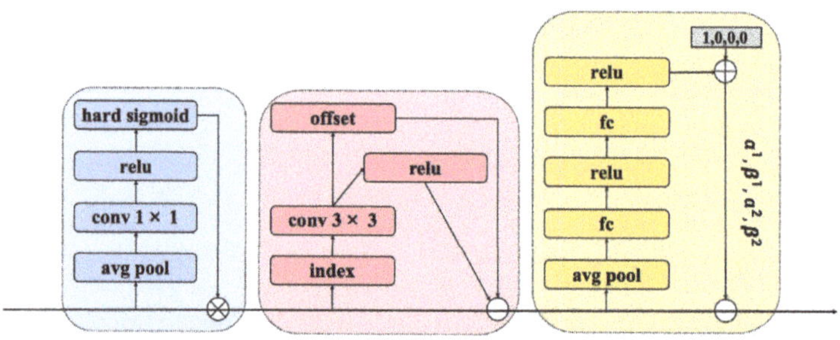

Figure 8. Illustration of the dynamic attention head proposed in this paper. The hard sigmoid is a simplified version of the sigmoid function, which accelerates computation through linear approximation.

Traditional multi-head self-attention aims to allow the model to simultaneously capture multiple different feature relationships. Specifically, each "head" independently executes self-attention operations, thereby focusing on different parts of the input features. This can be mathematically expressed as

$$\text{MultiHead}(Q, K, V) = \text{Concat}(\text{head}_1, \text{head}_2, \ldots, \text{head}_n)W^O, \tag{11}$$

where each head_i represents an independent self-attention operation as previously described, and W^O is the output linear transformation matrix. However, in the dynamic head design, instead of using a fixed number of "heads" for the self-attention operations, the number and weights of the "heads" are dynamically adjusted based on the content of the input features. Specifically, a weight coefficient is introduced for each "head", which adjusts dynamically based on the input features, allowing some "heads" to have higher weights when dealing with tiny focus features. Mathematically, the dynamic head can be expressed as

$$\text{DynamicHead}(Q, K, V) = \text{Concat}(\alpha_1 \times \text{head}_1, \alpha_2 \times \text{head}_2, \ldots, \alpha_n \times \text{head}_n)W^O, \tag{12}$$

where α_i is the weight coefficient for the ith "head", and it functions based on the input features. The design rationale behind the dynamic head originates from the observation that not all "heads" are equally important when processing tiny focus features; some "heads" might be more adept at capturing such features while others might overlook them. By introducing dynamic weights, the model can prioritize the more relevant "heads", achieving a more accurate capture of tiny focus features. Compared to the traditional multi-head self-attention, the dynamic head offers the following advantages:

1. Precise feature capture: Through dynamic weights, the model can place greater emphasis on those "heads" that are beneficial for capturing tiny focus features, thereby improving identification accuracy.
2. Enhanced model flexibility: The dynamic head is not limited to the detection of tiny focus features but is also applicable to other types of object detection tasks. This is because it can dynamically adjust the "head" weights based on input feature content, making the model more adaptive to the current task.

In conclusion, the dynamic head offers a novel and more potent feature extraction mechanism for chili pepper disease identification. It is believed that through this design, the model's accuracy and robustness can be further enhanced.

4.4.2. Advanced Focal Loss Function

In object detection tasks, especially with imbalanced data distributions, the classic cross-entropy loss may lead to a model preference for frequently occurring background classes at the expense of minority target classes, such as chili diseases [34]. To address this issue, an advanced focal loss function is proposed in this study. The original design of the focal loss function was intended to increase the weight of samples misclassified by the model, ensuring that these samples receive greater attention during training. It is mathematically defined as

$$\text{FL}(p_t) = -(1 - p_t)^\gamma \log(p_t), \tag{13}$$

where p_t is the model's predicted probability for the positive class, and γ is a tuning parameter used to control the rate of weight increase. However, the original focal loss function might still fail to capture some critical, hard-to-classify samples in certain cases. To further accentuate the model's focus on these samples, an advanced version of the focal loss function is introduced. A new parameter, α, is incorporated into the original focal loss function, combined with the sample's class imbalance. The specific form is

$$\text{AFL}(p_t) = -\alpha_t(1 - p_t)^\gamma \log(p_t), \tag{14}$$

where α_t is a coefficient related to the sample's class distribution, used to further enhance the weight of hard-to-classify samples. Such a design was chosen since in imbalanced data distributions, the hard-to-classify samples tend to be key and valuable. By introducing α_t, the model can place more emphasis on these samples during training, thereby improving the model's generalization capability. Mathematically, α_t can be defined as

$$\alpha_t = \frac{N_{\text{neg}}}{N_{\text{pos}} + N_{\text{neg}}}, \tag{15}$$

where N_{pos} and N_{neg} are the number of positive and negative samples, respectively. Compared to the original focal loss function, the advanced focal loss function offers the following advantages:

1. Enhanced ability to handle class imbalance: By introducing α_t, the weight of hard-to-classify samples can be further emphasized, ensuring that the model focuses more on these samples during training.
2. Improved generalization capability: In imbalanced data distributions, hard-to-classify samples are often key and valuable. By utilizing the advanced focal loss function, the model's focus on these samples during training can be accentuated, thereby enhancing its generalization capability.
3. Greater flexibility in loss adjustment: Compared to the original focal loss function, the advanced focal loss introduces a new parameter, α_t, providing flexibility to adjust the loss function according to specific task requirements, leading to improved training outcomes.

In summary, the advanced focal loss function offers a novel and more potent loss design for chili disease identification. It is believed that through this design, the accuracy and robustness of the model can be further enhanced.

4.5. Experiment Design

To provide a comprehensive and objective assessment of the proposed chili disease identification system, which is based on neural architecture search and knowledge graphs, a series of experiments have been designed. The subsequent sections detail the experimental design.

4.5.1. Experiment Platform

In this research, all experiments were conducted on the Linux operating system platform. To ensure efficient code execution and rapid model development, Python was chosen as the primary development language, owing to its extensive use in the fields of data science and machine learning. To build and test our model, we utilized several popular Python libraries. First, we employed the PyTorch library, version 1.8.0, which is an open-source deep learning framework that offers flexible and efficient model training and evaluation capabilities. Additionally, for data processing and analysis, we used NumPy (version 1.19.5) and Pandas (version 1.2.3), both of which provide a plethora of handy tools and functions. For the visualization of our model and the presentation of results, we employed Matplotlib (version 3.4.1) to generate high-quality graphics.

4.5.2. Dataset Partition and Baseline

The chili disease dataset was initially partitioned. Adhering to conventional data-splitting principles and aiming to ensure training stability, the dataset was divided into training, validation, and test sets at a ratio of 8:1:1. The training set is utilized for model training and parameter updates, the validation set for performance validation and hyper-parameter tuning, and the test set for the final evaluation of the model performance. To comprehensively evaluate the proposed model, several popular models in the object detection domain were chosen as baselines, including YOLOv5 [35], YOLOv8 [36], DETR [37], SSD [38], and EfficientDet [39]. These models have demonstrated remarkable performance in object detection tasks, thereby serving as suitable performance benchmarks. Notably, these models span various technological trends from real-time detection using the YOLO series to the Transformer-based DETR, offering a comprehensive perspective to evaluate the proposed approach.

4.5.3. Optimizer Selection and Hyperparameter Settings

The choice of optimizer plays a crucial role in influencing the training speed and final performance of the model. In these experiments, the Adam [40] optimizer was chosen due to its ability to adaptively adjust learning rates and its proven effectiveness across various tasks, combining the advantages of Momentum and RMSProp. The selection of hyperparameters is also a pivotal aspect of experimental design. Both grid search and random search strategies were employed to find the optimal hyperparameter combination, with various combinations validated on the validation set. Ultimately, a learning rate of 0.001, batch size of 32, and weight decay of 0.0005 were selected, as they exhibited the best performance on the validation set.

4.5.4. Ablation Study Design

To further validate the effectiveness and significance of each component within the proposed model, a series of ablation studies were conducted. These studies included variations such as: a model without the use of knowledge graphs; a model without the use of neural architecture search; the application of the original multi-head attention mechanism instead of the introduced dynamic head; and the utilization of cross-entropy loss instead of the advanced focal loss function proposed. Through these ablation studies, a deeper understanding of the role of each component within the model can be obtained, offering valuable insights for further research and improvements.

4.5.5. Experiment Metric

To objectively and comprehensively evaluate the performance of the proposed chili disease identification system, multiple evaluation metrics were employed, including precision, recall, mAP, and FPS [7].

1. Precision
   Precision measures the proportion of positive predictions that are actually correct. It reflects the accuracy of the model's predictions, indicating how many of the predicted positive samples are true positives.
   $$\text{Precision} = \frac{\text{TP}}{\text{TP} + \text{FP}} \quad (16)$$
   Here, TP represents the number of true positives, while FP indicates false positives. For chili disease identification, a high precision implies that the model has a low rate of false alarms when identifying diseases.
2. Recall
   Recall indicates the proportion of actual positive samples that are correctly predicted. It captures the model's capability to retrieve relevant instances, revealing how many of all positive samples are accurately predicted by the model.
   $$\text{Recall} = \frac{\text{TP}}{\text{TP} + \text{FN}} \quad (17)$$
   In this equation, FN denotes the number of false negatives. Recall is particularly important for chili disease identification as a high recall ensures that most diseases are detected, mitigating potential agricultural losses.
3. Mean Average Precision (mAP)
   mAP computes the average precision at varying levels of recall, commonly employed in object detection tasks. For each recall level, precision is calculated, and then an average of these precisions is taken.

$$\text{mAP} = \frac{1}{|\mathcal{R}|} \int_{r \in \mathcal{R}} P(r)\, dr \tag{18}$$

Here, $\mathcal{R}$ represents the set of recall values, and $P(r)$ indicates the precision at recall level r. mAP offers a holistic measure of the model's accuracy and recall capabilities.

4. Frames Per Second (FPS)
FPS serves as an indicator of the model's real-time capability, denoting the number of frames the model can process per second.

$$\text{FPS} = \frac{1}{\text{Time per frame}} \tag{19}$$

For chili disease identification, a high FPS suggests that the model can swiftly process images, offering timely disease detection results in practical applications.

In conclusion, the evaluation metrics employed in this study encompass the model's accuracy, recall capabilities, and real-time processing ability, providing a comprehensive and objective assessment standard. Particularly for chili disease identification, high accuracy and recall ensure timely and precise disease detection, while a high FPS guarantees real-time application, thus, offering farmers timely and effective disease control recommendations.

5. Results and Discussion

In machine learning and computer vision research, the evaluation and comparison of models serve as pivotal components for assessing their efficacy and robustness. The primary aim of this section is to evaluate and compare the performance of various object detection models on chili disease identification tasks, thereby offering theoretical and empirical foundations for practical applications. By employing consistent evaluation metrics—precision, recall, and mAP—an unbiased and objective assessment of the strengths and weaknesses of each model is achieved.

5.1. Detection Results

The experimental results from this study reveal that the model developed in this research achieved the best scores on the evaluation metrics of precision, recall, and mAP, being 0.95, 0.91, and 0.94, respectively, outperforming other baseline models, as shown in Tables 2 and 3.

Table 2. Detection results of different models.

Model	Precision	Recall	mAP
YOLOv5 [35]	0.89	0.87	0.88
YOLOv8 [36]	0.88	0.86	0.87
SSD [38]	0.86	0.83	0.85
DETR [37]	0.90	0.88	0.89
EfficientDet [39]	0.87	0.85	0.86
Ours	0.95	0.91	0.94

Table 3. Detection results of different chili disease types using our model.

Kind	Precision	Recall	mAP
Black Mold	0.93	0.90	0.92
Brown Spot	0.93	0.88	0.91
Black Spot	0.95	0.92	0.95
Leaf Blight	0.97	0.92	0.96
Early Blight	0.97	0.92	0.96

Among them, DETR demonstrated commendable performance, second only to the proposed model. This success can be attributed to its design based on the Transformer architecture, which is adept at capturing the contextual information of images, providing a more comprehensive perspective for chili disease detection. The YOLO series models (YOLOv5 and YOLOv8) also showcased noteworthy performance by predicting bounding boxes and categories in a single forward pass, ensuring the model's real-time capability and accuracy. However, the SSD model exhibited the most modest performance among all, possibly due to its inadequate handling of small or highly overlapping objects. EfficientDet, being a model focusing on efficiency, still holds considerable application value in scenarios with limited computational resources, despite its slightly inferior performance compared to the others. In summary, integrating knowledge graphs with deep learning techniques led to significant improvements in chili disease identification, and emerging detection models such as DETR also showcased vast potential.

5.2. Test on Different Edge-Platform

The objective of this experiment was to evaluate and compare the real-time performance of various object detection models across multiple edge computing platforms. These platforms encompassed common smartphone models such as Huawei P40 and iPhone 13, as well as microcomputers such as the Jetson Nano and Raspberry Pi. By assessing the frames per second (FPS) performance of these models on different hardware, insights were gained into the potential real-world performance of each model, especially in resource-constrained scenarios. The results are presented in Table 4.

Table 4. FPS comparison of different detection models on different hardware platform. Generally, we believe that if a model achieves a processing speed of 30FPS, it can be considered to have met the requirements for real-time monitoring [3]. On the Huawei P40, the implementation of the model in this paper was achieved using AI-related API interfaces provided by Google, with Java being the development language. On the iPhone, the development was carried out in the Swift language using Apple's Xcode software 14.0, as shown in Figure 9. For both the Jetson Nano and Raspberry Pi, since they run on a Linux system, their implementation is the same as that on servers. Both are developed using the Python language based on the PyTorch framework.

Model	Huawei P40	Jetson Nano	Raspberry Pi	iPhone 13
YOLOv5 [35]	31	49	11	28
YOLOv8 [36]	29	44	12	29
SSD [38]	13	27	-	-
DETR [37]	3	15	-	-
EfficientDet [39]	28	45	13	19
Ours	33	58	13	31

These findings are reflective of the inherent design nuances and optimization levels of each model. The YOLO series, renowned for its streamlined design and efficient forward computation, ensured commendable real-time performance across platforms. EfficientDet, on the other hand, sought a balance between efficiency and accuracy, rendering it slightly less optimal in certain resource-limited environments. While SSD aimed for efficiency during its design phase, it failed to match the expectations on some edge devices, possibly due to its multi-scale features and intricate default bounding box computations. As for DETR, its Transformer-based design excelled in capturing contextual information from images but at the cost of increased computational complexity, leading to a pronounced reduction in FPS on some edge devices.

Delving deeper into their mathematical constructs, the YOLO series fundamentally simplifies object detection to a regression problem, thereby eliminating the complexities of multi-stage computations and ensuring exemplary FPS. DETR, equipped with a Transformer structure, encompasses extensive matrix computations and self-attention mechanisms. This computational burden is particularly pronounced on edge devices, resulting in compromised performance. EfficientDet, meanwhile, endeavors to strike a balance between model size and computational intricacies. While it might occasionally fall short of YOLO's performance, its utility remains in resource-constrained settings.

Figure 9. Our method running on iPhone.

In conclusion, the varied performance of these models on diverse edge computing platforms stems from their unique design philosophies and underlying mathematical constructs. These insights offer valuable guidance for future research, aiding researchers and engineers in judicious model selection and optimization to cater to practical application needs.

5.3. Ablation Study on Different Dataset Augmentation Methods

The primary objective of this experiment was to evaluate and compare the effects of various dataset augmentation techniques on model performance. Dataset augmentation is commonly employed to expand the training set, enhance model generalization, and mitigate issues arising from insufficient data or overfitting. By systematically applying and combining various augmentation methods, such as flipping, cropping, resizing, and brightness adjustment, insights into their specific impacts on model performance can be gained. This paves the way for determining the optimal augmentation strategies for practical applications. The experimental results are presented in Table 5.

Table 5. Ablation experiment results of different dataset augmentation methods on our model.

Flipping	Cropping	Resize	Brightness	mAP
-	-	-	-	0.88
-	✓	✓	✓	0.93
✓	-	✓	✓	0.89
✓	✓	-	✓	0.93
✓	✓	✓	-	0.91
✓	✓	✓	✓	0.94

From the results table, it can be discerned that different augmentation techniques exert distinct effects on model performance. Specifically:

1. Without any augmentation, the model achieved an mAP of 0.88.
2. Employing solely cropping, resizing, and brightness adjustment, the model's mAP rose to 0.93, indicating that these three techniques significantly bolstered the model's performance.

3. With only flipping, resizing, and brightness adjustment, the mAP reached 0.89. This score, slightly higher than without any augmentation, is nonetheless inferior compared to the effect of cropping, suggesting that flipping might not always be as effective as cropping in certain contexts.
4. Retaining all methods but resizing, the mAP still remained at 0.93, possibly implying that, in the presence of other augmentations, the impact of resizing becomes less pronounced.
5. When excluding brightness adjustment but maintaining other methods, the mAP was 0.91, highlighting the contribution of brightness adjustment to model performance.
6. Integrating all augmentation techniques, the model achieved its best mAP of 0.94.

These findings underscore the pivotal role of data augmentation in enhancing model performance, particularly the cropping, resizing, and brightness adjustment methods, which seem to have a more pronounced effect on performance.

5.4. Ablation Study on Different Loss Functions

The primary objective of this study was to evaluate and compare the impact of different loss functions on model performance. Loss functions serve as a pivotal component in machine learning and deep learning training, determining how the parameters of the model are optimized and how features are learned during the training process. By contrasting various loss functions, insights can be garnered regarding their distinct roles and effects during model training, offering theoretical guidance for model selection and optimization in practical applications.

From the experimental data presented in Table 6, it can be inferred that varying loss functions considerably influence model performance. Specifically, the advanced focal loss outperforms in all metrics, achieving an mAP of 0.94, underscoring its efficacy for this task. The original focal loss also delivers commendable results with an mAP of 0.91, largely attributed to its design purpose of addressing class imbalance issues. Conversely, AP loss and DR loss present closely aligned performance, albeit marginally trailing behind focal loss, with mAPs of 0.88 and 0.87, respectively. Analytically, the focal loss accentuates the optimization of the model by emphasizing harder-to-classify samples, proving particularly advantageous for tasks potentially grappling with class imbalances. While AP Loss and DR Loss might exhibit promising results for certain tasks, they seem less effective than focal loss for this specific endeavor.

Table 6. Ablation experiment results of different loss functions on the proposed method.

Loss Function	Precision	Recall	mAP
AP Loss [41]	0.91	0.86	0.88
DR Loss [42]	0.90	0.85	0.87
Focal Loss [34]	0.93	0.87	0.91
Advanced Focal Loss	0.95	0.91	0.94

Mathematically dissecting the models reveals that the AP loss, a probability-based loss function, predominantly focuses on differentiation between positive and negative cases. It might render satisfactory outcomes when there is a balanced distribution of positive and negative samples. However, in scenarios involving class imbalances or other intricate factors, its efficacy might be overshadowed by other loss functions. DR Loss, used in conjunction with the Adam optimizer [40], incorporates gradient momentum and the second moment, contributing to a more stabilized optimization process. Nevertheless, this stability might come at the expense of performance. From a mathematical perspective regarding the focal loss, it augments the weight for samples mispredicted by the model. This means the model tends to pay more attention to challenging samples, which becomes crucial in situations with class imbalances.

5.5. Ablation Study on Dynamic Head Module

The aim of this experiment was to evaluate and compare the impacts of different attention mechanisms on model performance. Attention mechanisms play a pivotal role in deep learning, especially when processing sequential and image data. By contrasting various attention mechanisms, such as multi-head attention and the dynamic head module, insights into their distinct roles and effects during model training can be gleaned. A deep understanding of these mechanisms offers a theoretical foundation for the selection and optimization of models in practical applications. The results are presented in Table 7.

Table 7. Ablation experiment results of different attention mechanism on the proposed method.

Attention Mechanism	Precision	Recall	mAP
Multi-Head [26]	0.93	0.88	0.91
Dynamic Head Module	0.95	0.91	0.94

According to Table 7, it is evident that different attention mechanisms distinctly affect model performance. Specifically, the dynamic head module outperforms the multi-head attention mechanism across all metrics, achieving an mAP of 0.94, while the multi-head attention records an mAP of 0.91. Originating from the Transformer architecture, the multi-head attention mechanism processes information in parallel by segmenting the input into multiple distinct subspaces. Each subspace possesses its own weights, enabling the model to simultaneously attend to various information segments. While this mechanism aids the model in capturing a myriad of features and patterns within the data, it may also introduce some redundancy. In contrast, the dynamic head module presents a more flexible mechanism. It can dynamically adjust attention weights based on data characteristics, thus, capturing critical features with more specificity. This dynamism allows the model to better adapt to various data and scenarios, especially when the data exhibit intricate patterns or noise. This adaptability is a plausible reason why the dynamic head module surpasses the performance of the multi-head attention.

From a mathematical perspective, the multi-head attention mechanism processes multiple subspaces' information in parallel through matrix operations. Although this parallel processing boosts efficiency, it might introduce redundancy, leading to dispersed weights, which could compromise the model's performance. On the other hand, the dynamic head module pays closer attention to the main features within the data, dynamically adjusting weights to amplify the influence of these features, thereby enhancing the model's precision.

In conclusion, different attention mechanisms possess unique characteristics and outcomes when processing data. While the multi-head attention mechanism can process information across multiple subspaces in parallel, it might lead to weight dispersion, affecting the model's performance. Conversely, the dynamic head module, by dynamically adjusting weights, captures key features more specifically, enhancing the model's performance. Such insights provide valuable guidance for researchers and engineers when choosing and optimizing models.

5.6. Ablation Study on NAS and Knowledge Graph

The primary objective of the experimental design was to investigate the role and impact of neural architecture search (NAS) and knowledge graphs on the task of chili disease detection. By comparing both with the baseline model, it was discerned whether these mechanisms could enhance the model's performance, thereby elucidating their significance in crop disease detection.

As observed in Table 8, the baseline model, devoid of any attention mechanism, exhibited performances of 0.83, 0.85, and 0.84 in precision, recall, and mAP, respectively. This served as the benchmark for subsequent comparisons. With the exclusive utilization of NAS, all evaluation metrics displayed an increase. Such findings suggest that NAS can effectively optimize the model structure, thereby enhancing its performance. A notable advantage of NAS is its capability to automatically search for an optimal model architecture, hence identifying the most suitable model for a specific task. In this context, NAS likely pinpointed distinct features and patterns particularly apt for chili disease detection, leading to heightened accuracy and recall. In the scenario where only the knowledge graph was applied, there was an improvement in the model's performance, though potentially not as pronounced as with NAS. The primary role of the knowledge graph lies in its ability to consolidate domain knowledge, assisting the model in better understanding and interpreting data. Within the realm of chili disease detection, the knowledge graph might encompass diverse information pertinent to the disease, such as pathogens, symptoms, and growth environment. Such information can aid the model in more accurately identifying diseases. Notably, when the model integrated both NAS and the knowledge graph, all indicators experienced a significant surge. This underscores, to a certain extent, that NAS and the knowledge graph are complementary. Their concurrent application to the model can yield superior performance enhancements, indicating the considerable benefits of considering both model architecture optimization and domain knowledge integration for tasks such as chili disease detection.

Table 8. Ablation experiment results of NAS and knowledge graph.

Attention Mechanism	Precision	Recall	mAP
None (baseline)	0.83	0.85	0.84
Only NAS	0.90	0.88	0.88
Only Knowledge Graph	0.89	0.87	0.89
Both	0.95	0.91	0.94

From a mathematical standpoint, NAS primarily focuses on the optimization of the model's structure, ensuring the model's ability to capture the most valuable features from the data. Conversely, the knowledge graph emphasizes the model's semantic understanding, ensuring precise judgments in the intricate backdrop of crop diseases. Their combination equips the model with a robust discriminatory capability, making it exemplary in the task of chili disease detection. In conclusion, future crop disease detection tasks should contemplate the concurrent use of NAS and knowledge graphs to achieve heightened detection accuracy and robustness. Moreover, with the integration of more crop disease data and domain knowledge, the potential of the knowledge graph may further unfold, paving the way for significant breakthroughs in crop disease detection.

5.7. Limitations and Future Works

In the experiments conducted on edge computing platforms, a variety of common smartphones and microcomputers were covered. However, considering the rapid hardware updates, the hardware platforms used might not fully represent future devices. Additionally, for different application scenarios, a broader spectrum of hardware platforms might be considered. In this study, primary attention was given to the multi-head attention and dynamic head module attention mechanisms. Although the dynamic head module performed excellently in tests, it does not imply that it is suitable for all tasks or scenarios. The loss functions mentioned in the text, such as AP loss, DR loss, and focal loss, might not encompass all potential loss functions. Different loss functions might yield varied results under different tasks and data distributions.

Given the swift advancement in hardware technology, the future research should consider a more diverse range of new hardware platforms, including newly emerged chips and modules specifically designed for machine learning and microcomputers with higher computational capabilities. Apart from multi-head attention and dynamic head modules, many other attention mechanisms warrant exploration, such as axial attention and sparse attention. Delving deeper into these novel attention mechanisms could further enhance model performance. Regarding loss functions, future efforts might attempt to design new loss functions or combine and adjust existing ones to accommodate various tasks and data distributions.

6. Conclusions

In this study, a chili disease detection method based on deep learning is presented. Initially, the performance of various object detection models was evaluated on edge computing platforms. Through experiments, it was found that the dynamic head module and the multi-head attention mechanism exhibited distinct characteristics and performances in data processing. Notably, the dynamic head module, owing to its flexible nature, surpassed the multi-head attention mechanism in terms of performance. Furthermore, to optimize model performance, different data augmentation strategies and loss functions and their impact on model performance were explored. The experimental results indicated that when all the data augmentation methods were integrated, the model achieved the best mAP, reaching 0.94. Regarding the loss functions, the dynamic head module demonstrated higher precision and recall compared to the traditional multi-head attention mechanism.

Summarizing the core contributions of this study: First, a comprehensive deep learning framework for chili disease detection is introduced, encompassing every step from data preprocessing to model training and evaluation. Second, through a multi-faceted ablation study, various factors influencing model performance, such as data augmentation strategies, loss functions, and attention mechanisms, were revealed, offering valuable insights for future research in the field. Finally, this study not only presents an effective solution for chili disease detection but also provides insights and references for disease detection in other crops. In essence, this work offers a fresh perspective and approach to chili disease detection in modern agriculture, bridging the gap between traditional agricultural techniques and contemporary computer vision technologies. It is hoped that the findings of this study can be further expanded into practical agricultural production, contributing significantly to the advancement of modern agriculture.

Author Contributions: Conceptualization, B.X. and Q.S.; methodology, B.X. and J.W.; software, Y.L. and C.W.; validation, B.X., Q.S. and Y.L.; formal analysis, Q.S. and B.T.; resources, B.T. and Z.Y.; data curation, B.T., Z.Y., J.W. and C.W.; writing—original draft, B.X., Q.S., B.T., Y.L., Z.Y., J.W., C.W., J.L. and L.L.; writing—review and editing, L.L.; visualization, Z.Y. and J.W.; supervision, J.L.; project administration, J.L. and L.L.; funding acquisition, L.L. All authors have read and agreed to the published version of the manuscript.

Funding: This research received no external funding.

Conflicts of Interest: The authors declare no conflict of interest.

References

1. Ro, N.Y.; Sebastin, R.; Hur, O.S.; Cho, G.T.; Geum, B.; Lee, Y.J.; Kang, B.C. Evaluation of Anthracnose Resistance in Pepper (*Capsicum* spp.) Genetic Resources. *Horticulturae* **2021**, *7*, 460. [CrossRef]
2. Fidan, H.; Yildiz, K.; Sarikaya, P. Molecular detection of resistance-breaking strain Cucumber mosaic virus (rbCMV) (Cucumovirus; Bromoviridae) on resistant commercial pepper cultivars in Turkey. *J. Phytopathol.* **2023**, *171*, 234–241. [CrossRef]
3. Zhang, Y.; Wa, S.; Liu, Y.; Zhou, X.; Sun, P.; Ma, Q. High-accuracy detection of maize leaf diseases CNN based on multi-pathway activation function module. *Remote Sens.* **2021**, *13*, 4218. [CrossRef]
4. Zhang, Y.; Wang, H.; Xu, R.; Yang, X.; Wang, Y.; Liu, Y. High-Precision Seedling Detection Model Based on Multi-Activation Layer and Depth-Separable Convolution Using Images Acquired by Drones. *Drones* **2022**, *6*, 152. [CrossRef]
5. Zhang, Y.; He, S.; Wa, S.; Zong, Z.; Lin, J.; Fan, D.; Fu, J.; Lv, C. Symmetry GAN Detection Network: An Automatic One-Stage High-Accuracy Detection Network for Various Types of Lesions on CT Images. *Symmetry* **2022**, *14*, 234. [CrossRef]
6. Zhang, Y.; Liu, X.; Wa, S.; Liu, Y.; Kang, J.; Lv, C. GenU-Net++: An Automatic Intracranial Brain Tumors Segmentation Algorithm on 3D Image Series with High Performance. *Symmetry* **2021**, *13*, 2395. [CrossRef]
7. Zhang, Y.; Wa, S.; Zhang, L.; Lv, C. Automatic plant disease detection based on tranvolution detection network with GAN modules using leaf images. *Front. Plant Sci.* **2022**, *13*, 875693. [CrossRef]
8. Zeng, Y.; Zhao, Y.; Yu, Y.; Tang, Y.; Tang, Y. Pepper Disease Detection Model Based on Convolutional Neural Network and Transfer Learning. *IOP Conf. Ser. Earth Environ. Sci.* **2021**, *792*, 012001. [CrossRef]
9. Li, S.; Li, K.; Qiao, Y.; Zhang, L. A multi-scale cucumber disease detection method in natural scenes based on YOLOv5. *Comput. Electron. Agric.* **2022**, *202*, 107363. [CrossRef]
10. Abbas, A.; Jain, S.; Gour, M.; Vankudothu, S. Tomato plant disease detection using transfer learning with C-GAN synthetic images. *Comput. Electron. Agric.* **2021**, *187*, 106279. [CrossRef]

11. Sun, H.; Xu, H.; Liu, B.; He, D.; He, J.; Zhang, H.; Geng, N. MEAN-SSD: A novel real-time detector for apple leaf diseases using improved light-weight convolutional neural networks. *Comput. Electron. Agric.* **2021**, *189*, 106379. [CrossRef]
12. Wang, D.; Wang, J.; Ren, Z.; Li, W. DHBP: A dual-stream hierarchical bilinear pooling model for plant disease multi-task classification. *Comput. Electron. Agric.* **2022**, *195*, 106788. [CrossRef]
13. Peng, C.; Xia, F.; Naseriparsa, M.; Osborne, F. Knowledge Graphs: Opportunities and Challenges. *Artif. Intell. Rev.* **2023**, *56*, 13071–13102. [CrossRef] [PubMed]
14. Qiao, B.; Zou, Z.; Huang, Y.; Fang, K.; Zhu, X.; Chen, Y. A joint model for entity and relation extraction based on BERT. *Neural Comput. Appl.* **2022**, *34*, 3471–3481. [CrossRef]
15. Zhou, J.; Li, J.; Wang, C.; Wu, H.; Zhao, C.; Teng, G. Crop disease identification and interpretation method based on multimodal deep learning. *Comput. Electron. Agric.* **2021**, *189*, 106408. [CrossRef]
16. Zhu, D.; Xie, L.; Chen, B.; Tan, J.; Deng, R.; Zheng, Y.; Hu, Q.; Mustafa, R.; Chen, W.; Yi, S.; et al. Knowledge graph and deep learning based pest detection and identification system for fruit quality. *Internet Things* **2023**, *21*, 100649. [CrossRef]
17. Guan, L.; Zhang, J.; Geng, C. Diagnosis of Fruit Tree Diseases and Pests Based on Agricultural Knowledge Graph. *J. Phys. Conf. Ser.* **2021**, *1865*, 042052. [CrossRef]
18. Yu, C.; Wang, F.; Liu, Y.H.; An, L. Research on knowledge graph alignment model based on deep learning. *Expert Syst. Appl.* **2021**, *186*, 115768. [CrossRef]
19. Meng, X.; Yang, Y.; Qi, H.; Li, D.; Lu, Y.; Huang, G.; Zhang, J. Construction and Application of a Tree Knowledge Graph. In Proceedings of the 2021 IEEE/ACIS 19th International Conference on Computer and Information Science (ICIS), Shanghai, China, 23–25 June 2021. [CrossRef]
20. Zoph, B.; Le, Q.V. Neural architecture search with reinforcemenDenseNett learning. *arXiv* **2016**, arXiv:1611.01578.
21. Krizhevsky, A.; Sutskever, I.; Hinton, G.E. Imagenet classification with deep convolutional neural networks. *Adv. Neural Inf. Process. Syst.* **2012**, *25*, 1097–1105. [CrossRef]
22. Szegedy, C.; Liu, W.; Jia, Y.; Sermanet, P.; Reed, S.; Anguelov, D.; Erhan, D.; Vanhoucke, V.; Rabinovich, A. Going deeper with convolutions. In Proceedings of the IEEE Conference on Computer Vision and Pattern Recognition, Boston, MA, USA, 7–12 June 2015; pp. 1–9.
23. Huang, G.; Liu, Z.; Laurens, V.; Weinberger, K.Q. Densely Connected Convolutional Networks. In Proceedings of the IEEE Computer Society, 2016, Las Vegas, NV, USA, 27–30 June 2016.
24. Simonyan, K.; Zisserman, A. Very deep convolutional networks for large-scale image recognition. In Proceedings of the International Conference on Learning Representations (ICLR), Banff, AB, Canada, 14–16 April 2014.
25. He, K.; Zhang, X.; Ren, S.; Sun, J. Deep residual learning for image recognition. In Proceedings of the IEEE Conference on Computer Vision and Pattern Recognition, Las Vegas, NV, USA, 26 June–1 July 2016; pp. 770–778.
26. Vaswani, A.; Shazeer, N.; Parmar, N.; Uszkoreit, J.; Jones, L.; Gomez, A.N.; Kaiser, Ł.; Polosukhin, I. Attention is all you need. In Proceedings of the Advances in Neural Information Processing Systems 30 (NIPS 2017), Long Beach, CA, USA, 4–9 December 2017; Volume 30.
27. Devlin, J.; Chang, M.W.; Lee, K.; Toutanova, K. Bert: Pre-training of deep bidirectional transformers for language understanding. *arXiv* **2018**, arXiv:1810.04805.
28. Ji, S.; Pan, S.; Cambria, E.; Marttinen, P.; Philip, S.Y. A survey on knowledge graphs: Representation, acquisition, and applications. *IEEE Trans. Neural Netw. Learn. Syst.* **2021**, *33*, 494–514. [CrossRef]
29. Chen, Y.; Kuang, J.; Cheng, D.; Zheng, J.; Gao, M.; Zhou, A. AgriKG: An agricultural knowledge graph and its applications. In Proceedings of the Database Systems for Advanced Applications: DASFAA 2019 International Workshops: BDMS, BDQM, and GDMA, Chiang Mai, Thailand, 22–25 April 2019; Proceedings 24; Springer: Cham, Switzerland, 2019; pp. 533–537.
30. Qin, H.; Yao, Y. Agriculture knowledge graph construction and application. *J. Phys. Conf. Ser.* **2021**, *1756*, 012010. [CrossRef]
31. Chenglin, Q.; Qing, S.; Pengzhou, Z.; Hui, Y. Cn-MAKG: China meteorology and agriculture knowledge graph construction based on semi-structured data. In Proceedings of the 2018 IEEE/ACIS 17th International Conference on Computer and Information Science (ICIS), Singapore, 6-8 June 2018; pp. 692–696.
32. Blok, P.M.; Polder, G.; Peller, J.; van Daalen, T. *OPTIMA-RGB Colour Images and Multispectral Images (Including LabelImg Annotations)*; Wageningen University & Research: Wageningen, The Netherlands, 2022.
33. Hendrycks, D.; Gimpel, K. Gaussian error linear units (gelus). *arXiv* **2016**, arXiv:1606.08415.
34. Lin, T.Y.; Goyal, P.; Girshick, R.; He, K.; Dollár, P. Focal loss for dense object detection. In Proceedings of the IEEE International Conference on Computer Vision, Venice, Italy, 22–29 October 2017; pp. 2980–2988.
35. Zhu, X.; Lyu, S.; Wang, X.; Zhao, Q. TPH-YOLOv5: Improved YOLOv5 based on transformer prediction head for object detection on drone-captured scenarios. In Proceedings of the IEEE/CVF International Conference on Computer Vision, Montreal, BC, Canada, 11–17 October 2021; pp. 2778–2788.
36. Terven, J.; Cordova-Esparza, D. A comprehensive review of YOLO: From YOLOv1 to YOLOv8 and beyond. *arXiv* **2023**, arXiv:2304.00501.
37. Carion, N.; Massa, F.; Synnaeve, G.; Usunier, N.; Kirillov, A.; Zagoruyko, S. End-to-end object detection with transformers. In Proceedings of the European Conference on Computer Vision, Glasgow, UK, 23–28 August 2020; Springer: Cham, Switzerland, 2020; pp. 213–229.

38. Liu, W.; Anguelov, D.; Erhan, D.; Szegedy, C.; Reed, S.; Fu, C.Y.; Berg, A.C. Ssd: Single shot multibox detector. In Proceedings of the Computer Vision—ECCV 2016: 14th European Conference, Amsterdam, The Netherlands, 11–14 October 2016; Springer: Cham, Switzerland, 2016; pp. 21–37.
39. Tan, M.; Pang, R.; Le, Q.V. Efficientdet: Scalable and efficient object detection. In Proceedings of the IEEE/CVF Conference on Computer Vision and Pattern Recognition, Seattle, WA, USA, 13–19 June 2020; pp. 10781–10790.
40. Kingma, D.P.; Ba, J. Adam: A method for stochastic optimization. *arXiv* **2014**, arXiv:1412.6980.
41. Chen, K.; Lin, W.; Li, J.; See, J.; Wang, J.; Zou, J. AP-loss for accurate one-stage object detection. *IEEE Trans. Pattern Anal. Mach. Intell.* **2020**, *43*, 3782–3798. [CrossRef]
42. Qian, Q.; Chen, L.; Li, H.; Jin, R. Dr loss: Improving object detection by distributional ranking. In Proceedings of the IEEE/CVF Conference on Computer Vision and Pattern Recognition, Seattle, WA, USA, 13–19 June 2020; pp. 12164–12172.

agriculture

MDPI

Article

Artificial Intelligence-Based Fault Diagnosis and Prediction for Smart Farm Information and Communication Technology Equipment

Hyeon O. Choe [1] and Meong-Hun Lee [2,*]

[1] Department of Information and Communication Engineering, Sunchon National University, Suncheon-si 57922, Jellanam-do, Republic of Korea; wishind@scnu.ac.kr

[2] Department of Smart Agriculture Major, Sunchon National University, Suncheon-si 57922, Jellanam-do, Republic of Korea

* Correspondence: leemh777@scnu.ac.kr

Abstract: Despite the recent increase in smart farming practices, system uncertainty and difficulties associated with maintaining farming sites hinder their widespread adoption. Agricultural production systems are extremely sensitive to operational downtime caused by malfunctions because it can damage crops. To resolve this problem, the types of abnormal data, the present error determination techniques for each data type, and the accuracy of anomaly data determination based on spatial understanding of the sensed values are classified in this paper. We design and implement a system to detect and predict abnormal data using a recurrent neural network algorithm and diagnose malfunctions using an ontological technique. The proposed system comprises the cloud in charge of the IoT equipment installed in the farm testbed, communication and control, system management, and a common framework based on machine learning and deep learning for fault diagnosis. It exhibits excellent prediction performance, with a root mean square error of 0.073 for the long short-term memory model. Considering the increasing number of agricultural production facilities in recent years, the results of this study are expected to prevent damage to farms due to downtime caused by mistakes, faults, and aging.

Keywords: smart farming; sensors; RNN; LSTM; ontology; prediction

1. Introduction

In recent years, rapid progress has been made in agricultural technology in terms of enhancements in productivity and convenience, which are together referred to as smart farming. This has been facilitated by the convergence of various information and communication technologies (ICTs) [1]. In countries around the world, including South Korea, agriculture is undergoing technological evolution via smart convergence based on data collection, analysis, and prediction. Current agricultural practice is focused on developing differentiated technology related to "software and hardware platforms", "data intelligence", and "convergence of various technologies, such as artificial intelligence (AI), the cloud, and the Internet of Things (IoT)" to develop an intelligent smart farming industry [2]. Smart farms centered around facility horticulture provide services related to crop growth, environmental information management, system control, disease control, and growth algorithms customized to suit the requirements of local farmers. However, the applicability of these technologies to the current agricultural production process remains limited. A first-generation smart farming system operated based on ICT convergence was developed with a focus on labor reduction and convenience. However, it suffers from several problems, including difficulties in checking system operation and remote-monitoring-based control, cost-intensive CCTV operation for visual monitoring, and the high complexity of criteria for assessing abnormalities in sensing values. Besides such technical and financial problems, the application of advanced technologies to real-world local farms suffers from additional basic limitations [3–5].

The South Korean government's smart farm distribution project has considerably increased the number of smart farms across the country, but the aforementioned problems hinder the widespread adoption of smart farming. Unlike other sectors, an agricultural production system is extremely vulnerable to fault-induced downtime, which can lead to irrevocable damage to crops and farms, in addition to incurring maintenance and repair costs. Therefore, there is a compelling need for system prognostics, health management technology, and condition-based maintenance technology [6]. This will prevent damage to farms due to downtime induced by accidents, faults, and aging based on meticulous data monitoring of ICT equipment in agricultural production facilities. Several studies have investigated sensor-based fault detection by classifying various types of anomalous data in the field of sensor networks and proposed error detection methods for each data type [7]. Faults detected based on sensing values have been assessed corresponding to individual and multiple data within a given space. Particularly, a moving-average-based assessment technique was used to study time series, and anomalous data were assessed based on a geospatial understanding of the sensing values [3,4]. One study pointed out the problem of poor generalizability of the methods used for sensor outlier assessment based on the ambiguity of data forms and models [8]. Traditional outlier treatment uses Bayesian analysis of data sensed within a specific space; however, the utilization of only limit values within a specific dataset, such as temperature or humidity, is not suitable for facility horticulture smart farming [9,10].

To this end, this paper presents the design and implementation of an error detection system based on ontology and recurrent neural networks (RNNs). The system uses sensor and controller data generated during smart farming and implements an architecture to detect and diagnose the malfunction of sensors and controllers in the smart farming system.

The remainder of this paper is arranged as follows. In Section 2, an overview of previous studies on sensor-based fault detection is presented. In Section 3, the architecture and composition of the proposed system to detect faults in parts constituting the smart farming system are discussed. The system design of an AI model used for experimental verification is presented in Section 4. Subsequently, the model implementation and experimental analysis are presented in Section 5. Finally, Section 6 presents the conclusions of the study.

2. Materials and Methods

In this context, an AI algorithm and an ontological technique are developed to enhance model generalizability based on data type. The overall aim is to identify and predict malfunctions in smart farm ICT equipment by evaluating both internal and external control data, unlike the current prediction approach based solely on temperature and humidity data.

2.1. RNNs

RNNs are a type of artificial neural network (ANN) comprising a directed cycle composed of hidden nodes connected by directed edges. They have garnered significant attention recently, alongside convolutional neural networks (CNNs), owing to their suitability for processing sequential sensor data, such as voice and textual data [11].

As illustrated in Figure 1, RNNs can flexibly create various structures, which are capable of accepting inputs of any length, by manipulating their network architecture.

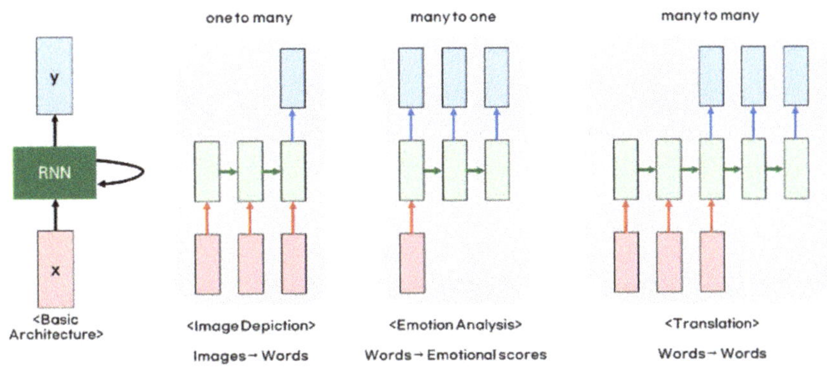

Figure 1. Architecture of the RNN algorithm ("one-to-many", "many-to-one", "many-to-many") [12].

Figure 2 depicts the basic architecture of RNNs. The green, red, and blue boxes represent the hidden states (h), input (x), and output (y), respectively. The current hidden state (h_t) is updated based on the previous hidden state (h_{t-1}). The current output (y_t) is updated by using h_t based on the equation given in the figure. The activation function of the hidden state is taken to be the nonlinear hyperbolic tangent function (tanh) [13–19].

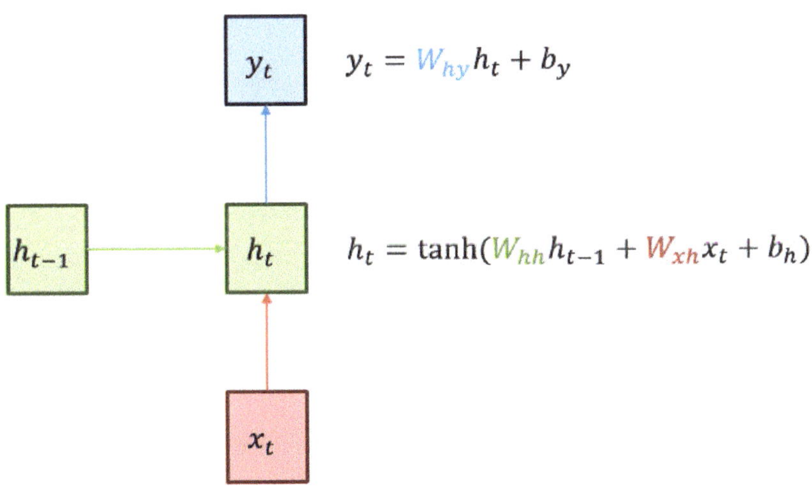

Figure 2. Basic architecture of RNNs [12].

2.2. Smart Farming Systems

Smart farming is defined as a convergence technology that incorporates ICT into existing agricultural, livestock, and fishery industries to improve their productivity. Smart farming enables the measurement and analysis of temperature, humidity, and sunlight using ICT and the remote control of the environment using mobile devices. In the smart farming system depicted in Figure 3, smart farm operations consist of defining the growth conditions such as temperature, humidity, and CO_2 level using growth environment maintenance/management software, and monitoring the growth environment by automatically collecting data related to temperature, humidity, solar radiation, and CO_2 levels. Moreover, the system enables convenient management of the environment, e.g., the automatic/remote operation of HVAC, window opening and closing, and the supply of CO_2 and nutrient feed. However, faults in ICT equipment can cause considerable damage to farms, making an effective fault diagnosis system essential [20–23].

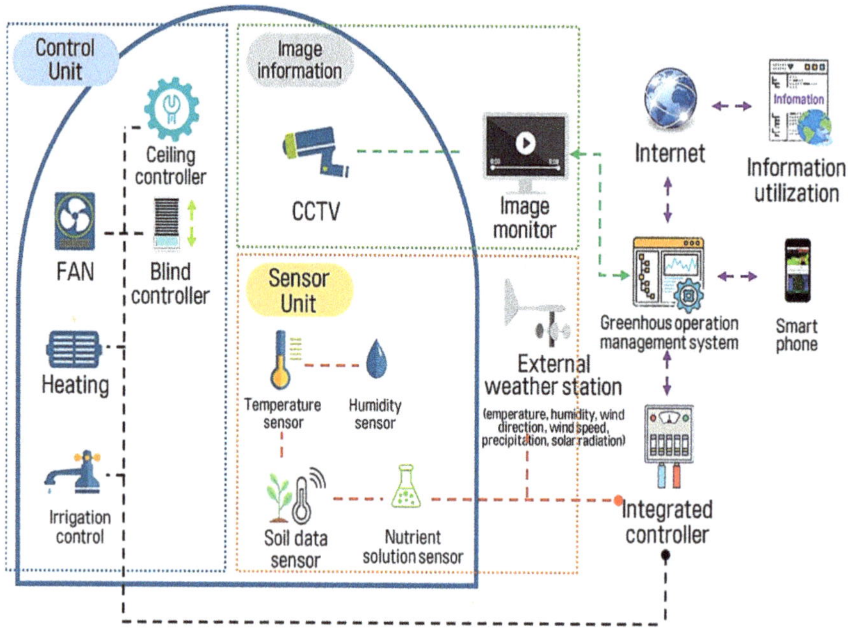

Figure 3. Smart farming architecture.

3. System Configuration

A testbed installed at a farm was used to implement the proposed fault detection system. The entire system constituting the smart farm comprises the cloud in charge of the IoT equipment installed at the smart farm, communication and control, system management, and the machine-learning- and deep-learning-based common framework constituting the fault diagnosis engine.

As depicted in Figure 4, the common framework for fault diagnosis consists of a framework to perform ontological and deep-learning-based fault diagnosis based on equipment thresholds, conditions, actuator conditions, and user-defined rules.

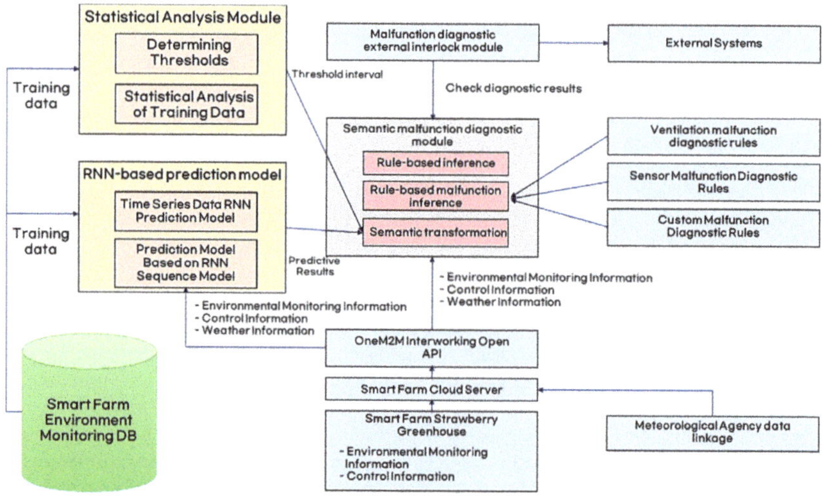

Figure 4. Block diagram of the fault diagnosis system.

Moreover, actual data are transmitted along the paths illustrated in Figure 5 within an integration module connecting the common interfaces to integrate the sensing data with the actuator control data received from the OneM2M Integration Open API cloud [24].

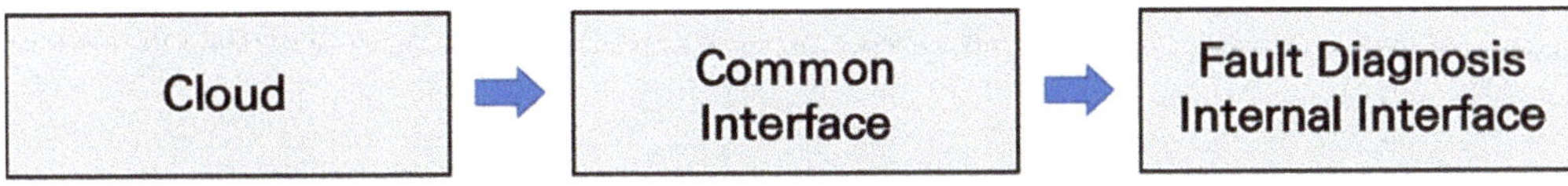

Figure 5. Data transfer paths of the actuator control interface integration module.

The statistical analysis module is a statistical value calculation module that diagnoses faults based on the difference between the measured value and the value predicted by the RNN-based prediction model. Time series, i.e., the RNN prediction model and the RNN sequence prediction model, are trained, and sensing values are predicted based on series of incoming sensing and control values.

The semantic fault diagnosis module consists of smart farm ontology, time ontology, geospatial ontology, event ontology, and fault diagnosis ontology. Fault diagnosis ontology is used to define the concepts of threshold and prediction data used to assess the faults detected by the statistical analysis module. Using the notification interface provided by the cloud, malfunction notifications are classified as faults and the results of fault diagnosis are notified to the user.

Figure 6 illustrates the software architecture, which consists of the implementation environment, a fault diagnosis engine layer, an interface layer, and a user interface layer.

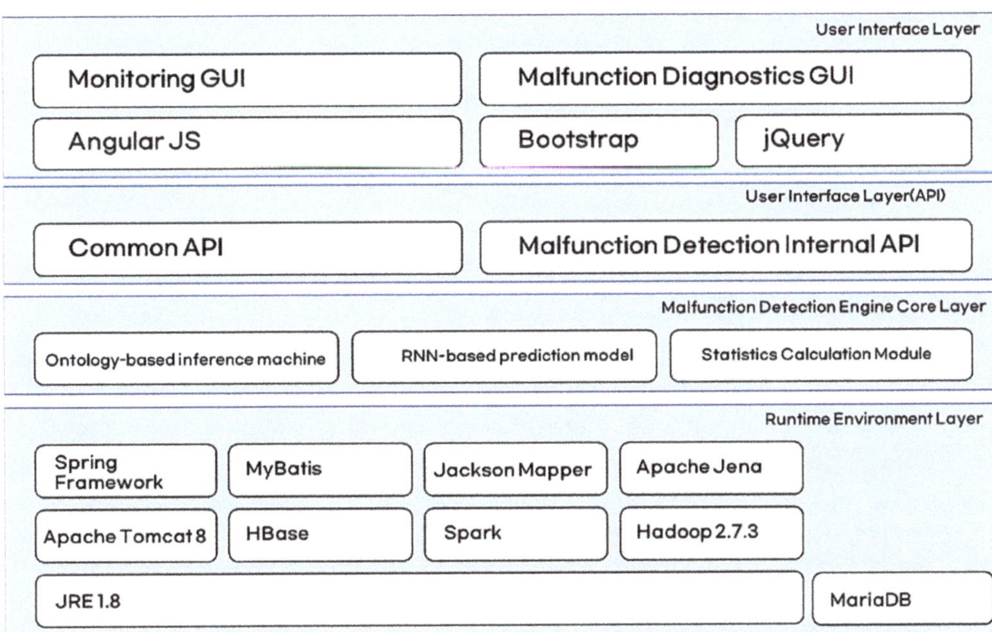

Figure 6. Software architecture.

Figure 7 illustrates the integration architecture within the fault diagnosis prediction system. Equipment control data and equipment-measured values transmitted through the common API and fault API are collected by the fault diagnosis module in real time via the HTTP RESTful API, stored in a queue for a certain length of time, and then diagnosed via the scheduler job.

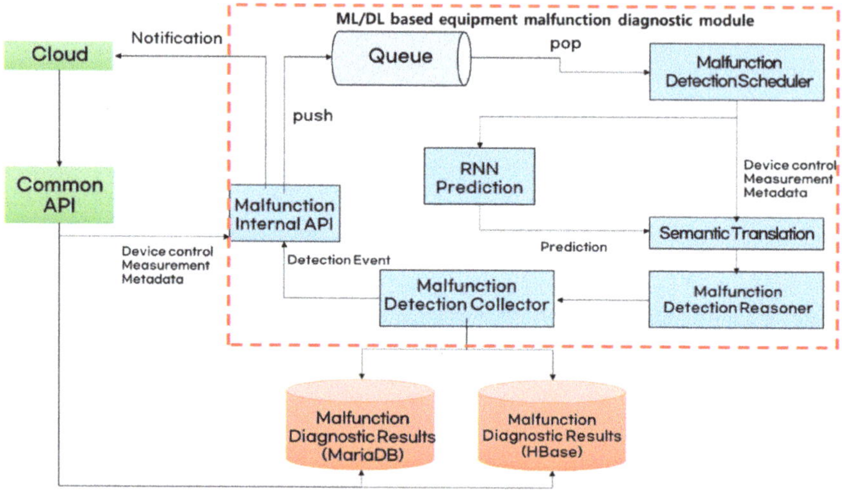

Figure 7. Integration architecture within the fault diagnosis prediction system.

To perform RNN prediction using the data collected by the scheduler job and add the RNN prediction results and the collected data source to the ontology, conversion-to-triples is performed during the semantic transformation process, and the converted triples are added to the semantic storage.

Inference rules registered in advance for ontology are applied to perform fault diagnosis as well as store and manage the inferred fault events in cloud notifications and RDBMS/HBase. Equipment fault diagnosis results are provided directly to RDBMS and HBase when requested in the fault API.

4. Model Design

The ontology for fault diagnosis is defined to enable fault diagnosis based on knowledge. The main environmental data for the smart farm are collected by the sensor network; thus, the ontology is defined based on the semantic sensor network ontology [25–29].

Figure 8 presents the ontology architecture for smart farm fault diagnosis. The semantic sensor network ontology is designed for general use, and includes the detection target of the sensor, the detection method, metadata, sensors, the sensor deployment system, and various attributes. However, its direct use for fault diagnosis is inefficient due to its complexity. Therefore, the smart farm sensor network is defined based on the stimulus–sensor–observation pattern and other patterns related to sensors and the sensor deployment system.

The W3C time ontology is used to describe the concept of time. For smart farm equipment deployment, a spatial concept suitable for smart farming is defined based on the open-source spatial data ontology.

Table 1 defines the concepts related to the fault diagnosis event and the conceptual relationships around the event.

Fault diagnosis methods are classified as (i) methods based on smart farm ontology and rules and (ii) fault event generation methods that use the ontology based on thresholds obtained by first analyzing the normal distribution of the difference between the sensor sequence values transmitted in real time using the RNN prediction model (abs (predicted value − measured value)) and the values generated during model training (abs (predicted value − measured value)) and, then, comparing these two values using the rules listed below.

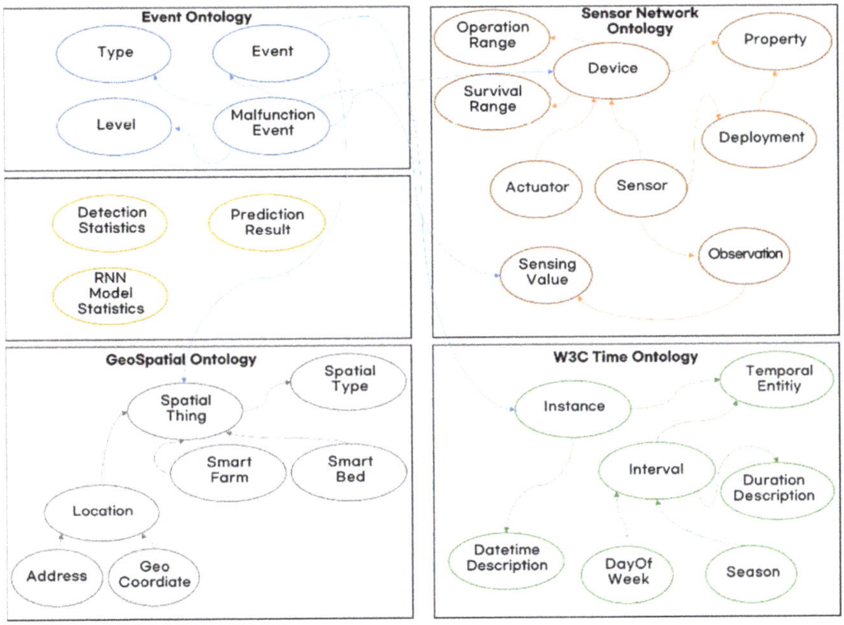

Figure 8. Ontology architecture for smart farm fault diagnosis.

Table 1. Concepts related to the fault diagnosis event.

Ontology	Description
Event Ontology	Event ontology defines the concept of an "event" representing the fault diagnosis result. Events are derived by ontological semantic transformation of the modeled equipment composition data and measurement values; storage is in the form of Resource Description Framework triples and user rule-based inference.
Detection Ontology	The sensing values exhibit specific patterns depending on various environmental factors, which enables the use of normal time series sensing values as training data to train the RNN model for the estimation of sensing values.
	To assess values exceeding a threshold as faults by comparing the predicted and measured values, it is necessary to define the concepts that the predicted values represent.
	Moreover, based on the predicted values, the concepts necessary for fault diagnosis are defined. For fault diagnosis, it is important to configure the threshold based on statistical analysis of normal data. Detection ontology is used to determine the threshold value for each equipment type.
Sensor Network Ontology	The sensor network ontology is composed of the concepts that represent the equipment used for smart farming (sensors, actuators, IoT nodes, network equipment, etc.) and their respective measurement values.
	The basic pattern comprises ontologies that are suitable for smart farming and based on the semantic sensor network ontology.
Geospatial Ontology	Geospatial ontology comprises spaces where smart farming equipment is installed and operated, e.g., farms, greenhouses, and zones. These spatial concepts share structural relationships.
Time Ontology	OWL-Time represents the OWL-2 DL ontology for the concept of time and is used to describe the temporal attributes of resources.
	Time ontology provides the words and phrases required to express topological (sequencing) relationships between moments and intervals related to temporal locations, including information about duration, date, and time.
	Temporal location and duration can be expressed using the traditional (Gregorian) calendar and clock, or other time reference systems, such as Unix time, geological time, or other types of calendars.

A fault event is added in the following cases:

Rule 1: The temperature sensing values obtained from the equipment differ from those obtained from other equipment of the same type installed in the same area.

Rule 2: A temperature sensor and a heater are adjacent among the equipment installed in the same area.

Rule 3: Temperature and humidity sensors are installed in the same area and the current temperature is 5 °C or less and the humidity is 10% or less.

Rule 4: The power measured by a power sensor is zero during the operation of some equipment.

Rule 5: When both the RNN-predicted value of an installed device and Many2OneModelStatistics for the device exist, a malfunction event is added if the VARIANCE of the RNN-predicted value is greater than the VARIANCE threshold of the Many2OneModelStatistics.

Figure 9 shows a hierarchical ontology configuration model, wherein the fault diagnosis processing flow involves a cyclic query of collected data via semantic transformation and their conversion into triples for the input (the optimal batch cycle is configured considering the quasi-real-time performance and the sensor collection cycle). The first request made after the data input is inferred automatically, and the inferred triples are extracted by transmitting the query to SparQL.

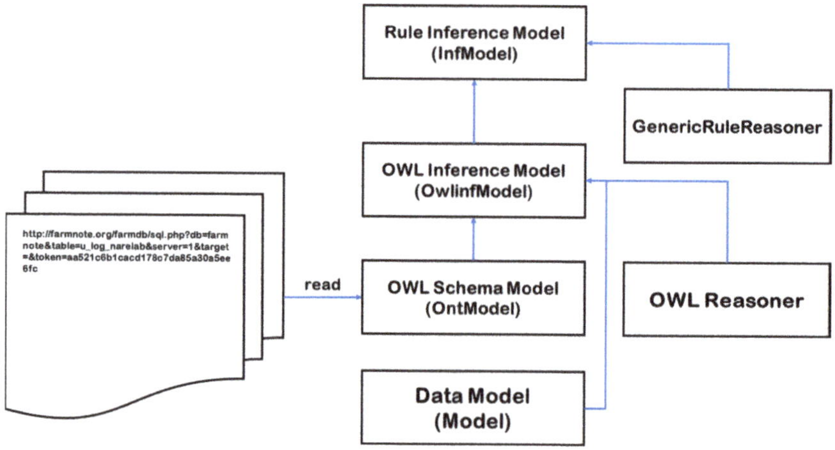

Figure 9. Hierarchical ontology configuration model. (http://farmnote.org/farmdb/sql.php?db=farmnote&table=u_log_narelab&server=1&target=&token=aa521c6b1cacd178c7da85a30a5ee6fc) (accessed on 11 June 2023).

The extracted inferred data are transferred to other integrated subsystems or stored in a database. Finally, all triples of the data model are deleted, as are all inferred triples, by rebinding the upper models (OWL inference model, rule inference model).

Table 2 provides an overview of the methods used to configure the fault detection inference rules. Inference rules for detecting an anomaly state of a sensor in a smart farm can be derived by applying these methods.

First, the sensing value is predicted using the RNN Many2One and RNN Seq2Seq models and compared with the measured value. If the difference between the predicted and measured values exceeds the pre-determined threshold, the event is considered a fault event. The threshold used at this time is determined by comparing the measured data used during model training with the values predicted by the RNN model.

Second, when the difference between the sensing values of sensors of identical type (e.g., temperature sensors) installed in the same zone/area exceeds the threshold (5 °C), the sensors in the zone/area are suspected to be faulty. As it is not possible to determine the exact faulty sensor, all sensors are deemed to be faulty.

Table 2. Methods for configuring the fault detection inference rules.

Category	Measure	Notes
Single-equipment (sensor) fault detection	Threshold exceeded	Single value; processing in real time.
	Improperly installed (location/process)	Metadata regarding sensor installation required (e.g., temperature sensors to be installed near ventilator and radiator at an appropriate distance).
	Sensor expired	Sensor expiration date metadata required.
	Disturbed communication and data collection	Storage and management of last collection date required.
Multi-equipment (sensor) fault detection	Detection by observing inter-sensor correlations	Collection of all correlated sensing values required (accumulation for a certain period of time before deployment) (e.g., high temperature and humidity).
	Detection by observing the difference from the nearest sensing value.	Occurrence of a difference exceeding the threshold of a given sensor.
Situational fault detection	Controller fault detection in an improper environment during the growth stage.	Data on growth stage required. Occurrence of an event in a suspected controller type, not a specific controller fault.
	Controller fault detection when the sensing value does not change for a long time during controller operation.	Controller activation data required.

Third, when high-temperature, high-humidity conditions beyond a certain threshold are different from the general conditions of high temperature and low humidity, which exerts a harmful effect on plant growth at smart farms, the sensor concerned is suspected to be faulty.

Finally, the minimum and maximum allowable ranges are checked as metadata of the measured values of a sensor to determine whether the sensor is in an environment where it can operate normally and reliably. The range may fall within the normal range, enabling normal operation (a range measurable by the sensor or ideal value to be specified in the normal plant factory environment). If the range checked falls outside the normal range, the corresponding sensors are deemed to be faulty. Table 3 provides an overview of the inference rules to be applied in different conditions.

Table 3. Inference rules to be applied in different conditions.

Rule ID	Description
rule_rnn_many2one	Anomaly state detection using the RNN Many2One model
rule_rnn_seq2seq	Anomaly state detection using the RNN Seq2Seq model
rule1	Fault inference based on a comparison with the threshold of a single sensor
rule2	Fault inference based on the correlations of two or more sensing values, which yields two events
rule3	Actuator fault inference based on actuator control data and power sensing values

5. Model Implementation and Experimentation

Diagnosis of malfunction using the above-defined ontology requires the consideration of the critical point of the smart farm sensor. As a result, if the sensor value is predicted to be abnormal and deviates significantly from the predicted value, it may be diagnosed as a malfunction.

Sensor value prediction was performed based on hourly data collected from a strawberry farm spanning one year obtained from the Korea Agency of Education, Promotion, and Information Service in Food, Agriculture, Forestry and Fishery. Data analysis reveals that various factors affect the temperature sensing values, as outlined in Table 4.

Table 4. Baseline training datasets [30].

Category	Datasets
Farm	Strawberry farm.
Environment	Indoor environment, outdoor environment, hydroponic environment, soil environment.
Sensing value	Temperature, humidity, CO_2, surface temperature, humidity, solar radiation, wind speed, precipitation (dry/wet).

The measurement items include measurement season, measurement time, outdoor temperature, solar radiation, clouds, fog, precipitation (dry or wet days), and amount of precipitation. The measurable and replaceable data applicable to a real prediction model are listed in Table 5.

Table 5. Training data affecting the temperature values.

Measurement Items	Alternative Items
Day (ref. 365 days)	Measurement season
Measurement time	Measurement time
Outdoor temperature	Outdoor temperature
Solar radiation	Solar radiation, clouds, fog
Precipitation	Dry day/wet day, amount of precipitation

As the data collected by the Korea Agency of Education, Promotion, and Information Service in Food, Agriculture, Forestry and Fishery provide information about the aforementioned items and temperature measurement values corresponding to each time slot, the data were extracted by sequence and item, and the training data were pre-processed. These datasets were used to train the many-to-one model by dividing the time series input sensing values into sequences. The errors corresponding to the differences between the predicted and measured values calculated during training followed a normal distribution, and a threshold was added (semantic transformation) to the ontology by setting appropriate confidence intervals. This value was input (semantic transformation) into the ontology, and faults were assessed by comparing it with the threshold value calculated by the model following the rules.

A system-wide configuration of constants and variables was applied to the RNN learning model, which was normalized and denormalized via MinMaxScaler and RevMinMaxScaler functions, respectively. The training and test datasets were separated by loading the farm file with the loadData function.

RNN cell/multi-RNN cell were defined to configure the RNN network after loading data from the "temp" directory corresponding to each farm into the main code, and a fully connected layer was defined to test the learning result. Based on the definition of the cost function, the cost was set to be minimized using the Adam optimizer, and several training iterations were performed by the (training) node following the system-wide configuration. Finally, the learning results were stored, and the learning model was tested using the farm data stored for testing.

Algorithm 1 presents the code responsible for calculating the predicted values based on an actual sensor sequence. It specifically focuses on the section that displays the outcomes of training the recurrent neural network (RNN) cell in the program.

In the main code, the code first examines the command line arguments and the file path for input data. It loads the data from the specified file. Afterwards, the RNN Cell/Multi-RNN cell is defined, the RNN network is configured, the fully connected layer is defined, the node is run, and learning is conducted as many times as globally set. Finally, the results are output as a graph.

Algorithm 1 Prediction algorithm based on sensor sequence

```
# ----------------------------------------------------------
# Program Start
# ----------------------------------------------------------
# Load training data
loadData()
print('Size of training data: ' + str(len(trainX)))
print('Size of test data: ' + str(len(inputX))
# Input placeholders
X = tf.placeholder(tf.float32, [None, seq_length, data_dim])
Y = tf.placeholder(tf.float32, [None, 1])
# Construct the RNN network
# RNN Cell (Available cells: Basic LSTM, LSTM, GRU, ...)
# cell = tf.contrib.rnn.BasicLSTMCell(
#       num_units=hidden_dim, state_is_tuple=True, activation=tf.tanh)
cell = tf.contrib.rnn.GRUCell(
num_units=hidden_dim, activation=tf.tanh)
# Multi-RNN Cells
cells = tf.contrib.rnn.MultiRNNCell([cell] * NUMBER_OF_RNN_CELL_LAYERS)
# Dynamic RNN (outputs: output, _states: previous states in the RNN network)
# If RNN cells and input data are given as arguments, the RNN cells are connected to form a network.
outputs, _states = tf.nn.dynamic_rnn(cell, X, dtype=tf.float32)
# Add fully connected layers to obtain prediction values
Y_pred = tf.contrib.layers.fully_connected(
outputs[:, −1], output_dim, activation_fn=None)
# Define the cost function (sum of the squares)
loss = tf.reduce_sum(tf.square(Y_pred − Y))
# Define the cost Tensor for Tensorboard
tf.summary.scalar("cost", loss)
# Summary
summary = tf.summary.merge_all()
# Define the optimizer
optimizer = tf.train.AdamOptimizer(learning_rate)
train = optimizer.minimize(loss)
# RMSE (Root Mean Square Error, the square root of the mean squared differences between actual and predicted values)
targets = tf.placeholder(tf.float32, [None, 1])
predictions = tf.placeholder(tf.float32, [None, 1])
rmse = tf.sqrt(tf.reduce_mean(tf.square(targets − predictions))
with tf.Session() as sess:
init = tf.global_variables_initializer()
sess.run(init)
# Create a summary writer
writer = tf.summary.FileWriter(TB_SUMMARY_DIR)
```

Algorithm 1 *Cont.*

```
writer.add_graph(sess.graph)
global_step = 0

# -----------------------------------------------------------
# Training Phase
# -----------------------------------------------------------
for i in range(iterations):
s, _, step_loss = sess.run([summary, train, loss], feed_dict={
X: trainX, Y: trainY})
#print("[step: {}] loss: {}".format(i, step_loss))

writer.add_summary(s, global_step=global_step)
global_step += 1

# Save the training results
saver = tf.train.Saver()
saver.save(sess, './model/malfunction_predict.pd')

# -----------------------------------------------------------
# Testing Phase
# -----------------------------------------------------------
# Perform predictions on test data and display the results using plots
test_predict = sess.run(Y_pred, feed_dict={X: inputX})
rmse_val = sess.run(rmse, feed_dict={
targets: sensingValueY, predictions: test_predict})
print("inputX RMSE: {}".format(rmse_val))

correct_prediction = test_predict - sensingValueY
accuracy = tf.reduce_mean(correct_prediction)

plt.figure(figsize=(20, 4))
plt.plot(RevMinMaxScaler(sensingValueY), 'b-', label='Sensing')
plt.plot(RevMinMaxScaler(test_predict), 'r-', label='Prediction')
plt.xlabel("Time Period")
plt.ylabel("Temperature")
plt.legend(loc='best')

# Perform predictions on the first test data and display the results using plots
test_predict = sess.run(Y_pred, feed_dict={X: firstTestX})
rmse_val = sess.run(rmse, feed_dict={
targets: firstTestY, predictions: test_predict})
print("firstTestX RMSE: {}".format(rmse_val))

correct prediction = test_predict - firstTestY
accuracy = tf.reduce_mean(correct_prediction)

plt.figure(figsize=(20, 6))
plt.plot(RevMinMaxScaler(firstTestY), 'b-', label='Sensing')
plt plot(RevMinMaxScaler(test_predict), 'r-', label='Prediction')
plt.xlabel("Time Period")
plt.ylabel("Temperature")
plt.legend(loc='best')

plt.show()

```

Figure 10 graphically depicts the tensor board cost with respect to the sensor values. The test was conducted by varying the number of training iterations from 500 to 1000, 5000, 10,000, and 100,000. After 5000 training iterations, no significant difference is observed.

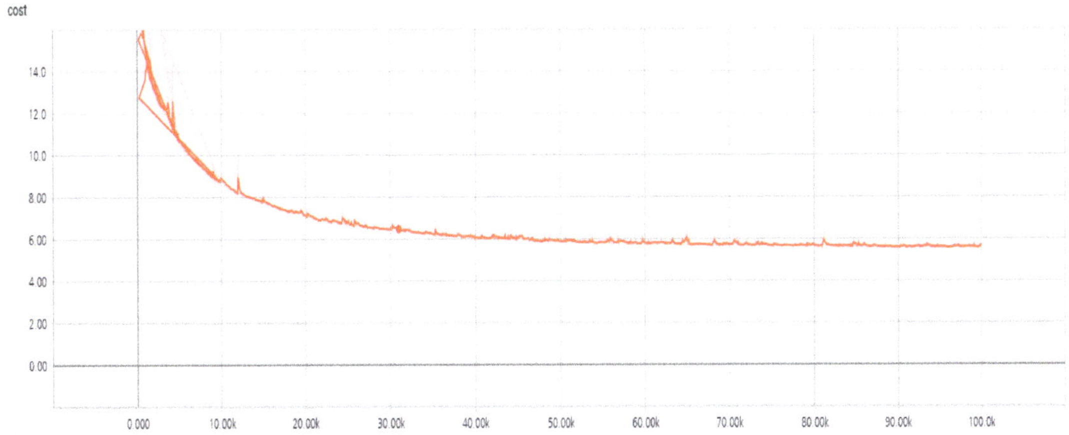

Figure 10. Tensor board cost graph.

The root mean squared error (RMSE), a widely used error metric, was used to compare the learning results. In the error measurement results listed in Table 6, the RNN cell exhibits an RSME value for the LSTM that is lower than that for the GRU by 0.003. The training data are time series data; thus, this can imply that the LSTM algorithm analyzes time series data more effectively than GRU.

Table 6. Errors corresponding to each learning model.

RNN Cell	RSME
LSTM	0.0732
GRU	0.0760

Table 7 presents the test results corresponding to different numbers of training iterations. The lowest mean error is achieved after 5000 iterations, and the mean error increases as the number of training iterations is increased to 10,000 and 100,000, presumably due to overfitting.

Table 7. Errors corresponding to different numbers of training iterations.

Number of Training Iterations	RSME
500	0.0806
1000	0.0785
5000	0.0784
10,000	0.0768
100,000	0.0928

Concrete accuracy is expressed numerically, as in Table 7; thus, it is difficult to determine the accuracy of these values. Figures 11–14 visually express the prediction accuracy by comparing actual data and predicted data, and it can be seen that the values predicted by the model follow the actual sensor values relatively well. A real-time smart farm equipment fault diagnosis experiment was conducted using the prediction model derived in this study.

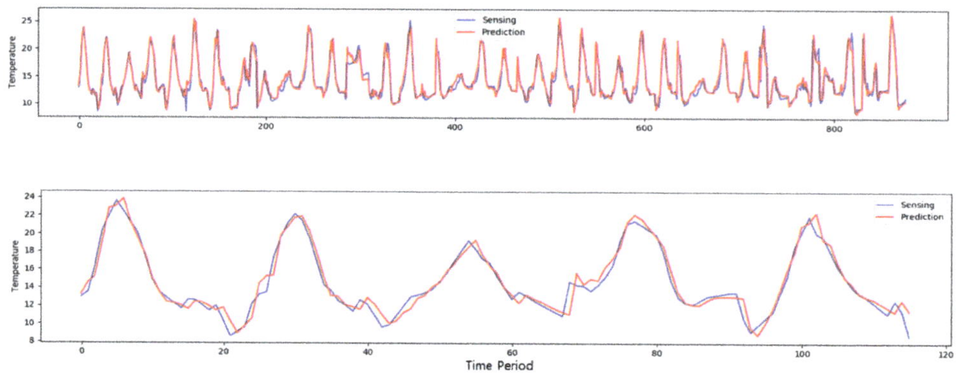

Figure 11. Test results obtained using the measured data (**top**: overall result value, **bottom**: magnified graph of some test results).

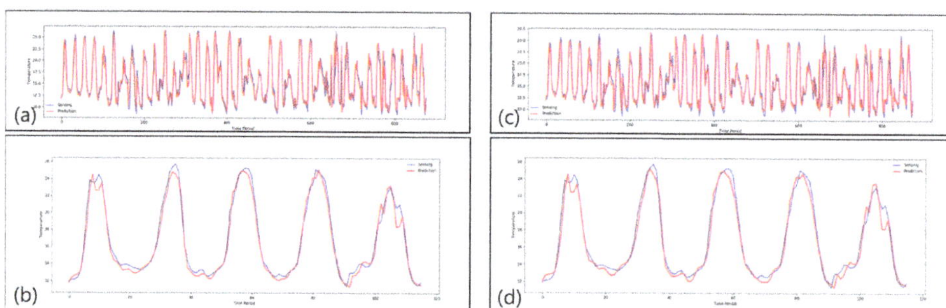

Parameter	Value
RNN Cell	LSTM
Sequence Length	5
Hidden Dimension	10
iterations	500

Parameter	Value
RNN Cell	GRU
Sequence Length	5
Hidden Dimension	10
iterations	500

Figure 12. (**Left**) LSTM model, 500 iterations; (**Right**) GRU model, 500 iterations. (a) LSTM model overall results, (b) LSTM model Enlarged graph, (c) GRU model Overall results, (**d**) GRU model Enlarged graph.

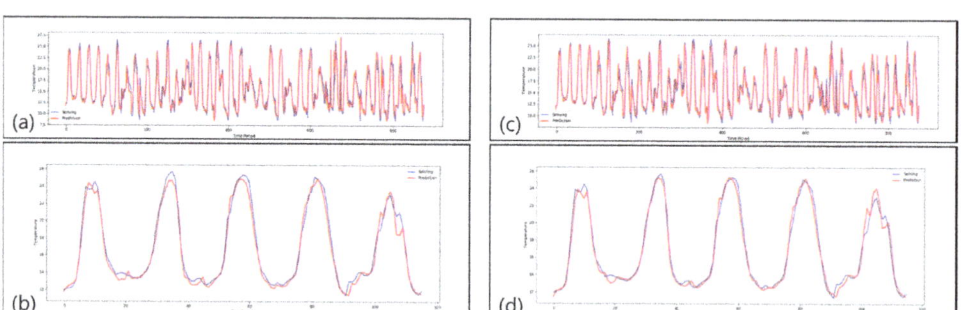

Parameter	Value
RNN Cell	GRU
Sequence Length	5
Hidden Dimension	3
iterations	5000

Parameter	Value
RNN Cell	GRU
Sequence Length	5
Hidden Dimension	10
iterations	5000

Figure 13. (**Left**) LSTM model, 5000 iterations; (**Right**) GRU model, 5000 iterations. (a) LSTM model overall results, (b) LSTM model Enlarged graph, (c) GRU model Overall results, (**d**) GRU model Enlarged graph.

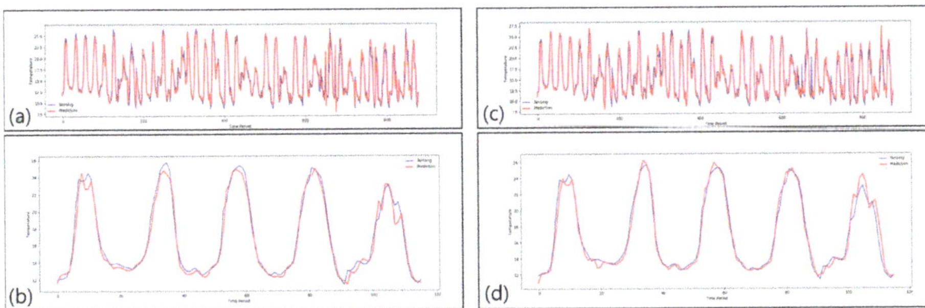

Parameter	Value
RNN Cell	GRU
Sequence Length	5
Hidden Dimension	5
iterations	1000

Parameter	Value
RNN Cell	GRU
Sequence Length	5
Hidden Dimension	5
iterations	10000

Figure 14. (**Left**) LSTM model, 1000 iterations; (**Right**) GRU model, 10,000 iterations. (a) LSTM model overall results, (b) LSTM model Enlarged graph, (**c**) GRU model Overall results, (**d**) GRU model Enlarged graph.

The observed RMSE value in the experiment is 0.062557. The graphs plotting the measured (blue) and predicted values (red) in Figure 11 are quasi-identical, indicating that the prediction model exhibits good performance.

Figure 15 depicts the smart farm equipment fault diagnosis test environment where the proposed model was evaluated experimentally. The equipment fault API integration function, equipment fault diagnosis unit function, power measurement equipment integration, and RNN model were evaluated. The sensor used in this experiment was a temperature sensor(Naretrends Co., Ltd., Buchon-si, Korea) used in smart farms.

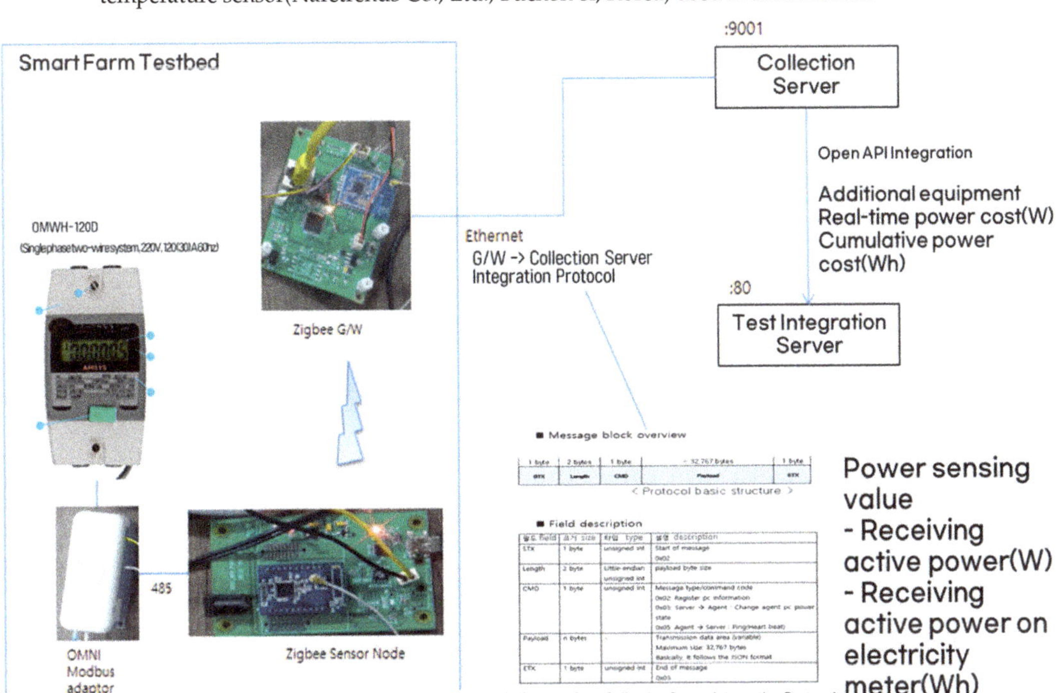

Figure 15. Configuration of the sensor node test environment.

Figure 16 shows a block diagram of the RNN-based fault diagnosis workflow. If fault diagnosis is conducted based on the threshold value after RNN-based prediction of sensor data, there is an integrated interface to submit external data. This is performed using the method "receiveDeviceMeasureNotification", and the path indicated is its URL path. The data received are not directly entered into the fault diagnosis ontology but are sent to the "Malfunction Detection Scheduler" queue. Fault detection inference can be conducted only after different types of sensor values have been received for a certain length of time. Subsequently, the semantic storage, which is loaded with ontology models, reasoners, and rules, is managed. Once instances are added to the semantic storage through the process of semantic transformation, an event query automatically leads to fault inference. In other words, a query triggers inference. Thus, adding triples does not automatically lead to inference, but inference starts as the need arises. A semantic transformation method such as "translatePredictionResult" is applied using a semantic transformer. The object "PredictionResult" is received as a factor and is transformed into a triple consisting of a subject, a predicate, and an object, which is then added to the ontology using the function "semanticStorageManager.addTriple". Then, the sensor values cyclically collected through the external interface stored in the "Malfunction Detection Scheduler" queue are subjected to semantic transformations, and fault event reporting is carried out through the process of query and fault inference. Here, "doCollect" issues an event query and triggers inference in the ontology, and the Cloud sends fault events to the website.

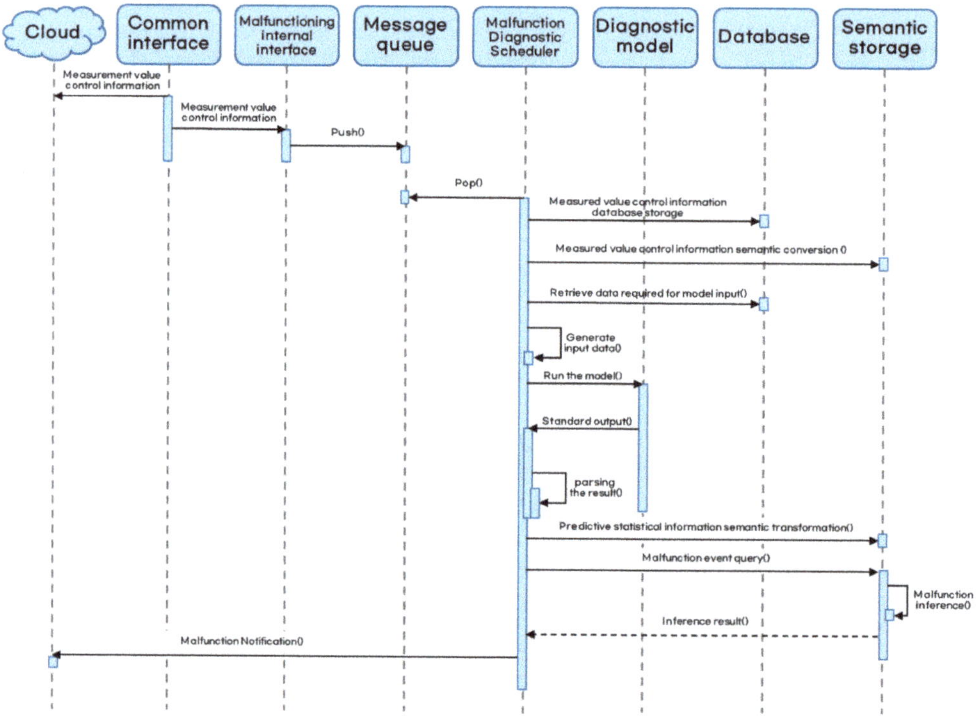

Figure 16. Block diagram of the RNN-based fault diagnosis workflow.

Figure 17 shows the RNN-based fault diagnosis program. The sensor values are predicted using the prediction model, the predicted data are subjected to ontology inference, and the diagnosis results are presented to the user. Once the greenhouse is selected as the area and test data are entered, the predicted and measured values are presented via real-time monitoring to notify the user of threshold crossings.

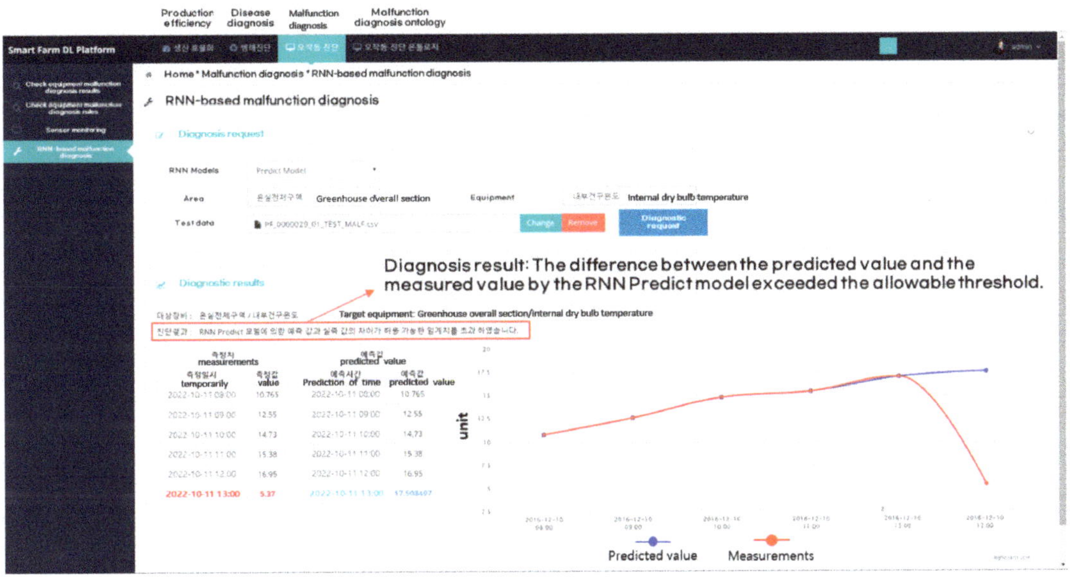

Figure 17. RNN-based fault diagnosis program interface.

The proposed system and previously provided fault detection systems have long been used in other fields, and accurate comparisons between them are difficult because they employ different methods. However, the continuous analysis of sensor outliers detected by the proposed system is expected to improve fault detection in smart farming systems.

6. Conclusions

In this study, a fault detection technique for smart farming equipment was designed and implemented with the aim of preventing damage to farms due to downtime caused by mistakes, faults, and aging of ICT devices. This is particularly significant owing to the widespread use of such devices in agricultural production facilities.

To this end, a model was designed based on RNN algorithms. The model was trained using hourly data obtained from the Korea Agency of Education, Promotion, and Information Service in Food, Agriculture, Forestry and Fishery which were collected at a strawberry farm over a one-year span. The data were extracted in sequence, and necessary items were preprocessed as training data. The learning results were tested by considering 500, 1000, 5000, 10,000, and 100,000 training iterations. The RSME of the optimized model was 0.07, confirming that it exhibits a high prediction power in an environment in which ICT equipment operation is difficult.

In farms operating large-scale, modernized, high-tech greenhouses, growth management optimization with respect to facilities and crop characteristics is essential, and efficient operation of sensors and controllers is fundamental. Further, to ensure effective smart farm operation, human interference should be minimized. Furthermore, damage to smart farm equipment and crops can be minimized by detecting malfunctions and arranging for prompt replacements. The technology discussed in this study is a key element in the construction of smart farms capable of self-reliant operation and fault detection.

Moreover, this study enables the prevention of disputes between farms and companies, empowers device PL insurance through linkage with insurance companies, and can be used as a source of base data suitable for agricultural research.

It is also expected to maximize the output of the smart farm system industry by predicting the remaining useful life and promoting the use of sensors and data to monitor the status of smart farm equipment or mechanical systems, the use of secure diagnostic technology to detect signs of failure, and the use of condition-based maintenance technology to maintain normal operation. It is expected to enhance the reliability of the service and secure global competitiveness for smart farm companies.

In future work, we intend to predict malfunction faults by collecting vibration, current, and image values obtained using actuators used in greenhouses.

Author Contributions: Writing—original draft, H.O.C.; Project administration, M.-H.L. All authors have read and agreed to the published version of the manuscript.

Funding: This work was supported by the Korea Institute of Planning and Evaluation for Technology in Food, Agriculture and Forestry (IPET) and the Korea Smart Farm R&D Foundation (KosFarm) through the Smart Farm Innovation Technology Development Program funded by the Ministry of Agriculture, Food and Rural Affairs (MAFRA) and the Ministry of Science and ICT (MSIT), Rural Development Administration (RDA) (421021-03). This work was supported by Korea Institute of Planning and Evaluation for Technology in Food, Agriculture and Forestry (IPET) through Short-term Advancement of Smart Agricultural Technology in Open Fields Project, funded by Ministry of Agriculture, Food and Rural Affairs (MAFRA) (322031-03).

Institutional Review Board Statement: Not applicable.

Data Availability Statement: Data are contained within the article.

Conflicts of Interest: The authors declare no conflict of interest. The funders had no role in the design of the study; in the collection, analyses, or interpretation of data; in the writing of the manuscript, or in the de-cision to publish the results.

References

1. Kaaya, J. Role of Information Technology in Agriculture. *Proc. FOA Conf.* **1999**, *4*, 315–328.
2. Sharma, A.B.; Golubchik, L.; Govindan, R. Sensor Faults: Detection Methods and Prevalence in Real-World Datasets. *ACM Trans. Sens. Netw.* **2010**, *6*, 23. [CrossRef]
3. Mourad, M.; Bertrand-Krajewski, J.L. A Method for Automatic Validation of Long Time Series of Data in Urban Hydrology. *Water Sci. Technol.* **2002**, *45*, 263–270. [CrossRef] [PubMed]
4. Ahn, H.; An, S.-Y.; Kim, J.-Y.; Lee, C.W. Humidity Sensor Prediction Model Using Environmental Data Analysis in Greenhouse Smartfarm. In Proceedings of the Symposium of the Korean Institute of Communications and Information Sciences, Jeju, Republic of Korea, 18–21 June 2021; pp. 159–160.
5. Lee, S. Cloud-Based Smart Farm Technology. *J. Korean Inst. Commun. Sci.* **2016**, *34*, 51–57.
6. Jeffery, S.R.; Alonso, G.; Franklin, M.J.; Hong, W.; Widom, J. Declarative Support for Sensor Data Cleaning, Lecture Notes in Computer Science. In Proceedings of the of 4th International Conference on Pervasive Computing, Dublin, Ireland, 7–10 May 2006; pp. 83–100. [CrossRef]
7. Ni, K.; Ramanathan, N.; Chehade, M.N.H.; Balzano, L.; Nair, S.; Zahedi, S.; Kohler, E.; Pottie, G.; Hansen, M.; Srivastava, M. Sensor Network Data Fault Types. *ACM Trans. Sens. Netw.* **2009**, *5*, 25. [CrossRef]
8. Hong, S.-J.; Kim, J.H.; Kim, T.-H. An Integrated Verification System for an RNN Inference Processor. In Proceedings of the Korean Society of Electronics Engineers Conference, Jeju, Republic of Korea, 28 June–2 July 2022; pp. 1918–1921.
9. Kim, Y.-D.; Jung, G.-H. Bayesian Methodology for Machine Learning. *J. Korean Inst. Commun. Sci.* **2016**, *33*, 60–64.
10. Seo, D.-S.; Park, Y.-G.; Park, J.-Y.; Kim, Y.-J. *A Study on Smart Farm Operation Status and Development Direction*; Ministry of Agriculture, Food and Rural Affairs, National Library of Korea: Sejong, Republic of Korea, 2016; Volume 3.
11. Joonyong, K.; Rack, P.K. Analysis of Accuracy and Loss Performance According to Hyperparameter in RNN Model. *J. Converg. Inf.* **2021**, *11*, 31–38.
12. Introduction to the Architecture of Recurrent Neural Networks (RNNs). Available online: https://pub.towardsai.net/introduction-to-the-architecture-of-recurrent-neural-networks-rnns-a277007984b7 (accessed on 11 June 2023).
13. Cho, S.M.; Kim, W. Automatic Document Title Generation with RNN and Reinforcement Learning. *J. Inf. Technol. Appl. Manag.* **2020**, *27*, 49–58.
14. Shin, S.H.; Lee, M.K.; Song, S.K. A Prediction Model for Agricultural Products Price with LSTM Network. *J. Korean Contents Assoc.* **2018**, *18*, 416–429.
15. Kim, D.-H. Livestock Price Prediction Study Using LSTM Based on Big Data. Ph.D. Thesis, Chonbuk National University, Cheongju, Republic of Korea, 2021.
16. Oh, H.-W.; Huh, J.-D. *IoT-Based Smart Factory Failure Prediction Analysis Technology for Productivity and Quality Improvement*; IEEE: Piscataway, NJ, USA, 2020; pp. 33–43.

17. Han, J.H.; Choi, D.-J.; Park, S.-U.; Hong, S.-K. DT-CNN Based Motor Failure Prediction Considering Outlier Data. *J. Inst. Control Robot. Syst.* **2020**, *26*, 932–939. [CrossRef]
18. Ryu, M.; Cha, S.-H. Developing a Knowledge Graph Based on Ontology Learning. *J. Korean Assoc. Comput. Educ.* **2022**, *25*, 51–57.
19. Park, J.; Song, M.; Ahn, S. Developing the Fault Diagnostics and Prognostics Model of a Rotating Machinery. *JKORMS* **2020**, *45*, 25–38. [CrossRef]
20. Lee, S.; Lee, J. Monitoring Procedure of Autocorrelated Processes Using the Deep Learning-Based LSTM Model. *J. Korean Data Inf. Sci. Soc.* **2022**, *33*, 237–248.
21. Elnahrawy, E.; Nath, B. Cleaning and Querying Noisy Sensors. In Proceedings of the of 2nd ACM International Conference on Wireless Sensor Networks and Applications, San Diego, CA, USA, 19 September 2003; pp. 78–87.
22. Tolle, G.; Polastre, J.; Szewczyk, R.; Culler, D.; Turner, N.; Tu, K.; Burgess, S.; Dawson, T.; Buonadonna, P.; Gay, D.; et al. A Macroscope in the Redwoods. In Proceedings of the 2nd International Conference on Embedded Networked Sensor Systems, San Diego, CA, USA, 2–4 November 2005; ACM Press: New York, NY, USA, 2005; pp. 51–63.
23. Bae, N.J. *Implementation of Ontology-Based Context-Aware Control Service Model for Plant Factory Environment*; Suncheon University: Suncheon, Republic of Korea, 2014.
24. Park, M.Y.; Kim, M.J.; Park, Y.M.; Song, J.S. Development of Internet of Things Platform Based on oneM2M Standards to Provide Data Augmentation Features for Artificial Intelligence Services. In Proceedings of the Symposium of the Korean Institute of Communications and Information Sciences, Gyeongju, Republic of Korea, 16–18 November 2022; pp. 1683–1684.
25. Ji, K.; Kwon, Y. Hadoop MapReduce Performance Optimization Analysis by Calibrating Hadoop Parameters. *J. Korean Inst. Inf. Technol.* **2021**, *19*, 9–19. [CrossRef]
26. Choi, M.; Gyeong, N.-J.; Lim, J.S. Design of Smart Farming System with Kafka and Spark Streaming. In Proceedings of the Korea Contents Association Comprehensive Conference, Limassol, Cyprus, 7 September 2022; pp. 405–406.
27. Lembo, D.; Stantarelli, V.; Savo, D.F.; Giacomo, G.D. Graphol: A graphical language for ontology modeling equivalent to OWL 2. *Futur. Internet* **2022**, *14*, 78. [CrossRef]
28. Jeon, Y.J.; Lee, H. A Study on the Ontology Modeling by Analyzing RiC-CM v0.2. *J. Korean Rec. Manag. Assoc.* **2020**, *20*, 139–158.
29. Kim, K.Y.; Jongmo, K.; Park, G.-D.; Sohn, M.M. Value of Information Assessment Framework Using Fuzzy Inference Ontology. *J. Korean Soc. Manag. Eng.* **2020**, *25*, 117–135.
30. Smartfarm Datamart. Available online: https://data.smartfarmkorea.net/structuredData/selectContHortiCultureDataViewLists.do (accessed on 11 June 2023).

agriculture

MDPI

Article

Data-Driven Analysis and Machine Learning-Based Crop and Fertilizer Recommendation System for Revolutionizing Farming Practices

Christine Musanase [1,*], Anthony Vodacek [2], Damien Hanyurwimfura [1], Alfred Uwitonze [1] and Innocent Kabandana [1]

[1] African Centre of Excellence in Internet of Things, College of Science and Technology, University of Rwanda, Kigali P.O. Box 4285, Rwanda

[2] Chester F. Carlson Center for Imaging Science, Rochester Institute of Technology, Rochester, NY 14623, USA

* Correspondence: musanasechristine@gmail.com

Abstract: Agriculture plays a key role in global food security. Agriculture is critical to global food security and economic development. Precision farming using machine learning (ML) and the Internet of Things (IoT) is a promising approach to increasing crop productivity and optimizing resource use. This paper presents an integrated crop and fertilizer recommendation system aimed at optimizing agricultural practices in Rwanda. The system is built on two predictive models: a machine learning model for crop recommendations and a rule-based fertilization recommendation model. The crop recommendation system is based on a neural network model trained on a dataset of major Rwandan crops and their key growth parameters such as nitrogen, phosphorus, potassium levels, and soil pH. The fertilizer recommendation system uses a rule-based approach to provide personalized fertilizer recommendations based on pre-compiled tables. The proposed prediction model achieves 97% accuracy. The study makes a significant contribution to the field of precision agriculture by providing decision support tools that combine artificial intelligence and domain knowledge.

Keywords: precision agriculture; Internet of Things; artificial intelligence; crop recommendation; fertilizer recommendation

1. Introduction

Agriculture is a vital part of the global economy, providing food, fibe, and other essential products for human consumption [1]. However, the agricultural sector faces significant challenges in meeting the growing demand for food as the global population increases, climatic conditions change, and the significant concern subjected to the critical role of soil and fertilizers in achieving optimal crop yields and maintaining soil health [2]. It is well known that the right soil type and precise applicability of fertilizers are critical factors that can significantly enhance crop growth and overall agricultural sustainability. However, the conventional agricultural system has long been plagued by a shortfall in intelligent recommendations; as such, systems are often based on general guidelines, historical knowledge, and limited experimentation [3]. The traditional systems do not consider the specific needs of individual crops and fields, which often leads to the inefficient allocation of resources, increased costs for farmers, and sub-optimal environmental outcomes [4]. In light of these challenges, the agricultural sector stands at the threshold of transformation. There is an urgent need to revolutionize farming practices.

Precision agriculture (PA), a subset of smart agriculture, is a promising solution to these challenges, as it can help improve agricultural practices' efficiency and sustainability [5]. PA involves using advanced technology and data-driven techniques to optimize agricultural practices [6,7]. Key components of precision agriculture often include the use of sensors, GPS (global positioning system) technology, drones, and data analytics [8,9].

This transition towards a more effective, data-driven, and user-friendly approach is essential to improve farming operations' efficiency, productivity, and sustainability by making more precise and informed decisions based on real-time data [10]. In particular, this shift is made possible by the advent of Crop and Fertilizer Recommendation Systems (CFRS), which harness the power of technology such as the Internet of Things (IoT), data analytics, and Artificial Intelligence (AI) to provide specialized guidance to farmers [11,12]. Such system holds promising scope to optimize soil and fertilizer interactions, increase agricultural productivity, and promote sustainable practices. Moreover, CFRS can empower farmers with valuable insights, reduce uncertainties, and mitigate the risks associated with traditional farming methods.

In the modern era, the integration of IoT and AI in agriculture has witnessed significant growth in recent years [13,14]. IoT is reshaping how we collect, process, and utilize data in real time through a network of interconnected sensors and devices embedded in the agricultural landscape. IoT-based systems are deployed to collect real-time sensory data on various factors affecting crop growth, such as temperature, humidity, soil pH, and nutrient levels [15]. This wealth of real-time data and AI analytics forms the foundation upon which the CFRS operates to make informed decisions on crop selection and fertilizer application. The study aimed to bridge the gap between the physical and digital realms by integrating IoT technology with AI into modern agriculture. The proposed CFRS reported in this paper not only facilitates the recommendation of suitable crops to specific agricultural land but also offers fertilizer recommendations based on the soil condition for the crops during both pre-sowing (before plantation) and post-sowing until the crop reaches maturity. With such a comprehensive perspective, the proposed CFRS (Crop and Fertilizer Recommendation Systems) holds significant potential for addressing the challenges faced by smallholder farmers and for countries subject to food insecurity such as Rwanda due to low productivity, declining soil organic matter, and adverse topography.

Rwanda has been heavily involved in promoting crop intensification programs to increase the agricultural productivity of high-value food crops and achieve food security and self-sufficiency [16]. However, smallholder farmers, who play a vital role in the agricultural sector, face challenges that limit their crop productivity. These challenges include sub-humid conditions that cause frequent crop failures, the prevalence of acidic soils, declines in soil organic matter due to high population density, and the country's topography that makes agricultural systems vulnerable [17]. A study by NISR found that agriculture employs nearly 72% of Rwandans and contributes nearly 33% to the country's GDP [18]. This mismatch between labor force participation and GDP contribution shows how serious the problems in the agricultural sector are. Moreover, the socioeconomic impacts are severe, with persistent poverty and alarming malnutrition rates, especially among children. Therefore, by addressing both crop and fertilizer recommendations in an integrated way, the proposed CFRS has a transformative impact. It can potentially revolutionize the current agriculture system in Rwanda by optimizing the selection of suitable crops on particular farming land, in-creasing crop yields, reducing fertilizer waste, and promoting sustainable practices. Although several studies reported in the literature focus on developing sustainable agriculture systems in the Rwandan context, many of these approaches utilize data analysis machine learning (ML) techniques and deep learning models to build predictive models addressing various challenges to crop production [19–21]. It has been identified that most of the existing works are subjected to predicting yields of different crops, assessing the soil quality, reviewing agricultural crop policies and single crop recommendations, and understanding the impact of climate anomalies on crops. However, the scope of the existing works reported in the literature is often narrowly defined, either addressing a single aspect of farming or not integrating crucial components of fertilizer suggestion and soil conditions. This isolation often results in a lack of comprehensive insights, limiting the existing approaches' real-world applicability and accuracy. The subsequent section highlights this research manuscript's prime aim and core contribution.

1.1. Contributions of the Study

The prime aim of this study is to help farmers choose the best crops to grow and the right amount of fertilizer to use. The study develops a comprehensive model that integrates CFRS holistically, considering various relevant factors to provide more precise and specialized recommendations to farmers. The key contributions of this study are highlighted as follows:

- Data Processing and Profiling: Comprehensive data on Rwanda's major crops have been collated and analyzed, focusing on structured content and data integrity.
- Correlation Analysis: The study examines how inter-variable correlation can enhance predictive modeling, leading to better decision-making.
- Crop Recommendation: The study implements a neural network model to recommend crops. This model has been thoroughly trained and tested and is more effective than other prominent ML models.
- Fertilizer Recommendation: The modeling of fertilizer recommendation adopts a simple logical function that supports the foundational understanding that each soil and crop combination has specific nutrient requirements. The predetermined rules grounded in real agricultural practices were adopted, allowing farmers to understand the why behind a recommendation, fostering trust and encouraging adoption.
- Practical Application with IoT: Real-world testing was conducted with IoT sensors, where proposed CFRS is applied to collect data to offer actionable insights.
- The study also presents a conceptual architecture for deploying the proposed CFRS on a cloud server to provide real-time, effective, and data-driven agricultural recommendations, including monitoring soil conditions and nutrient dynamics over time.

By combining IoT and AI to framing practices, this paper offers a solution to the rising global food demand facing challenges, such as population growth and changing climate condition: an advanced system that not only suggests which crops to grow but also offers the right fertilizer to use based on various data inputs. The novelty of the proposed work stems from multiple factors. Firstly, the proposed system is more comprehensive than existing approaches, which often restrict recommendations to one or two crops. It caters to various crops predominantly cultivated in Rwanda and considers soil conditions at various crop growth stages to recommend the proper fertilizer application. Secondly, the proposed system also includes effective data processing and profiling to ensure the completeness and reliability of the data used for recommendation modeling. The proposed schemes are designed to be lightweight, ensuring rapid and cost-effective computations.

1.2. Outline of the Study

The remaining part of the proposed manuscript is organized as follows. Section 2 briefly discusses the related work, demonstrating the current research status and differentiating the proposed work from the existing one. Further, Section 3 elaborates on a system design following the implementation procedures adopted in each module of the proposed CFRS. Next, Section 4 discusses the performance metrics adopted in the experimental process, result analysis, and performance discussion. This section also presents the use case scenario of the proposed system concerning real-world deployment scenarios. Finally, Section 5 concludes the work and core findings reported in this paper.

2. Related Works

The application of precision agriculture has been the focus of extensive research and development efforts, with scholars across the globe developing various strategies and technologies to optimize farming practices. This section reviews the research works done in the context of soil quality prediction, crop recommendation, and fertilizer recommendation.

The researchers in Rivera and Bonilla [22] trained neural network and generalized linear model (GLM) models on a dataset of soil samples from different regions with varying properties such as texture, organic matter content, and pH to predict soil quality. The experimental results showed that the neural network model outperformed the GLM model regarding prediction accuracy. The study provides two models that can be used to predict aggregate stability when direct measurements are unavailable, which can help improve the comprehensiveness of soil surveys. Suchithra and Pai [23] utilized extreme learning machines (ELM) to optimize agriculture practices through soil testing and classification. By analyzing soil test report values, village-wise soil fertility indices for essential nutrients are categorized. The Gaussian radial basis function emerges as the top performer, with over 80% accuracy in most classifications. The presented approach can reduce fertilizer waste, enhance profitability, and improve soil health and environmental quality for sustainable agriculture in India. In [24], Chambers showed that the type of ML model used can affect the accuracy of soil property predictions and that local farms tend to have more accurate predictions than farms in different locations. Principal component analysis (PCA) with 50 components was found to be beneficial. Wu et al. [25] showed that the Generalized Regression Neural Network (GRNN) model can effectively estimate soil nutrients for Dacrydium pectinatum communities in China. The GRNN model, along with the k-nearest neighbor (KNN) and support vector machine (SVM) model, is utilized to assess soil nutrient content and quality grades. The work of Rose et al. [26] emphasized the significance of ML classifiers and statistical approaches in predicting soil fertility and regulating ecosystems with reduced human intervention. In [27], Rajamanickam used Decision trees, KNN, and SVM algorithms to predict soil fertility based on macro- and micronutrient data, achieving 99% accuracy with the decision tree algorithm.

In [28], Rajamanickam and Mani addressed the impact of climate anomalies on crops and environmental challenges on agriculture practices. The authors have proposed a probabilistic neural network for the soil fertility prediction approach, providing higher accuracy and reduced processing time. Katarya et al. [29] discuss various artificial intelligence (AI) techniques for improving crop yields in agriculture. These techniques are based on the paradigm of precision agriculture (PA), specifically crop recommender systems. The specific approaches discussed include K-nearest neighbor (KNN), similarity-based classifiers, ensemble learning, and neural networks. The authors introduce a model that considers external factors such as meteorological data, temperature, and soil profile to recommend optimal crops for cultivation. This can lead to improved yields and more efficient use of resources. Klerkx et al. [30] provided a comprehensive overview of the emerging field of digital agriculture, covering a wide range of sub-fields, including the adoption of digital technologies on farms, the impact of digitalization on farmer identity and skills, ethics in digital agriculture, the effects of digitalization on agricultural knowledge and innovation systems, and the economics of digital agriculture. The study maps the contributions of 17 special issue articles to these clusters. It offers insights into the links between digital agriculture and farm diversity, new economic and institutional arrangements, and the governance of digital agriculture. Shadrin et al. [31] developed a low-power embedded system with AI capabilities for continuous analysis of plant leaf growth. The system uses a GPU to run a recurrent neural network (LSTM) on board, enabling autonomous operation for 180 days on a standard Li-ion battery. This study opens up new possibilities for intelligent monitoring in agriculture, and the authors have shared the Tomato Growth dataset with the research community.

Kumar et al. [32] investigate how wireless sensor networks (WSNs) can be used in precision agriculture to improve crop yields and quality. They highlight a variety of WSN applications, such as pest and disease control, animal tracking, and crop strength assessment, which have the potential to significantly boost crop production. Talaviya et al. [33] discuss the importance of AI in addressing agricultural challenges posed by rising population and food demand. They review AI applications in agriculture, including irrigation, weeding, and spraying using sensors, robots, and drones, with a focus on soil moisture sensing, automated weeding techniques, and drone applications for spraying and crop monitoring. Kamilaris et al. [34] introduce a smart farming framework that uses IoT platforms to process diverse sensor data streams in real time. The Agri-IoT framework supports reasoning across heterogeneous data streams, enabling seamless integration of sensors, services, processes, farmers, and online information sources. It provides a comprehensive and adaptable solution for the agri-food industry, bridging the gap between external factors and the food supply chain. Rekha et al. [35] develop an IoT framework to help farmers improve their farming methods and increase crop yields. The framework uses wireless sensor networks to collect data and a decision support system to provide farmers with personalized advice on irrigation, fertilization, and other practices. The advice is delivered in the farmers' regional language through an Android app, making it easy for them to follow and improve their farming practices, which can lead to increased income.

Rehman et al. [36] propose a smart farming approach that uses real-time sensor data and machine learning to improve agricultural practices. By integrating these two technologies, their approach enhances precision agriculture and overcomes the limitations of traditional smart farming methods. Priya et al. [37] suggested using deep learning algorithms to predict the best crops to grow based on factors such as soil moisture, humidity, temperature, pH, soil type, and land type. This crop recommendation system helps farmers make informed decisions to improve productivity, especially in the face of changing weather patterns. Biradar et al. [38] highlight the potential of IoT and data mining to develop intelligent systems for more efficient water management in agriculture. Sensor networks can provide a cost-effective way to monitor and control water use, leading to improved crop yields and food security. Akhter et al. [10] demonstrate how IoT, WSN, data analytics, and machine learning can be used to revolutionize apple disease prediction in apple orchards. They also explore the challenges of implementing these technologies in traditional farming practices. Ref. [39] studied how IoT can revolutionize traditional irrigation scheduling on a flood-irrigated subtropical lemon farm. They also explore the challenges of implementing these technologies in traditional farming practices. Gupta et al. [40] show that using the right algorithms on sensor data can recommend the best crops to grow, leading to higher yields and better-quality produce. Vi-vekanandhan et al. [41] introduce an adaptive neuro-fuzzy inference system (ANFIS) technique for analyzing agricultural plant growth based on soil, water level, temperature, and moisture conditions.Their smart irrigation system is effective in monitoring and improving crop growth.

Hence, it can be seen that there is much research work presented on PA applications and ML-based predictive modeling to benefit agriculture systems. However, each method is associated with its own advantages, and they have limitations too. It has also been analyzed that very little work is done in the context of Rwanda's agricultural system. The literature is rich with studies on various ML applications in smart agriculture. However, such approaches are subjected to theoretical discussion, not implementations. The theoretical discussions are valuable, and validation of these models in real-world agricultural settings is essential to assess their feasibility and effectiveness. The potential of integrating the IoT and AI in agriculture has been explored in previous research. However, in isolation, most of these studies either focus on crop or fertilizer recommendations. It has also been noticed that the existing studies lack details about the data source system implementations, even if they do not specify the features used and on what basis they selected a particular learning model. All these gaps are addressed by the proposed system discussed in the next section.

3. Materials and Methods

The development of the proposed recommendation models is carried out using python programming language in Anaconda distribution installed on windows 10 machine. Five-pin soil transmitter (Type485) sensors from (Shandong Renke Control Technology Co., Ltd., Jinan City, China), were used to collect data across agricultural fields.

This section presents the design of the proposed CFRS as a support system for precision agriculture and sustainable farming practices. Building a robust and efficient CFRS for Rwanda's agricultural system requires a suitable dataset that includes crop information, soil properties, and nutrients. However, no standard dataset for building CFRS in the Rwandan context is available. The first steps in building the proposed CFRS system are to collect and prepare the dataset. This study has developed a sophisticated data modeling and feature extraction approach to effectively train the learning model for crop recommendation based on current soil attributes. Figure 1 shows the schematic architecture of the proposed system, which includes various computing modules such as dataset collection and selection, data profiling, data preprocessing, neural network-driven crop recommendation, and decision logic implementation for fertilizer recommendation.

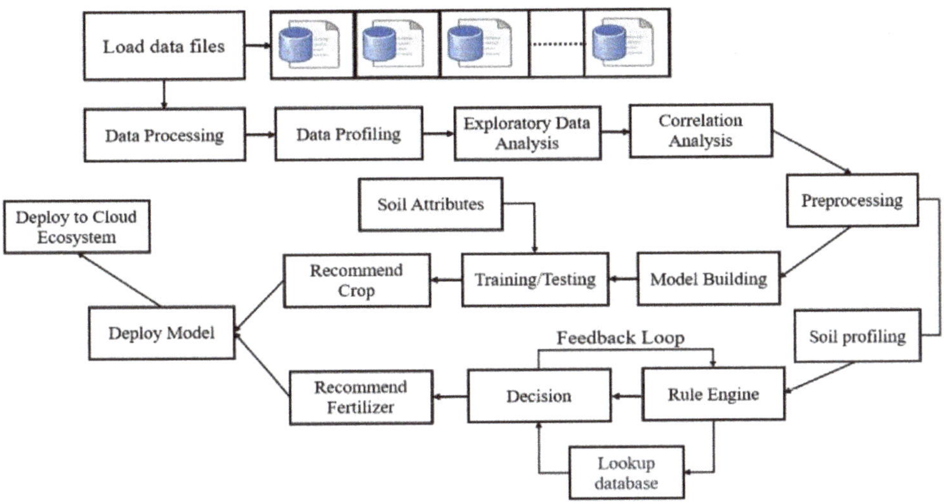

Figure 1. Schematic architecture of the proposed CRFS workflow.

The proposed system architecture, shown in Figure 1, involves a highly synchronized approach to data analytics and computational intelligence. The top layer of the system design is responsible for collecting data from different sources. These data are then analyzed and mapped to construct a final crop dataset in the next layer. Exploratory data analysis and correlation analysis are then performed to understand the nature of the data and obtain significant insights into which preprocessing techniques to apply to ensure the dataset's completeness. This process is crucial for ensuring the dataset's quality, making it suitable for training the learning model for crop recommendation. The study used the common practice of splitting the dataset into training and testing sets in an 80:20 ratio. The neural network model was configured and optimized for the specific problem and input data. The trained model was then validated on the testing dataset, which consisted of different soil attributes.

The next part of the proposed system integrates a rule-based fertilizer recommendation system. This phase first performs soil profiling using the preprocessed dataset from the crop recommendation system and builds a lookup table based on scientific evidence and expert knowledge. The proposed system is based on the ideology that while many tasks can benefit from the predictive capabilities of machine learning, there is undeniable value in domain-specific, expert-driven rules. The proposed rule-based fertilizer recommendation system is based on the principle that each soil and crop combination has specific nutrient requirements. It is transparent and easy to understand, as it is based on established knowledge in agriculture. This allows farmers to understand the reasoning behind the recommendations, which can build trust and lead to improved crop yields and more sustainable farming practices. The system also includes a feedback loop mechanism to help update and improve the rules and recommendations over time. The proposed study discusses the conceptual architecture of the system and underscores its real-world applicability and feasibility for deployment in a cloud environment.

3.1. Data Processing and Profiling

In order to ensure the effectiveness of the crop recommendation system, diverse data are collected, including information on soil properties, nutrient levels, and crop performance from various agricultural regions. The data collection phase of the proposed system involved compiling crop datasets from various sources, primarily the crop recommendation dataset and soil fertility data obtained from Kaggle [42,43], and soil nutrient balance (NPK) dataset of Rwanda obtained from [44]. Combining data from different sources is difficult because the data can be in different formats and sizes. This study uses a sophisticated data modeling process called data profiling to harmonize and integrate the data to address this challenge. First, the system selects the major crops cultivated in Rwanda, such as maize, potatoes, beans, tomatoes, coffee, cassava, sweet potatoes, sorghum, onion, kidney beans, and banana. The selection criteria are based on familiarity with Rwanda's agricultural system, sustainability concerns, and empirical evidence from previous studies [45,46]. This process results in a comprehensive crop data collection relevant to the study objectives.

Data profiling is the process of examining a dataset to understand its structure, content, and quality. The first step in this process is to review the columns of the dataset. In this case, the dataset included columns for nitrogen (N), phosphorus (P), potassium (K), temperature, humidity, pH, rainfall, and a label column. These columns represent essential factors that affect crop growth. Next, the study selected the columns that are relevant to the study, which are N, P, K, pH, and major crop. Focusing on these variables simplified the dataset while retaining the most critical information. Finally, the study filtered the dataset to include only rows corresponding to the significant crops of interest. To prevent bias, the data frame was randomly shuffled. The data frame index was reset to maintain data integrity. Finally, a new data frame was created with only the selected variables. Algorithm 1 shows how data profiling was used to clean and streamline the dataset after preprocessing. This helped to identify the key features in the data and ensure that the analysis focused on the variables that were most relevant to the study objectives.

Algorithm 1 Dataset Profiling for data integration and harmonization

Input: $\mathcal{D} = d_1, d_1, \cdots, d_n$ (Sets of datasets sourced from different origins)
Output: df (harmonized dataset for CFRS)
Procedure:
1. Initialization:
$\mathcal{C} = \{'Maize'.'Potatoes','Beans', \cdots\}$ (Set of major crops of interest)
2. Data Selection:
$\mathcal{D}' \leftarrow \cup_{i=1}^{n} \{x | x \epsilon d_i \wedge crop(x) \in \mathcal{C}\}$
// Union of all datasets retaining only records related to crops in $\mathcal{C}$
$\mathcal{D}' \leftarrow \mathcal{D}' \{x | column(x) \notin RelevantColumns\}$ (Substraction of irrelevant columns from $\mathcal{D}'$)
3. $\forall d \in \mathcal{D}'$: Convert units of 'N', 'P', 'K' to standard units, if not already
4. Data Randomization:
Shuffle the order of records in $\mathcal{D}'$
Reset indices of $\mathcal{D}'$
5. Data Integration:
$df \leftarrow \cup_{i=1}^{n} d_i'$ (Union of all dataset $\mathcal{D}'$)
6. Review & Validation:
Conduct exploratory data analysis on df
7. Data pre-processing: (Post data profiling, detailed in next sub-section)
8. $\mathcal{D}' \leftarrow \mathcal{D}' \{ x | xhasmissingvalues\} (Removingrecordswithmissingvalues)$
9. Identify correlations and rectify outliers in $\mathcal{D}'$
10. Remove duplicates in df (ensures completeness of the dataset)
End

3.2. Preprocessing

Preprocessing is a critical step in data-driven predictive modeling because the quality of the data and the useful information that can be extracted from them directly affects the model's ability to learn the underlying patterns in the training data. This phase of the proposed system involves exploratory data analysis (EDA) and correlational analysis. EDA is a process of analyzing datasets to identify their main characteristics, often using visual methods. Under EDA, the study calculates descriptive statistics such as mean, median, mode, minimum, maximum, range, quartiles, variance, and standard deviation to understand the relationships between the variables.

Table 1 shows different crops' average NPK (kg/ha) and pH requirements. The average NPK requirements for bananas are 100.19 kg of nitrogen, 80.89 kg of phosphorus, and 50.04 kg of potassium per hectare. The average pH of soil where bananas are grown is 6.07, considered moderately acidic. Similarly, beans require an average of 75.09 kg of nitrogen, 25.32 kg of phosphorus, and 34.73 kg of potassium per hectare. The average pH for growing beans is 5.89, considered highly acidic. Cassava requires an average of 74.77 kg of nitrogen, 34.92 kg of phosphorus, and 59.79 kg of potassium per hectare. The average pH of cassava fields is 5.92, which is also highly acidic. Similar interpretations can be made for other crops to understand their fertilizer requirements and soil suitability for more efficient crop production.

Table 1. A sample visualization of crops with mean values.

	Major_CROP	Nitrogen	Phosphorus	Potassium	pH
0	banana	110	70	70	6.00
1	beans	80	25	35	6.25
2	cassava	135	70	50	6.24
3	coffee	70	25	50	6.00
4	kidneybeans	75	25	35	6.26
5	maize	135	70	50	6.26
6	onion	50	25	35	6.49
7	potato	90	50	70	5.75
8	sorghum	110	60	50	6.48

Figure 2 provides a comparative analysis of different crops based on their nitrogen requirements in Kg/ha. The analysis demonstrates that the crop Maize requires the most nitrogen, followed by onion and sorghum. Kidney beans require the least nitrogen. Bananas have a wide range of nitrogen needs, appearing in the highest and lowest categories. Similarly, Figure 3 gives a comparative analysis of different crops based on their phosphorus requirements in Kg/ha.

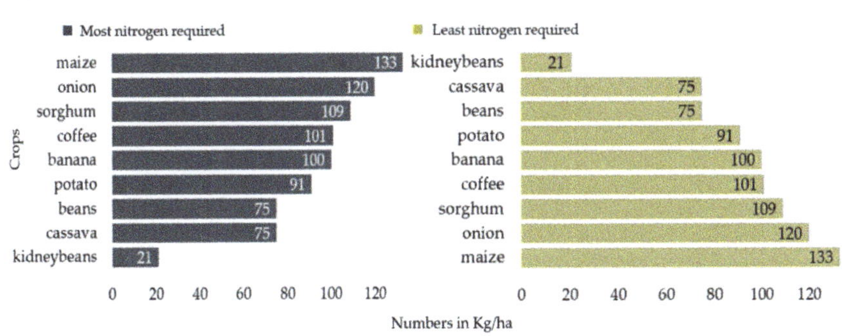

Figure 2. Analysis regarding most and least nitrogen-requiring crops.

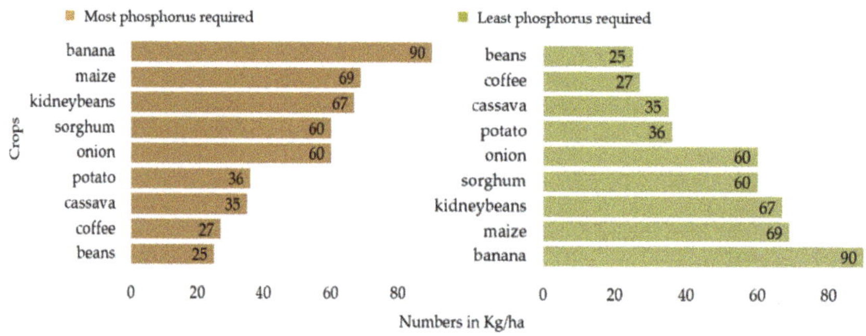

Figure 3. Analysis concerning most and least phosphorus-requiring crops.

The insight from Figure 3 exhibits that bananas require the most phosphorus, followed by maize and kidney beans. Beans and coffee require relatively less phosphorus. Onions have diverse phosphorus requirements, ranking in the highest categories. A closer analysis of Figure 4 reveals that onions, followed by cassava and potatoes, require the most potassium, while kidney beans, maize, and sorghum require little potassium. This analysis emphasizes the varied nutrient profiles of different crops across categories.

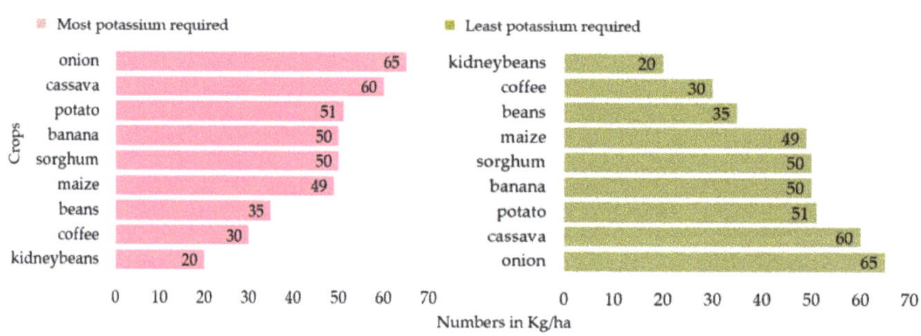

Figure 4. Analysis for most and least potassium-requiring crops.

The analysis shown in Figure 5 offers a holistic view of the data's distribution and relationships between different data points color-coded based on different crops under major crops. In this visualization, the diagonal part shows frequent distribution of the data points. In contrast, the scatter representation of the data points showcases the distinction among the different crops concerning different nutrient requirements. It can be seen that nitrogen (N) and phosphorus (P) are correlated; it seems that crops that need high amounts of N may also require high amounts of P.

A closer analysis of the above-mentioned Figure 5 also reveals that specific data points are densely packed, indicating that such crops have similar requirements for those paired nutrients. Moreover, a few data points are also found that lie far away from others, which can be potentially an outlier, which, if not addressed, may introduce ambiguity in predictive learning. To eliminate the outliers, the proposed study uses the standard score method, which measures how many standard deviations an element is from the mean. About 99.7% of the data in a normal distribution fall within three standard deviations from the mean. Therefore, a score greater than three is used as an indicator of an outlier. Mathematically, the outlier score (S) for a data point x is computed as follows:

$$S = \frac{x - \mu}{\sigma}, \tag{1}$$

where x is the data point, μ refers to the mean value of the dataset, and σ denotes the standard deviation of the dataset. So, if the value of S of a data point falls outside a predefined threshold, the data point is tagged as an outlier. Figure 6 presents a correlation heatmap to gain insight into correlation among different data points.

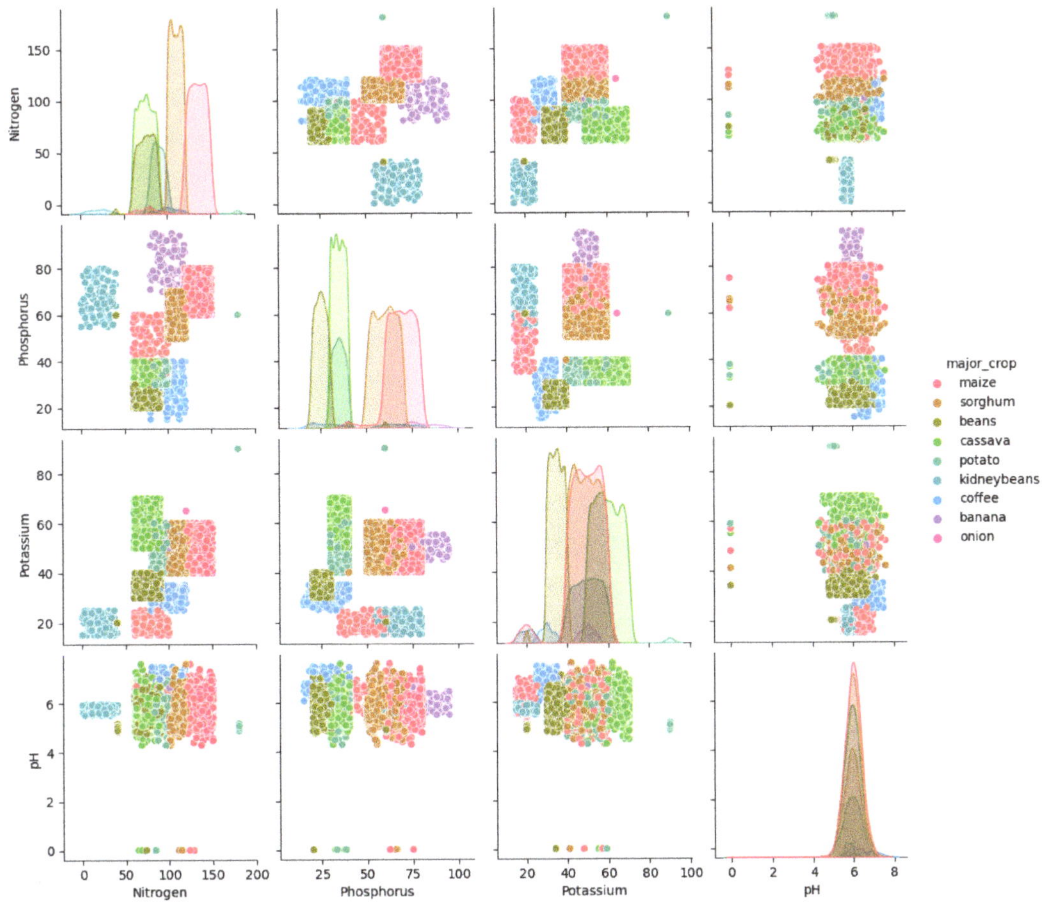

Figure 5. Pair-wise relationship visualization.

Figure 6 shows a correlation plot of different crop nutrients and factors. The values in the plot range from −1 to 1, where −1 indicates a strong negative correlation, 1 indicates a strong positive correlation, and 0 indicates no correlation. Nitrogen (N) has a strong positive correlation with phosphorus (P), meaning crops that need much nitrogen also need much phosphorus. This correlation is 0.77, which is considered high. Nitrogen's correlations with other elements and factors are weaker. For example, it has a weak positive correlation with potassium (K) at 0.15 and a similar trend with pH at 0.04. This analysis shows a slight but positive relationship between nitrogen and potassium and between nitrogen and pH. The major crop type has a moderate positive correlation with nitrogen (N) levels, with a coefficient of 0.52. This shows a medium-strength relationship between the crop type and the amount of N it needs. Phosphorus (P) has weaker positive correlations with potassium (K) and pH, with correlation coefficients of 0.13 and 0.01, respectively. This examination suggests a slight but positive relationship exists between P and K and between P and pH. The overall analysis shows that the amount of N a crop needs is most strongly correlated with the amount of P it needs. It is also correlated with the crop type but to a lesser extent. The correlations between N and other elements and factors are even weaker.

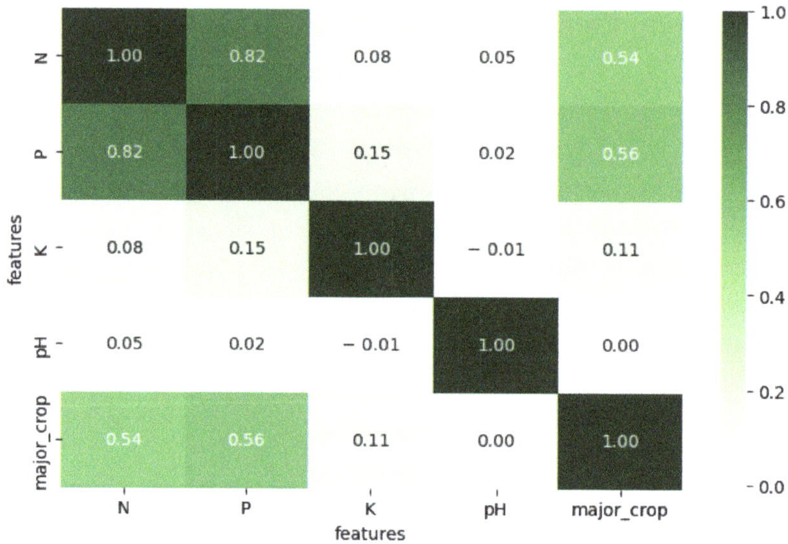

Figure 6. Correlation analysis.

3.3. Crop Recommendation System

The machine learning model applied in the proposed study offers the potential to learn intricate patterns and relationships between multiple variables, making it an effective tool for crop recommendations. This model has the ability to map a high-dimensional input space to outputs and adaptively learn from the data. The proposed study built a neural network and trained it on the prepared crop dataset to recommend suitable crops.

3.3.1. Data Attributes

It is well recognized in agronomic studies that many environmental and geographical factors, such as humidity, temperature, rainfall, altitude, soil type, and more, can significantly impact crop growth, yield, and health. These factors, individually or in tandem, play a role in determining the suitability of a specific crop in a given geographical region.

In the proposed study, the study has considered the learning model around the essential soil nutrients N (nitrogen), P (phosphorus), K (potassium), and soil quality represented by pH. These are foundational factors that majorly influence crop recommendations and are consistent indicators across various datasets. The major crop type serves as the output of our model. While it would be ideal to incorporate all influential environmental and geographical parameters for a holistic recommendation, the proposed study is focused primarily on the soil's health, nutrients, and quality. The reasons being:

- Data Availability & Consistency: Our dataset is a compilation from various sources, with the majority being oriented towards N, P, K, pH, and crops. Some datasets did encompass parameters such as rainfall and temperature, but to maintain consistency and avoid introducing data biases, it was essential to have uniform features across all data points.
- Complexity in Data Collection for Other Factors: Gathering a comprehensive dataset that includes all geographical and climatic factors is a herculean task. It requires expert interventions, prolonged data curation processes, and introduces the risk of human errors.

- Interrelation with pH: The soil's pH, which measures its acidity or alkalinity, can act as a proxy for some environmental factors. For instance, consistent rainfall can influence soil pH; likewise, soil pH can reflect certain climatic conditions such as humidity and temperature. By considering pH, the model indirectly captures some of the environmental conditions' impacts on the soil.

3.3.2. Model Training

The machine learning model for crop recommendations was trained using a supervised learning approach. Initially, we preprocessed the data to handle any missing values, outliers, and to normalize the features. The dataset consists of approximately 10,440 samples, which were then split into a training and testing set considering a split ratio of 70:30, where 80 percent of data, i.e., 8352 samples, are kept for training and the remaining 20%, i.e., 2088 samples, are kept for testing the trained model. The study also considers a validation set, which is 10% of the training dataset, to ensure that the model is not overfitting or underperforming during its training. Therefore, training data consists of 7308 data samples and four predictors (N, P, K, pH) and a single response variable named major_crop.

Given the complexity and nonlinear relationships between soil attributes and crop types, the study employed a neural network, a subset of machine learning models, that are adept at capturing intricate patterns and dependencies in the data. The employed neural network architecture consists of three layers: an input layer, two hidden layers, and an output layer, as shown in Figure 7.

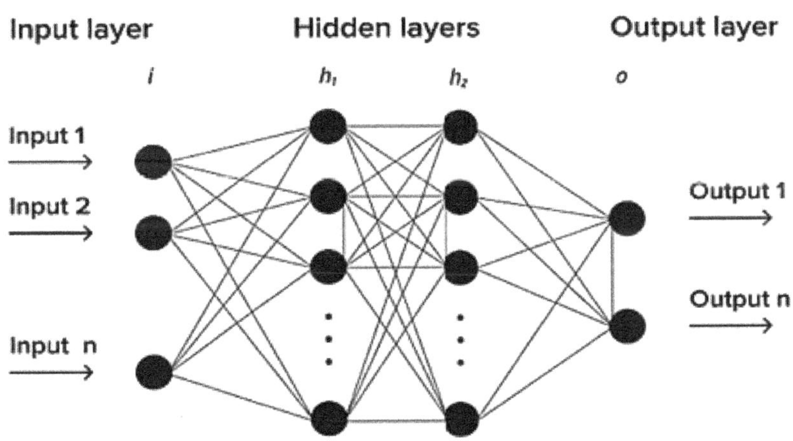

Figure 7. Neural network architecture used in the proposed study.

The input layer serves as a placeholder for input data to the model. The study experimented with multiple architectures, adjusting the number of hidden layers and nodes within each layer to optimize performance. Based on the empirical analysis, the study found the consideration of two hidden layers with 64 and 32 neurons, respectively. The output layer contains nodes corresponding to the number of unique crop types in the dataset. The activation function used here is typically the SoftMax function, which provides a probability distribution over the potential crop types. The model was trained using backpropagation, a standard method for training neural networks. This involves iteratively adjusting the model weights to minimize the difference between the predicted output and the actual target values. The training process of the neural network is discussed in Algorithm 2.

Algorithm 2 Neural Network Model For Crop Recommendation

Input: x_{train} Training data (N,P,K,pH)y_{train} labels (crop),
Number of predictors η, Number of unique crop classes κ,
Learning rate α, Number of epochs e, Batch size s, Adam
hyperparameters $\beta_1, \beta_2, \epsilon$
Output: Trained Neural network model
Procedure:
1. Initialization:
Define input layer with η
2. Initialize first hidden layer 64 with weights $W_1 \in \mathbb{R}^{64xn}$ and biases $b_1 \in \in \mathbb{R}^{64}$
3. Initialize first hidden layer 64 with weights $W_2 \in \mathbb{R}^{64xn}$ and $\in \mathbb{R}^{32}$
4. Training: For epoch = 1 to e:
5. Shuffle the training data.
6. Divide x_{train} and y_{train} into batches of size s
For each batch:
Compute: $Z_1 = W_1X + b_1$ and $A_1 = ReLU(Z_1)$
Compute: $Z_2 = W_1X + b_1$ and $A_2 = ReLU(Z_2)$
Compute: $Z_o = W_oA_2 + b_o$ and $A_o = softmax(Z_o)$
Compute Loss L
Compute gradient of loss w.r.t. final output $\frac{\partial L}{\partial A_o}$
compute: $\frac{\partial L}{\partial Z_o}$ = $A_o - y_{train}$
$\frac{\partial L}{\partial W_o} = \frac{\partial L}{\partial Z_o} A_2^T$
$\frac{\partial L}{\partial b_o} = \frac{\partial L}{\partial Z_o}$
Using chain rule and considering ReLU derivatives:
$\frac{\partial L}{\partial A_2} = W_2^T \frac{\partial L}{\partial Z_o}$
$\frac{\partial L}{\partial Z_2} = \frac{\partial L}{\partial A_2} g^{'(Z_2)}$ // where g′ is the ReLU derivative
Update learnable parameters W and b using Adam optimizer with learning rate α
7. Validate the model using testing dataset
End

The input layer has 'n' neurons, corresponding to the number of predictors. Mathematically, this can be represented as $X \in \mathbb{R}^n$ being the input vector, where $X = [N, P, K, pH]^T$. The first hidden layer has 64 neurons, such that $W_1 \in \mathbb{R}^{64xn}$ is the weight matrix connecting the input layer to the first hidden layer, and $b_1 \in \mathbb{R}^{64xn}$ is the bias vector for the first hidden layer. The weighted sum $Z_1 \in \mathbb{R}^{64xn}$ of the first hidden layer is $W_1X + b_1$. The ReLU activation function is applied element-wise to the weighted sum, $A_1 = ReLU(Z_1)$, where $A_1 \in \mathbb{R}^{64xn}$. The second hidden layer has 32 neurons, such that $W_2 \in \mathbb{R}^{32xn}$ is the weight matrix connecting the first hidden layer to the second hidden layer, and $b_2 \in \mathbb{R}^{32xn}$ is the bias vector for the second hidden layer. The weighted sum $Z_2 \in \mathbb{R}^{32xn}$ of the second hidden layer is $W_2A_1 + b_2$. The ReLU activation function is applied element-wise to the weighted sum $A_2 = ReLU(Z_2)$, where $A_2 \in \mathbb{R}^{32}$. The next output layer has k neurons, representing the number of unique crop classes, such that $W_o \in \mathbb{R}^k$ is the weight matrix connecting the second hidden layer to the output layer, and $b_o \in \mathbb{R}^k$ is the bias vector for the output layer. The weighted sum $Z_o \in \mathbb{R}^k$ of the output layer can be calculated, $Z_o = W_oA_2 + b_o$. Here, the softmax activation function is applied to the weighted sum to obtain the final output probabilities for each class.

The model is compiled using the Adam optimizer with a learning rate of 0.0001. The loss function employed is the sparse categorical cross-entropy. The model is trained using the fit( ) method, which takes in the training data. The training data are split into a training subset and a validation subset. In this case, 20% of the training data are used for validation during training. The model is trained for 200 epochs. The training dataset is divided into batches, and in each iteration, the model updates its weights and biases based on the gradients calculated from a batch of size 64. The weight and bias updates are performed using the backpropagation algorithm and the Adam optimizer. The algorithm is responsible for training the neural network optimally for crop recommendation by learning latent attributes and complex patterns of the data points from the training dataset.

3.4. Fertilizer Recommendation System

A fertilizer recommendation system is a specialized tool aimed at aiding farmers in making informed decisions about the suitable amounts of fertilizers to use for their crops. This system aims to increase agricultural productivity while minimizing adverse environmental impacts. Many farmers might not have a comprehensive understanding of their soil's current nutrient levels. Without soil testing, it is challenging to know which nutrients are deficient, which are abundant, and which are at optimum levels. The cost of professional soil testing is prohibitive for many small-scale farmers, especially in developing countries. Even where affordable tests are available, the infrastructure to understand and act on the results might be lacking.

This study proposes a rule-based fertilizer recommendation system to guide farmers on the most beneficial types and amounts of fertilizers for specific crops. The system's foundation is rooted in the well-established scientific principles of soil chemistry and plant biology. Recognizing that different crops have varied nutrient requirements and that these needs are influenced by soil pH, the system seeks to bridge the knowledge gap by providing specific fertilizer recommendations based on these factors. The fertilizer recommendation system considers various parameters such as soil quality determined based on pH level, crop type, and specific nutrient requirements of each crop. By analyzing these factors, the system provides recommendations for the optimal amounts of N, P, and K, the primary nutrients needed by crops. It is to be noted that soil pH is a crucial parameter because it affects the solubility of nutrients, which has a high impact on plant growth. A pH of 7 is considered neutral, while anything below 7 is acidic and anything above is alkaline. The soil's pH can influence the crop's health, yield, and disease resistance. Different crops prefer different pH ranges. The study first builds a lookup table for determining soil qualities using their pH measure, as shown in Table 2.

Table 2. Soil quality based on different pH value [27,29].

pH Value	Soil Quality
<4.5	Strongly acidic
4.5–5.5	Highly acidic
5.6–6.5	Moderately acidic
6.6–7.0	Slightly acidic
7.0	Neutral
7.1–8.0	Slightly alkaline
8.1–9.0	Moderately alkaline
9.1–10.0	Strongly alkaline
>10.0	Very strongly alkaline

The proposed system considers the values of pH and associated quality indicator as the primary input source to the rule-based system, in which a set of logical conditions are established. The study further focuses on the building reference database consisting of recommended ranges of fertilizer (N, P, K) and suitable pH for different crops. A sample visualization of recommended fertilizers and pH for the crops under consideration is shown in Table 3.

Table 3 provides the data needed to build fertilizer recommendations. The first step is to conduct a comprehensive soil analysis. Different crop types have different nutrient needs, so it is essential to understand each crop's specific N-P-K requirements at different growth stages. Using expert knowledge and proven scientific data, the system establishes rules to ensure reliable recommendations. The system then cross-references the user's input with fertilizer data to determine if the soil pH falls within the acceptable range for the chosen crop. If the soil pH is within the acceptable range, the system recommends the optimal amounts of N, P, and K based on the nutrient needs specified in the fertilizer table. If the soil pH is not within the acceptable range, the system advises the user to adjust the soil pH and suggests potential soil amendments. The system also offers alternative crops that could thrive in the existing soil pH range, along with the N, P, and K fertilizer recommendations for those crops. This approach allows the system to make precise and comprehensive fertilizer recommendations without requiring the explicit training of a predictive model on a dataset. The rationale behind this method is the universal nature of the recommended nutrient values for each crop type, thus providing a straightforward measure of the required fertilizer based on the difference between the current and recommended N, P, and K values. The utility of this rule-based approach lies in its ability to provide transparency and personalized fertilizer recommendations.

Table 3. Ideal nutrient levels and pH ranges for effective crop cultivation.

Crop	Nitrogen (N)	Phosphorus (P)	Potassium (K)	Suitable pH Range
Maize	120–150	60–80	40–60	5.5–7.0
Sorghum	100–120	50–70	40–60	5.5–7.5
Cassava	60–90	30–40	50–70	5.0–6.5
Beans	60–90	20–30	30–40	5.5–7.0
Potato	80–100	40–60	60–80	5.0–6.5
Coffee	60–80	20–30	40–60	5.5–6.5
Banana	100–120	60–80	60–80	5.0–7.0
Kidney beans	60–90	20–30	30–40	5.5–7.0
Onion	40–60	20–30	30–40	6.0–7.0

The proposed system can also reduce farmers' costs by precluding the overuse or underuse of fertilizers. An additional environmental advantage of this system is its potential to minimize the detrimental impact of excessive fertilizer usage, such as water pollution caused by fertilizer runoff. Finally, the system equips farmers with the information needed to make informed decisions about crop rotation based on soil's pH suitability for various crops.

The system starts by receiving user inputs on the crop type and the soil's pH value. It then cross-references this with a precompiled fertilizer table containing the ideal pH ranges and nutrient (N, P, K) requirements for various crops. Depending on the pH level input, the system first classifies the soil into categories ranging from "Strongly acidic" to "Very strongly alkaline". It then checks if the input pH falls within the appropriate range for the chosen crop. If the pH level is suitable, the system provides fertilizer recommendations specific to the crop, considering the nutrient needs from the fertilizer table. The recommendations regarding the required amounts of N, P, and K are given. If the soil pH is not right for the chosen crop, the system will tell you how to adjust it to the correct range. It will also suggest ways to improve the pH using soil amendments. In addition, the system will list alternative crops that could grow well in the current soil pH range. It will also provide N, P, and K fertilizer recommendations for each of these crops.

4. Result and Discussion

The development of the proposed recommendation models is carried out using python programming language in Anaconda distribution installed on windows 10 machine.This section presents the performance analysis of the proposed predictive model for crop recommendation. The accuracy of the predictions was assessed using various metrics, such as accuracy, precision, recall, F1-score, and the ROC curve. These metrics provide a comprehensive view of the model's performance, ensuring that the recommendations it provides are both precise and reliable. The performance metric accuracy is the proportion of the total number of predictions that were correct. It is given as follows:

$$Accuracy = \frac{TP + TN}{TP + TN + FP + FN} \quad (2)$$

where *TP* denotes True Positives, *TN* is the True Negative, *FP* is False Positives, and *FN* refers to False Negatives.

Precision is the proportion of positive identifications that were actually correct. It is computed as follows:

$$Precision = \frac{TP}{TP + FP}. \quad (3)$$

Recall (or Sensitivity) is the ability of a model to find all the relevant cases within a dataset. The recall is given by:

$$Recall = \frac{TP}{TP + FN}. \quad (4)$$

F1 Score is the harmonic mean of precision and recall, aiming to find a balance between both. The formula for computing *F1 Score* is given as follows:

$$F1\ Score = 2 \times \frac{Precision \times Recall}{Precision + Recall} \quad (5)$$

4.1. Training Performance Analysis

The training phase of a machine learning model is crucial because it determines how well the model learns the patterns in the training data. The training accuracy is a measure of how well the model can predict the labels of the training data. If a model has poor training accuracy, it indicates that there may be problems with the dataset, the model architecture, or the hyperparameters. Training accuracy is also an important metric for ensuring that the model generalizes well, meaning that it can make accurate predictions on new data that it has not seen before. This is because a model with high training accuracy following high validation accuracy has learned the underlying patterns in the data, not just the specific examples in the training set. Figure 8 shows the training performance of a neural network over 200 epochs.

From Figure 8, the training accuracy reaches up to 99%, indicating that the model has learned the underlying patterns in the training data very well. However, it is worth noting that a training accuracy of 100% is not always desirable, as it can lead to overfitting, where the model becomes too specific to the training data and performs poorly on unseen data. Therefore, the study considers validation of the model during the training processes. It can be seen that the validation accuracy reaches up to 95%, which is slightly lower than the training accuracy. This can be due to the fact that the model may have overfit to the training data or that the validation set may have some variability that the model could not capture. However, a validation accuracy of 95% is still a good level of accuracy, indicating that the model can generalize well to new, unseen data. Overall, the high training and validation accuracy suggest that the model is a good fit for the data and has the potential to make accurate predictions.

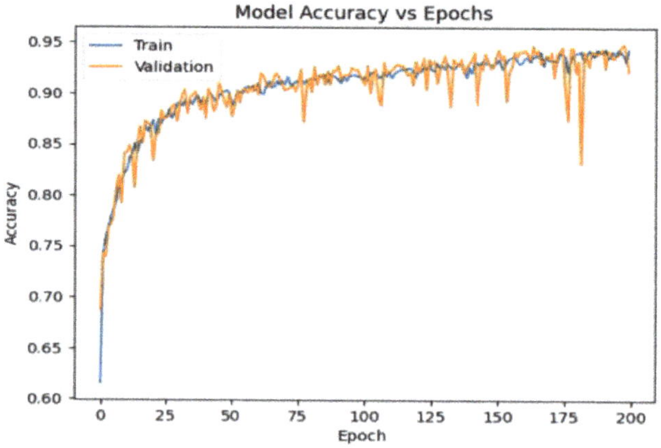

Figure 8. Training and validation curve.

4.2. Model Performance on Test Data

This section evaluates the performance of the model on the test dataset for crop recommendations. The metrics considered for this analysis are precision, recall, and F1-score across different crop classes and overall accuracy, as shown in Table 4. The support column indicates the sample size for each class, offering context to the reported statistical outcomes.

Table 4. Performance analysis of the proposed model.

Class	Label	Precision	Recall	F1-Score	Support
0	Maize	1.00	0.80	0.89	23
1	Sorghum	1.00	0.98	0.99	300
2	Cassava	0.90	0.95	0.93	435
3	Beans	1.00	1.00	1.00	23
4	Potato	1.00	1.00	1.00	11
5	Coffee	0.99	0.99	0.99	566
6	Banana	1.00	1.00	1.00	3
7	Kidney beans	0.84	0.74	0.78	208
	Overall Accuracy	0.97			

As shown in Table 4, for class 0, the model correctly predicted this class with 100% precision, meaning that every time the model predicted class 0, it was correct. The re-call of 80% indicates that the model was able to identify 80% of the actual instances of class 0 in the data. The F1-score, which is the harmonic mean of precision and recall, is 0.89. There were 25 instances of class 0 in the test set. Additionally, the model performed very well on Class 1, with a precision and recall of 1.00 and 0.98, respectively. This means that the model correctly identified 98% of the Class 1 instances, and none of the predictions were false positives. The F1-score of 0.99 for Class 1 indicates a strong balance between precision and recall. The model's performance on Class 2 was slightly lower, with a precision of 0.90 and recall of 0.95. This means that the model correctly identified 95% of the Class 2 instances, but there were some false positives. The F1-score of 0.93 for Class 2 still indicates a good balance between precision and recall.

The model performed perfectly on Classes 3, 4, and 6, achieving a precision, recall, and F1-score of 1.00 for all three classes. However, it is important to note that these classes were very underrepresented in the dataset, with only 20 instances for Class 3, 4 instances for Class 4, and 1 instance for Class 6. This means that the model's perfect performance on these classes may be due to their simpler nature or distinctive features, which made them easier to distinguish. Class 5 was also well-represented in the dataset, with 557 instances.

The model performed very well on this class, with a precision and recall of 0.99. This means that the model correctly identified 99% of Class 5 instances and made very few false positives. Class 7 was the only class where the model's performance was not as good. The model had a precision of 0.84 and recall of 0.74, meaning that it correctly identified 74% of Class 7 instances and made 16% false positives. The F1-score of 0.78 indicates that there was a moderate balance between precision and recall for this class. Class 8 was the largest class in the dataset, with 543 instances. The model performed very well on this class, with a precision, recall, and F1-score of 0.99. This means that the model correctly identified 99% of Class 8 instances and made very few false positives.

The Receiver Operating Characteristic (ROC) score is a widely used metric to evaluate the performance of classification models.Based on the outcome statistics shown in Figure 9, the Receiver Operating Characteristic (ROC) score, which measures a trade-off be-tween the True Positive Rate (sensitivity) and False Positive Rate (specificity). The ROC score ranges from 0 to 1, where a score of 1 denotes a perfect classifier, and a score of 0.5 indicates a model that is no better than random chance.

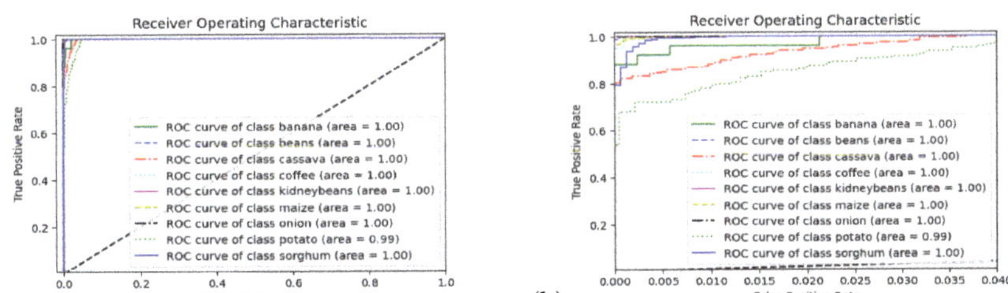

(a) (b)

Figure 9. Analysis of ROC curve; (**a**) ROC plot; (**b**) Magnified view of overlapping ROC regions.

The study also considers a comparison of the performance of the proposed neural network model with other different supervised classifiers as crucial in determining its suitability for specific tasks. A comparative analysis was conducted in Figure 10 to assess the proficiency of the proposed neural network and three other different models, namely, SVM (Support Vector Machines), Decision Tree, and XGBoost. The evaluation considered three vital metrics—weight precision, recall, and F1-Score—to gauge the performance nuances of each model. Based on the outcome analysis, the proposed neural network model outperformed all other models, including SVM, Decision Tree, and XGBoost, in a comparative evaluation. The neural network achieved a precision of 99.18%, while XGBoost came in second with a precision of 97.36%. The Decision Tree performed similarly to XGBoost, with a precision of 97.36%. SVM performed the worst, with a precision of 93.93%. This analysis shows that the neural network model is exceptionally capable of accurately identifying correct classifications. The proposed neural network model has a high recall of 98.66%, which means that it can accurately identify a large portion of the actual positive examples in the dataset. XGBoost is also effective, but its recall is slightly lower at 97.03%.

The F1-score, which measures the balance between precision and recall, is even higher for the neural network model at 98.98%. XGBoost is a close second with 97.17%, followed by the Decision Tree with 96.87%. Based on the comprehensive evaluation, the proposed neural network model distinctly outperforms the other models. Its consistently high scores across all metrics not only underline its accuracy in predictions but also its capacity to maintain a balance between precision and recall.

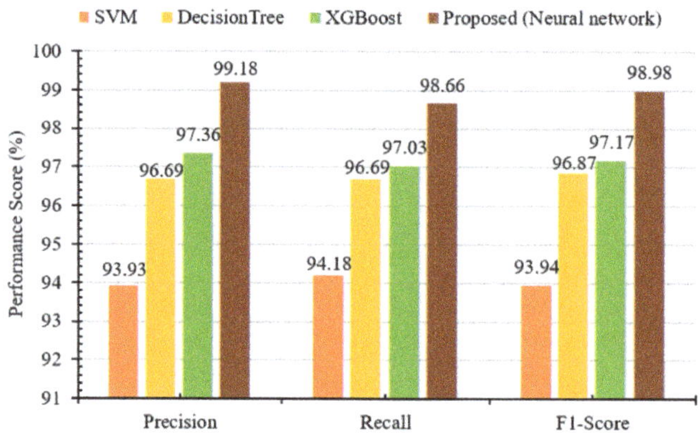

Figure 10. Comparative analysis with different predictive models.

4.3. Real-World Testing

This section discusses how we deployed and tested our crop recommendation system in real time and the results we obtained when using it with live data. The most important part of evaluating our system is the data it uses. For this study, we collected data from a set of IoT sensors (Five-pin soil transmitter (Type485) sensors, Shandong Renke Control Technology Co. Ltd., Jinan City, Shandong Province, China), that we strategically deployed across agricultural fields, as shown in Figure 11. The solar panels capture sunlight, convert it to electricity, and then tranfer that electricity to a battery. The battery powers a printed circuit board (PCB), which provides electrical connection and mechanical support to the electrical components of a circuit (sensor, GSM, microcontroller). The transmitter steel needle (sensor) was inserted horizontally into the soil in a pit that was dug with a diameter of >20 cm vertically to detect soil properties, which are transmitted to the cloud using GSM sim 900,and a microcontroller was integrated into a system to manage the device function.

(a) (b) (c)

Figure 11. Visual depiction of IoT sensor setup; (**a**) sensors, battery and solar; (**b**) sensor assembled; (**c**) sensor deployed in cropland.

Figure 11a shows a visual representation of the IoT soil sensor, battery, and solar as a power source. Figure 11b shows sensors were meticulously positioned within croplands, silently working to gather valuable agronomic data. Figure 11c showcases an IoT sensor, equipped with solar panels, ready for data collection. The IoT sensors measure N, P, and K in parts per million (ppm), which is a common unit used in soil testing. In this case, ppm represents absolute concentrations, not percentages. Therefore, the study converted the raw data from the IoT sensors from ppm to kg/ha before feeding it to the trained model for the proposed crop recommendation system. The conversion between the two depends on the depth of soil sampled and its bulk density. To convert from ppm to kg/ha, the study followed the standard convention that 1 ppm is equivalent to 1 mg/kg. So, if we had a reading of 50 ppm N, that means there are 50 mg of nitrogen per kg of soil. Hence, in a hectare, we have 50 mg/kg $\times$ 2,600,000 kg = 130,000,000 mg of nitrogen, which is

130 kg of nitrogen. The study converted all of the data collected by the IoT sensor node in real time from ppm to kg/ha in this manner. The transformed data were then fed to the trained model for the proposed crop recommendation system. The results of the study were promising, justifying the scope of the proposed work. This suggests that the proposed crop recommendation system can be used to accurately recommend crops for farmers based on the real-time data collected from IoT sensors.

4.4. Scope and Limitations

The proposed crop recommendation and rule-based fertilizer system holds immense promise, with a wide scope and diverse applicability. By soil data attributes and machine learning algorithms, it considers N, P, K, and soil quality pH, which are crucial for crop growth based on soil conditions. A balanced pH ensures optimal nutrient uptake, promoting healthy crop growth. By ensuring that these core parameters are in their ideal ranges, a significant portion of crop health following fertilizer recommendation yield optimization is addressed. By factoring in these vital soil attributes, the system ensures a tailored approach to crop cultivation, rooted in the specific conditions and needs of the soil. By focusing on these four key parameters, the study reduces the complexities often associated with integrating multiple environmental factors. While environmental factors such as rainfall and temperature can vary significantly across regions and seasons, the importance of N, P, K, and pH remains consistent for crop growth globally. This gives the proposed system a universal applicability, making it relevant across diverse geographical areas. However, despite the advantages of proposed system, it has limitations too when considering the implementation at a large scale in the agriculture sector. The potential challenges and limitations are highlighted as follows.

Data Generalization: The system is based on specific soil properties, such as N, P, K, and pH. However, when scaled up, these properties can vary widely across different regions, which could limit the accuracy of the recommendations. Additionally, it is difficult to create a comprehensive dataset that includes all possible soil, crop, environmental, and geographic conditions. Inaccuracies can also arise from inconsistent or incomplete data.

Exclusion of Environmental Factors: The model does not consider factors such as rainfall, humidity, and temperature. While N, P, K, and pH are important, the absence of these environmental variables may not provide a complete view of what is needed for all regions.

Infrastructure Challenges: Large-scale implementation may require extensive infrastructure, including IoT sensors, data transmission systems, and more.

Maintenance and Updation: The model will need to be updated regularly as soil conditions, crop varieties, and farming practices change. This will be challenging on a large scale.

Economic Implications: While the system might reduce some costs, the initial setup, training, and maintenance can be economically taxing for small-scale farmers or in regions with limited funding.

Rule-Based Fertilization Challenges: For rule-based fertilization recommendation models, the fixed set of rules might not adapt quickly to changing conditions. It can also become complex to update the rules for too many crops. If there are too many rules, it can potentially lead to contradictions.

4.5. Use Case Scenario and Conceptual Architecture of IoT Farm

Figure 12 shows sensors that are deployed in cropland to monitor and collect detailed data on soil nutrient levels (nitrogen, phosphorus, and potassium), as well as other important parameters. The IoT gateway connects the sensors to the cloud and securely and quickly transfers the data to a cloud database. This robust and scalable storage solution archives the collected data. Being cloud-based, it ensures that the data are accessible, redundant, and secure. The proposed system is deployed over the cloud and leverages the power of machine learning to analyze the real-time data using predictive algorithms to generate two key outputs: crop recommendations and fertilizer recommendations. The user can access both the real-time monitoring data and the generated recommendations through a tailored interface, enabling informed decision-making. Therefore, the proposed system's conceptual architecture showcases an end-to-end solution, from data collection in croplands to actionable insights delivered to the user. This synergy ensures that farmers are equipped with the best tools and information to drive productivity and sustainability in their work.

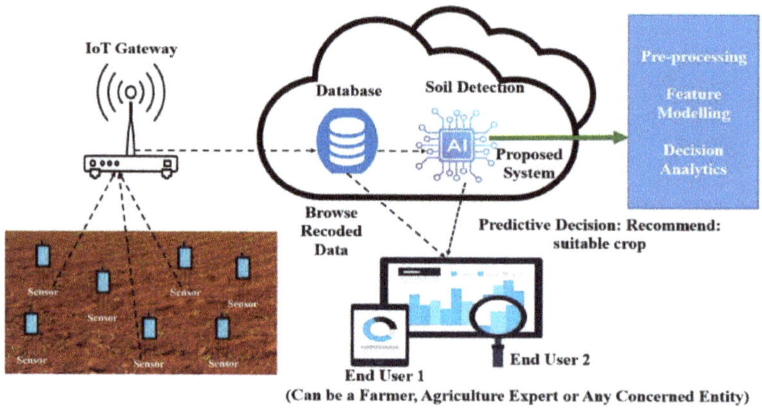

Figure 12. A schematic illustration of an IoT-assisted agriculture farm.

5. Conclusions

This study has introduced a novel crop and fertilizer recommendation system (CFRS) that is personalized specifically for Rwanda's agricultural landscape. The system uses machine learning and data analysis to give farmers insights that can help them make informed decisions about crop selection and fertilizer use. In rigorous comparative analyses, the neural network outperformed other popular machine learning models, demonstrating its precision, balance, and proficiency. The system has the potential to improve crop yield and quality, while also promoting cost-effective agricultural practices and reducing environmental impact. However, the system is not without its limitations. It needs to consider more environmental and geographical factors, and the data modeling and feature extraction process needs to be refined and expanded. Future work will focus on incorporating additional environmental and geographical factors, such as rainfall, temperature, humidity, and altitude, into the system. The study will also adopt more sophisticated deep learning algorithms and collaborate with environmental scientists, agronomists, and technologists to develop a more comprehensive, adaptive, and impactful system.

Author Contributions: Methodology, C.M. and A.V.; Formal analysis, C.M. and A.V.; Investigation, C.M. and I.K.; Writing—original draft, C.M.; Writing—review & editing, A.V., D.H. and A.U.; Visualization, C.M. and I.K.; Supervision, A.V., D.H. and A.U. All authors have read and agreed to the published version of the manuscript.

Funding: This research was funded by the African Center of Excellence in Internet of Things (ACEIoT), College of Science and Technology, University of Rwanda.

Institutional Review Board Statement: Not applicable.

Data Availability Statement: Data presented in tables are available upon request to the corresponding author.

Acknowledgments: This work was supported by Seed Technology Engineering and Science Group Ltd.

Conflicts of Interest: The authors declare no conflict of interest.

References

1. Pawlak, K.; Kołodziejczak, M. The Role of Agriculture in Ensuring Food Security in Developing Countries: Considerations in the Context of the Problem of Sustainable Food Production. *Sustainability* **2020**, *12*, 5488. [CrossRef]
2. Norton, G.W.; Alwang, J. Changes in Agricultural Extension and Implications for Farmer Adoption of New Practices. *Appl. Econ. Perspect. Policy* **2020**, *42*, 8–20. [CrossRef]
3. Khan, N.; Ray, R.L.; Sargani, G.R.; Ihtisham, M.; Khayyam, M.; Ismail, S. Current Progress and Future Prospects of Agriculture Technology: Gateway to Sustainable Agriculture. *Sustainability* **2021**, *13*, 4883. [CrossRef]
4. Chen, Y.; Kuang, J.; Cheng, D.; Zheng, J.; Gao, M.; Zhou, A. AgriKG: An agricultural knowledge graph and its applications. In Proceedings of the Database Systems for Advanced Applications: DASFAA 2019 International Workshops: BDMS, BDQM, and GDMA, Chiang Mai, Thailand, 22–25 April 2019; Proceedings 24; pp. 533–537.
5. Sanjeevi, P.; Prasanna, S.; Kumar, B.S.; Gunasekaran, G.; Alagiri, I.; Anand, R.V. Precision agriculture and farming using Internet of Things based on wireless sensor network. *Trans. Emerg. Telecommun. Technol.* **2020**, *31*, e3978. [CrossRef]
6. Tantalaki, N.; Souravlas, S.; Roumeliotis, M. Data-driven decision making in precision agriculture: The rise of big data in agricultural systems. *J. Agric. Food Inf.* **2019**, *20*, 344–380. [CrossRef]
7. Sharma, A.; Jain, A.; Gupta, P.; Chowdary, V. Machine learning applications for precision agriculture: A com-prehensive review. *IEEE Access* **2020**, *9*, 4843–4873. [CrossRef]
8. Bucci, G.; Bentivoglio, D.; Finco, A. Precision agriculture as a driver for sustainable farming systems: State of art in literature and research. *Calitatea* **2018**, *19*, 114–121.
9. Cisternas, I.; Velásquez, I.; Caro, A.; Rodríguez, A. Systematic literature review of implementations of precision agriculture. *Comput. Electron. Agric.* **2020**, *176*, 105626. [CrossRef]
10. Akhter, R.; Sofi, S.A. Precision agriculture using IoT data analytics and machine learning. *J. King Saud Univ.-Comput. Inf. Sci.* **2021**, *34*, 5602–5618. [CrossRef]
11. Hossain, M.A.; Siddique, M.N.A. Online Fertilizer Recommendation System (OFRS): A Step Towards Precision Agriculture And Optimized Fertilizer Usage By Smallholder Farmers In Bangladesh: Online fertilizer recommendation. *Eur. J. Environ. Earth Sci.* **2020**, *1*. [CrossRef]
12. Bhat, S.A.; Huang, N.-F. Big Data and AI Revolution in Precision Agriculture: Survey and Challenges. *IEEE Access* **2021**, *9*, 110209–110222. [CrossRef]
13. Singh, R.K.; Berkvens, R.; Weyn, M. AgriFusion: An Architecture for IoT and Emerging Technologies Based on a Precision Agriculture Survey. *IEEE Access* **2021**, *9*, 136253–136283. [CrossRef]
14. Shaikh, T.A.; Rasool, T.; Lone, F.R. Towards leveraging the role of machine learning and artificial intelligence in precision agriculture and smart farming. *Comput. Electron. Agric.* **2022**, *198*, 107119. [CrossRef]
15. Ruan, J.; Jiang, H.; Zhu, C.; Hu, X.; Shi, Y.; Liu, T.; Rao, W.; Chan, F.T.S. Agriculture IoT: Emerging trends, cooperation networks, and outlook. *IEEE Wirel. Commun.* **2019**, *26*, 56–63. [CrossRef]
16. MINAGRI. *Crop Intensification Program*; Ministry of Agriculture and Livestock: Kigali, Rwanda, 2007.
17. Giller, K.E.; Corbeels, M.; Nyamangara, J.; Triomphe, B.; Affholder, F.; Scopel, E.; Tittonell, P. A research agenda to explore the role of conservation agriculture in African smallholder farming systems. *Field Crop. Res.* **2011**, *124*, 468–472. [CrossRef]
18. Available online: https://www.nri.org/images/documents/development-programmes/gender_soc_dif/publications/farming_for_impact_-_full_report.pdf (accessed on 18 September 2023).
19. Rugimbana, C. Predicting Maize (*Zea mays*) Yields in Eastern Province of Rwanda Using Aquacrop Model. Doctoral Dissertation, University of Nairobi, Nairobi, Kenya, 2019.
20. Ngaruye, I.; von Rosen, D.; Singull, M. Crop yield estimation at district level for agricultural seasons 2014 in Rwanda. *Afr. J. Appl. Stat.* **2016**, *3*, 69–90. [CrossRef]
21. Breure, M.S.; Kempen, B.; Hoffland, E. Spatial predictions of maize yields using QUEFTS—A comparison of methods. *Geoderma* **2022**, *425*, 116018. [CrossRef]
22. Rivera, J.I.; Bonilla, C.A. Predicting soil aggregate stability using readily available soil properties and machine learning techniques. *Catena* **2020**, *187*, 104408. [CrossRef]
23. Suchithra, M.S.; Pai, M.L. Improving the prediction accuracy of soil nutrient classification by optimizing extreme learning machine parameters. *Inf. Process. Agric.* **2020**, *7*, 72–82. [CrossRef]
24. Chambers, O. Machine Learning Strategy for Soil Nutrients Prediction Using Spectroscopic Method. *Sensors* **2021**, *21*, 4208. [CrossRef]
25. Wu, C.; Chen, Y.; Hong, X.; Liu, Z.; Peng, C. Evaluating soil nutrients of Dacrydium pectinatum in China using machine learning techniques. *For. Ecosyst.* **2020**, *7*, 30. [CrossRef]
26. Rose, S.; Nickolas, S.; Sangeetha, S. Machine Learning and Statistical Approaches used in Estimating Parameters that Affect the Soil Fertility Status: A Survey. In Proceedings of the 2018 Second International Conference on Green Computing and Internet of Things (ICGCIoT), Karnataka, India, 16–18 August 2018; IEEE: New York, NY, USA, 2018; pp. 381–385.
27. Rajamanickam, J. Predictive model construction for prediction of soil fertility using decision tree machine learning algorithm. *Infocomp J. Comput. Sci.* **2021**, *20*, 49–55.
28. Rajamanickam, J.; Mani, S.D. Kullback chi square and Gustafson Kessel probabilistic neural network based soil fertility prediction. *Concurr. Comput. Pract. Exp.* **2021**, *33*, e6460. [CrossRef]

29. Katarya, R.; Raturi, A.; Mehndiratta, A.; Thapper, A. Impact of Machine Learning Techniques in Precision Agriculture. In Proceedings of the 2020, 3rd International Conference on Emerging Technologies in Computer Engineering: Machine Learning and Internet of Things, ICETCE, Jaipur, India, 7–8 February 2020; pp. 1–6.
30. Klerkx, L.; Jakku, E.; Labarthe, P. A review of social science on digital agriculture, smart farming and agriculture 4.0. New contributions and a future research agenda. *Njas-Wagening. J. Life Sci.* **2019**, *90*, 100315. [CrossRef]
31. Shadrin, D.; Menshchikov, A.; Somov, A.; Bornemann, G.; Hauslage, J.; Fedorov, M. Enabling Precision Agriculture through Embedded Sensing with Artificial Intelligence. *IEEE Trans. Instrum. Meas.* **2019**, *69*, 4103–4113. [CrossRef]
32. Kumar, S.A.; Ilango, P. The Impact of Wireless Sensor Network in the Field of Precision Agriculture: A Review. *Wirel. Pers. Commun.* **2018**, *98*, 685–698. [CrossRef]
33. Talaviya, T.; Shah, D.; Patel, N.; Yagnik, H.; Shah, M. Implementation of artificial intelligence in agriculture for optimization of irrigation and application of pesticides and herbicides. *Artif. Intell. Agric.* **2020**, *4*, 58–73.
34. Kamilaris, A.; Gao, F.; Prenafeta-Boldu, F.X.; Ali, M.I. Agri-IoT: A semantic framework for Internet of Things-enabled smart farming applications. In Proceedings of the 2016 IEEE 3rd World Forum on Internet of Things (WF-IoT), Reston, VA, USA, 12–14 December 2016.
35. Rekha, P.; Rangan, V.P.; Ramesh, M.V.; Nibi, K.V. High yield groundnut agronomy: An IoT based precision farming framework. In Proceedings of the 2017 IEEE Global Humanitarian Technology Conference (GHTC), San Jose, CA, USA, 19–22 October 2017; pp. 1–5.
36. Rehman, A.; Liu, J.; Li, K.; Mateen, A.; Yasin, Q. Machine Learning Prediction Analysis using IoT for Smart Farming. *Int. J. Emerg. Trends Eng. Res.* **2020**, *8*, 1–30.
37. Priya, P.K.; Yuvaraj, N. An IoT Based Gradient Descent Approach for Precision Crop Suggestion using MLP. *J. Phys. Conf. Ser.* **2019**, *1362*, 012038. [CrossRef]
38. Biradar, H.B.; Shabadi, L. Review on IOT based multidisciplinary models for smart farming. In Proceedings of the 2017 2nd IEEE International Conference on Recent Trends in Electronics, Information & Communication Technology (RTEICT), Bangalore, India, 19–20 May 2017; pp. 1923–1926.
39. Zia, H.; Rehman, A.; Harris, N.R.; Fatima, S.; Khurram, M. An Experimental Comparison of IoT-Based and Traditional Irrigation Scheduling on a Flood-Irrigated Subtropical Lemon Farm. *Sensors* **2021**, *21*, 4175. [CrossRef]
40. Gupta, A.; Nagda, D.; Nikhare, P.; Sandbhor, A. Smart Crop Prediction using IoT and Machine Learning. *Int. J. Eng. Res. Technol.* **2021**, *9*, 18–21.
41. Vivekanandhan, V.R.; Sakthivel, S.; Manikandan, M. Adaptive neuro fuzzy inference system to enhance the classification performance in smart irrigation system. *Comput. Intell.* **2021**, *38*, 308–322. [CrossRef]
42. Sharma, S. Crop Recommendation Dataset [Data Set]. 2021. Available online: https://www.kaggle.com/datasets/siddharthss/crop-recommendation-dataset (accessed on 18 September 2023).
43. Jaiswal, R. Soil Fertility Dataset [Data Set]. 2023. Available online: https://www.kaggle.com/datasets/rahuljaiswalonkaggle/soil-fertility-dataset (accessed on 18 September 2023).
44. Uwiragiye, Y.; Ngaba, M.J.Y.; Zhao, M.; Elrys, A.S.; Heuvelink, G.B.; Zhou, J. Modelling and mapping soil nutrient depletion in humid highlands of East Africa using ensemble machine learning: A case study from Rwanda. *CATENA* **2022**, *217*, 106499. [CrossRef]
45. Perez-Guzman, K.; Imanirareba, D.; Jones, S.K.; Neubauer, R.; Niyitanga, F.; Naramabuye, F.X. Sustainability implications of Rwanda's Vision 2050 long-term development strategy. *Sustain. Sci.* **2023**, *18*, 485–499. [CrossRef]
46. Chapter III Research Resltlts: Rwandan Peasants and Their Living Standards. Available online: https://www.ide.go.jp/library/English/Publish/Reports/Jrp/pdf/127_6.pdf (accessed on 17 September 2023).

agriculture

MDPI

Article

Comprehensive Analysis of Model Errors in Blueberry Detection and Maturity Classification: Identifying Limitations and Proposing Future Improvements in Agricultural Monitoring

Cristhian A. Aguilera [1,*], Carola Figueroa-Flores [2], Cristhian Aguilera [3] and Cesar Navarrete [3]

[1] Facultad de Ingeniería, Arquitectura y Diseño, Universidad San Sebastián, Lago Panguipulli 1390, Puerto Montt 5501842, Chile
[2] Departamento de Ciencias de la Computación y Tecnologías de la Información, Facultad de Ciencias Empresariales, Universidad del Bío-Bío, Chillan 3800708, Chile; cfigueroa@ubiobio.cl
[3] Departamento de Ingeniería Eléctrica y Electrónica, Facultad de Ingeniería, Universidad del Bío-Bío, Concepción 4051381, Chile; cristhia@ubiobio.cl (C.A.); cesar.navarrete1401@alumnos.ubiobio.cl (C.N.)
* Correspondence: cristhian.aguilera@uss.cl

Abstract: In blueberry farming, accurately assessing maturity is critical to efficient harvesting. Deep Learning solutions, which are increasingly popular in this area, often undergo evaluation through metrics like mean average precision (mAP). However, these metrics may only partially capture the actual performance of the models, especially in settings with limited resources like those in agricultural drones or robots. To address this, our study evaluates Deep Learning models, such as YOLOv7, RT-DETR, and Mask-RCNN, for detecting and classifying blueberries. We perform these evaluations on both powerful computers and embedded systems. Using Type-Influence Detector Error (TIDE) analysis, we closely examine the accuracy of these models. Our research reveals that partial occlusions commonly cause errors, and optimizing these models for embedded devices can increase their speed without losing precision. This work improves the understanding of object detection models for blueberry detection and maturity estimation.

Keywords: blueberry detection; maturity estimation; edge computing; smart agriculture; computer vision; machine learning

1. Introduction

In modern agriculture, accurately determining the number and maturity of blueberries is essential for identifying the ideal harvest time. With noticeable variations in maturity levels among blueberry clusters [1], obtaining accurate and timely information is vital in enhancing productivity, reducing costs, and maximizing profits. This challenge has guided research efforts toward automating such assessments, offering a more data-driven and efficient strategy for determining the optimal harvesting period.

Recognizing the importance of this problem, agricultural sector researchers have achieved substantial advancements, particularly in Deep Learning applications, with convolutional neural networks (CNNs) leading these developments. Known for their exceptional ability to process complex visual data, CNNs excel in a variety of intricate tasks such as object recognition, image classification, and instance segmentation, all of which are highly valuable in numerous agricultural applications (e.g., [2–9]). These applications underscore the versatility and adaptability of CNNs in meeting the diverse challenges faced by the agricultural sector.

In blueberry detection and maturity estimation, considerable advancements have been made, leveraging a fusion of machine learning and computer vision techniques (e.g., [10,11]). Innovative approaches, including hyperspectral imaging, partial least squares regression, and Deep Learning models, have been employed to extract color and texture features for maturity classification [12,13]. Notable contributions include pipelines designed by Gonzalez et al. [14] and Ni et al. [1], which use CNN models for classifying blueberry traits, including maturity estimation.

More novel approaches such as the work conducted by Mu et al. [15] significantly enhanced the accuracy and efficiency of blueberry quality detection, leveraging Deep Learning for classification tasks. Obsie et al. [16] demonstrated the viability of various machine learning algorithms in developing predictive models for blueberry yield prediction. MacEachern et al. [17] successfully trained models to identify wild blueberry ripeness stages, achieving high mean average precision (mAP) values for two and three types of ripeness, alongside an impressive runtime inference compared to previous approaches.

While advancements in Deep Learning for agriculture are significant, a critical area of research remains in assessing their real-world viability, especially on embedded devices like those in agricultural robots and drones. These autonomous systems require careful consideration of processing power, size, weight, and connectivity, typical of edge computing environments. Furthermore, cost-effectiveness in processing is crucial for widespread adoption. Although edge computing devices are affordable and compact, they present challenges in balancing speed and precision. Effectively applying these methods for tasks such as blueberry detection and maturity estimation in practical settings is an evolving field.

Additionally, the practical application of these advancements warrants further exploration. Most existing research involves image acquisition in controlled environments, which only partially represents the complexities of real-world conditions. This discrepancy between laboratory and field settings and the computational challenges addressed in this study highlight the need for more research to bridge the gap between theoretical models and their practical implementation in agriculture.

Contributions of the Study

This study evaluates various state-of-the-art models for detecting and estimating blueberry maturity across multiple devices, assessing their real-world application potential. Utilizing a dataset specially curated for this purpose, the performance of these models is examined, employing the Type-Influence Detector Error (TIDE) method for a detailed analysis of prevalent issues. This approach identifies critical areas needing improvement and facilitates a discussion on future research directions, potentially leading to more refined and efficient methods in agricultural technology.

The main contributions of this work are as follows:

- A new and publicly available dataset of blueberries for object detection tasks covering various maturity stages has been created, captured, and labeled. While other datasets include object detection labels, this incorporates maturity classification.
- This study offers novel insights into the performance of current models, particularly in terms of runtime and error analysis. State-of-the-art blueberry detection and maturity estimation models have been thoroughly evaluated, with the errors identified and quantified using the TIDE method. These insights are crucial for guiding future research efforts.
- An essential contribution of our study is the exploration of model adaptability in edge computing environments, explicitly examining their computational demands and performance. We show that some object detection models can operate in real-time on edge devices while maintaining their ability to detect and classify blueberry maturity effectively. Although runtime information for these models on embedded devices is known, the impact of optimization techniques required for these devices on their capability for blueberry detection and maturity estimation was previously unexplored.
- The code associated with the evaluations is made available to promote research reproducibility and encourage further advancements in this field. The code can be accessed at https://github.com/ngunsu/bb2023 (accessed on 13 November 2023).

2. Materials and Methods

2.1. Image Dataset

In this study, a set of blueberry images was collected from a plantation situated in Quillón, a town in the Ñuble Region of Chile. This region is known for its mild microclimate, exhibiting an average annual temperature of 14.9 °C. Typical January temperatures fluctuate between 27 and 30 °C, while the annual precipitation ranges from 700 to 1000 mm, with most rainfall occurring between April and September. The images were captured on three separate occasions, from late October to early December 2021, during sunny days when the temperatures exceeded 20 °C.

2.1.1. Image Acquisition

The image acquisition process employed a Nikon Coolpix B700 and a Basler acA2440-20gc camera, both firmly mounted on a SOLIGOR WT-330A tripod to ensure stability during the capture process. Figure 1 displays the configuration of this imaging system. The Nikon Coolpix B700, equipped with a 60× optical Zoom-NIKKOR glass lens (4.3–258 mm), captures images with a resolution of 5184 × 3888 pixels. Meanwhile, the Basler acA2440-20gc, using a Fujinon HF9HA-1B Lens (9 mm 1.5MP 2/3″ f/1.4 C-Mount), captures images at 2448 × 2048 pixels.

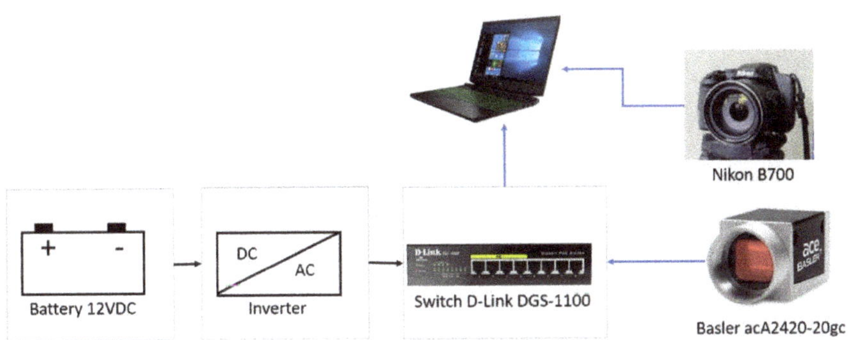

Figure 1. Camera acquisition setup showcasing two cameras: the Nikon Coolpix B700 and the Basler acA2440-20gc. The Nikon Coolpix B700 is directly connected to a notebook, while the Basler acA2440-20gc connects to the notebook via an ethernet switch. The notebook facilitates the capture process. Power to the ethernet switch, vital for the Basler camera's operation, is supplied by a 12-volt battery.

Using the previously described camera setup, approximately 500 images were captured from different locations within the plantation. The cameras' automatic illumination settings were used for each image capture, and any specialized adjustments were deliberately avoided. This approach was intentionally chosen to create challenging conditions where the variable lighting could affect the blueberries' coloration. This approach aims to generate edge-case scenarios, thereby providing a comprehensive evaluation of our models' performance under diverse and demanding environmental conditions.

2.1.2. Image Labeling

Image labeling is crucial in creating datasets for machine learning applications. In this context, labels serve as the ground truth that a machine learning model aims to learn. For our study, labeling involved meticulously outlining each blueberry with rectangular bounding boxes and categorizing them according to their maturity levels. The categories were defined as follows: berries with green or reddish tones were classified as *unripe*, those with light purple to darker red hues were *pint*, and berries showing blue or dark purple coloration were labeled as *ripe*.

Label Studio 1.5.0 was employed to label the images to facilitate this process. Three individuals participated in this detailed labeling process, carefully drawing rectangular bounding boxes around the blueberries and assigning the appropriate ripeness category. Throughout the labeling process, several images were discarded if a labeler could not reliably assess the ripeness of any blueberries in the image. This quality control step resulted in a refined dataset of 265 images, representing a wide range of ripeness stages. Figure 2 shows examples of these annotations, highlighting the diversity of ripeness stages included in the dataset.

(**a**) Original image (**b**) Labeled image

Figure 2. Representative image from the dataset. The image on the left presents the original capture, while the image on the right displays the same capture with manually added labels illustrating blueberry locations and their respective maturity categories.

2.1.3. Dataset Splits

Of the 265 images labeled, 85% were allocated for training and 15% for testing purposes. The training data were also subdivided into two sets: a primary training set and a validation set. This subdivision followed the same 15–85% ratio, with 85% of the images used for training and 15% for validation. Each split was conducted through random partitioning to ensure variety and unpredictability in the data distribution.

Crucially, the datasets were manually reviewed to guarantee that no clusters of blueberries were duplicated across the sets, thus maintaining distinct and unique image sets for training, validation, and testing. Table 1 displays the final distribution and number of images across these sets, illustrating the breakdown of the dataset for the different phases of the machine learning process.

Table 1. Distribution of images across training, validation, and testing subsets within the image dataset.

Data	Training	Validation	Test	Total
Images	190	33	42	265

Table 2 presents a comprehensive breakdown of the generated labels for each image, based on its usage in training and evaluation, as well as its maturity level (see Figure 3). It is important to note that the dataset exhibits a slight imbalance, with the *pint* class having fewer instances than other classes. This imbalance can be attributed to the date of the image captures.

Table 2. Classification and distribution of labels corresponding to blueberry maturity stages.

Class	Train	Validation	Test	Total
unripe	2825	539	680	4044
pint	431	66	170	667
ripe	3271	556	628	4455

Figure 3. Examples of blueberry labeling: (**a**) unripe, (**b**) pint, and (**c**) ripe.

2.2. Model Training and Evaluation

2.2.1. Model Training

Training and evaluation were conducted on three distinct object detection architectures for identifying and classifying the maturity of blueberries: YOLOv7 (including YOLOv7-tiny, YOLOv7-w6, and YOLOv7-default) [18], Mask-RCNN [19], and RT-DETR-L [20]. The primary goal of this study was to assess the impact of different models, with their varying number of parameters and runtime speeds, on the detection and classification accuracy within our dataset.

A crucial step in training these models was the hyperparameter tuning phase, where a range of adjustments was explored. Fine-tuning strategies were also implemented, using pre-trained models from the COCO dataset to enhance the models' performance. This involved rigorously evaluating various learning rates, a critical factor in how quickly a model learns during training and, thus, affects its overall performance. Furthermore, several data augmentation techniques were used to artificially increase the dataset's size. Given the small size of our training set, this approach was especially advantageous. Data augmentation, involving image transformations like rotation or vertical mirroring, allowed us to generate multiple samples from a single image, thereby improving the model's learning efficiency and generalization capability. The optimal hyperparameters for each model found through grid search are listed in Table 3.

The computational constraints of the GPU setup required minor image size modifications to ensure successful model training. All experiments were conducted on a computer system with a 12th Gen Core i7 CPU, 32GB of RAM, a 1TB SSD, and an NVIDIA RTX3080TI 10GB GPU.

Table 3. Hyperparameters utilized for training each object detector in our experimental analysis, with a uniform training duration of 350 epochs for all models. All models were pre-trained on the COCO dataset.

Model	Image Size	lr	Augmentation
YOLOv7-tiny	640 × 640	0.01	Translation, Scale, Rotation, Vertical Flip, Horizontal Flip, Copy Paste, Mosaic
YOLOv7-default	640 × 640	0.001	Translation, Scale, Rotation, Vertical Flip, Horizontal Flip, Copy Paste, Mosaic
YOLOv7-w6	1280 × 1280	0.010	Translation, Scale, Rotation, Vertical Flip, Horizontal Flip, Copy Paste, Mosaic
RT-DETR-L	640 × 640	0.001	Translation, Vertical Flip, Horizontal Flip, Mosaic
Mask-RCNN	[800, 1333] × [800, 1333]	0.010	Vertical Flip, Horizontal Flip

2.2.2. Model Evaluation

The average mean precision is a standard metric for evaluating object detection models, assessing both the accuracy of the detected objects and the model's confidence in these detections. Among its variations, mAP50 is widely used, where detection is considered accurate if the intersection-over-union (IoU) between the predicted bounding box and the ground truth is at least 50%. The calculation of mAP50 involves sorting all detections by their confidence scores, determining each detection as a true positive or a false positive based on the IoU threshold, and then calculating precision and recall at each threshold level. The final mAP50 score is an average of these precision values, taken at the points where recall changes, across all classes in the dataset.

Similarly, mAP75 follows the same calculation process but with a stricter IoU threshold of 75%, providing a more rigorous evaluation of the model's accuracy. This metric is especially relevant in our study, where precision in the localization of objects is crucial.

Additionally, the precision, recall, and F1 score of the models are evaluated. Precision measures the proportion of correct identifications made by the model, while recall measures the proportion of actual positives that were correctly identified. The F1 score is a measure that combines precision and recall, providing a balance between them. These metrics can be mathematically represented as follows:

$$\text{Precision} = \frac{\text{True Positives}}{\text{True Positives} + \text{False Positives}} \tag{1}$$

$$\text{Recall} = \frac{\text{True Positives}}{\text{True Positives} + \text{False Negatives}} \tag{2}$$

$$\text{F1 Score} = 2 \times \frac{\text{Precision} \times \text{Recall}}{\text{Precision} + \text{Recall}}. \tag{3}$$

These metrics offer a comprehensive evaluation of the model's performance in object detection tasks.

Regarding runtime, the same procedure is followed for each model. First, a warmup phase of 100 runs is started, which initializes the GPU. This step ensures the system is fully operational before beginning the measurements. After the warmup, an additional 100 runs are conducted, and the inference times are meticulously recorded. The runtime is then determined by calculating the average time across these runs.

2.3. Edge Computing

Edge computing emerged as a solution for situations when the acquired data must be processed on the spot without the possibility of being transmitted to a remote server. Therefore, it is a suitable solution for technology's deployment in rural areas, which usually lack Internet connectivity. Furthermore, edge computing devices' compact size and energy efficiency make them ideal for integration into drones or mobile robots, rendering them fit for real-world applications. This attribute expands the technology's accessibility, balancing computational power, energy consumption, and cost-effectiveness. For all the above considerations, we adopted edge computing as a compelling approach for real-time blueberry detection and classification.

Selecting appropriate devices for edge computing is crucial, particularly when aligned with the specific requirements of an application. This study focuses on NVIDIA's Jetson line, particularly the Jetson AGX Xavier and Jetson AGX Orin models. These models were selected for their superior technical features and software compatibility. Essentially, both Jetson devices are compact computers equipped with integrated GPUs and capable of being powered by batteries. This setup enables the execution of Deep Learning models with relatively lower costs than traditional desktop setups. Specifically, we chose the Jetson AGX Xavier and Orin models for their proficiency in efficiently running advanced AI models, such as RT-DETR and YOLOv7, which are integral to our research. This efficiency marks a significant improvement over earlier models like the Jetson Nano, which is limited by its outdated software capabilities. The technical specifications of the Jetson AGX Xavier and Orin are detailed in Table 4, underscoring their suitability for our research.

Table 4. Technical specifications of the NVIDIA Jetson devices evaluated in this study.

Device	Specifications
Jetson AGX Xavier	CPU: 8-core NVIDIA Carmel ARM v8.2 64-bit CPU @ 2.26 GHz GPU: 512-core Volta GPU with Tensor Cores DLA: 2 × NVDLA engines RAM: 16 GB 256-bit LPDDR4x@137 GB/s Storage: 32 GB eMMC 5.1 onboard Power: 9 V 20 V DC
Jetson AGX Orin	CPU: 2 × 12-core NVIDIA Arm® Carmel CPU@2.75 GHz GPU: 2 × NVIDIA Ampere architecture Tensor Cores and 2 × NVIDIA Volta architecture Tensor Cores DLA: 2 × NVDLA engines Memory: 128 GB/s 256-bit LPDDR4x | 200 GB/s 2048-bit LPDDR5 Storage: 1 × 10GbE, 1 × 5GbE, 1 × 2.5GbE, 1 × 1GbE Power: 9 V 36 V DC

It is crucial to emphasize that although the Jetson devices—namely the AGX Xavier and AGX Orin—are powerful mini-computers, they are primarily designed for inference tasks rather than for the training phase of machine learning models. Consequently, in our research, these embedded systems will be utilized exclusively for evaluating the runtime performance of machine learning models that have already been trained. The training phase of these models will be conducted on more robust desktop GPUs.

2.4. Optimizing Model Runtime

Deploying Deep Learning models on embedded devices often necessitates a post-processing phase to optimize them for efficient runtime performance. Typically, these models are designed for desktop-grade GPUs, and their performance on embedded devices is comparatively lower due to the limitations of these devices' internal GPUs, such as reduced memory capacity and fewer GPU cores. Therefore, it is necessary to adapt the models for these devices to enhance their runtime speed once they are trained. Several techniques are employed for this purpose. For instance, knowledge distillation, as described in [21], involves training a compact model to emulate the behavior of a larger, more complex model, making the smaller version more suitable for embedded devices. Another prevalent technique is quantization, which accelerates network inference by utilizing lower precision computations, like 16-bit, 8-bit, or even 1-bit precision models [22].

However, this study adopts a more direct optimization approach using TensorRT 8.5.0.2 [23], a software tool developed by NVIDIA. TensorRT effectively reduces the model size and enhances runtime performance through quantization, converting 32-bit floating-point computations to 16-bit or 8-bit formats. This adaptation increases the runtime speed as GPUs process these calculations more quickly. TensorRT also implements layer and tensor fusion, combining operations to run faster and fully utilizing GPU capabilities. TensorRT takes a trained model as the input and produces a new, more efficient version for inference. This type of optimization is essential for applications that require high-speed processing.

While TensorRT offers substantial benefits, assessing its impact on the network's performance is crucial. The trade-off between runtime efficiency and model accuracy is a significant factor in this assessment. Consequently, our article focuses on an in-depth analysis of TensorRT's optimization effects, particularly in real-time edge detection and maturity estimation of blueberries on edge devices.

2.5. The Type-Influence Detector Error

The Type-Influence Detector Error (TIDE) [24] analysis is a tool to examine the types of errors made by an object detector and how these errors affect the mean average precision metric (mAP). It provides a detailed perspective on specific categories of errors and their contribution to the detector's overall performance. Essentially, TIDE quantifies the impact of each error type on the total mAP (denoted as dAP), offering an estimate of potential mAP improvement if a particular error was effectively addressed.

TIDE analysis gives information about the following:

1. Misclassification errors (Cls), which occur when the detected object is incorrectly classified;
2. Localization errors (Loc), which arise when the algorithm accurately classifies an object but inaccurately localizes it, underscoring the need for enhancements in object detection algorithms;
3. Combined misclassification and mislocalization errors (Both);
4. Duplication errors (Dup), which occur when an object is detected multiple times;
5. Background errors (Bkg), which occur when the algorithm wrongly identifies parts of the background as objects;
6. Missed errors (Miss), which represent overlooked ground truth tags by the algorithm;
7. False positive (FP) errors, depicting instances where the algorithm mistakenly identifies non-objects as objects;
8. False negative (FN) errors result when the algorithm fails to detect an existing object.

3. Results

3.1. Blueberry Detection and Maturity Estimation

Building on the findings in [17], the empirical evaluation has been broadened to encompass a range of more recent and diverse models. Table 5 presents the results of our trained models for blueberry detection and maturity estimation. This analysis adheres to the methodology described in [17], particularly employing a stringent mean average precision (mAP) criterion of IOU 75%. The findings highlight that the mAP scores for most models range between 0.3 and 0.5, suggesting a moderate accuracy level. Among the models, MASK-RCNN stands out for its superior accuracy, though it is also the slowest in runtime. Additionally, the analysis reveals a consistent precision level across all maturity levels, indicating that no single class disproportionately contributes to errors despite the imbalance in the dataset.

Table 5. This table offers a comparative assessment of various YOLOv7 configurations, RT-DETR-L, and Mask-RCNN in detecting blueberries at three stages of maturity—ripe, pint, and unripe. The performance metrics are evaluated on an NVIDIA RTX3080TI GPU, using mAP75.

Model	Class	Precision	Recall	F1	mAP75	Runtime (ms)
YOLOv7-tiny	Ripe	0.547	0.387	0.443	0.330	3.308
	Pint	0.568	0.433	0.489	0.364	
	Unripe	0.485	0.323	0.388	0.231	
	All	0.533	0.380	0.443	0.309	
YOLOv7-default	Ripe	0.626	0.456	0.528	0.435	8.059
	Pint	0.641	0.508	0.567	0.432	
	Unripe	0.605	0.415	0.492	0.348	
	All	0.624	0.460	0.530	0.405	
YOLOv7-w6	Ripe	0.598	0.500	0.544	0.445	19.551
	Pint	0.631	0.494	0.554	0.431	
	Unripe	0.591	0.457	0.516	0.381	
	All	0.607	0.484	0.539	0.419	
RT-DETR-L	Ripe	0.547	0.454	0.496	0.416	11.551
	Pint	0.575	0.348	0.434	0.314	
	Unripe	0.518	0.396	0.449	0.309	
	All	0.547	0.399	0.462	0.346	
Mask-RCNN	Ripe	0.612	0.490	0.543	0.447	34.301
	Pint	0.680	0.574	0.622	0.558	
	Unripe	0.582	0.488	0.530	0.426	
	All	0.625	0.518	0.565	0.477	

Figure 4 presents the detection results from four distinct models applied to the same image. This comparative analysis reveals that across this particular sample, the detection bounding boxes generated by each model are similar. Notably, the results from Mask-RCNN align more closely with the actual contours of the blueberries. Furthermore, the RT-DETR model uniquely identifies one blueberry that the other models overlooked. Of particular interest is that Mask-RCNN is the only detector that accurately identifies the pint berries in this sample, demonstrating its superior precision in distinguishing between different maturity stages of the blueberries.

3.2. The TIDE Analysis

Table 6 presents our TIDE analysis, which is based on the results from the previous subsection, explicitly targeting mAP75. This table highlights that the predominant error is the models' inability to detect all blueberry instances, leading to many false negatives. This issue underscores the need for the enhanced localization of blueberries, especially those partially obscured by plant foliage or too small for the network to detect accurately. Additionally, localization error (Loc) is the second primary source of inaccuracies. This can be attributed to the natural clustering of blueberries, where, often, a single detection may encompass parts of adjacent blueberries, leading to skewed bounding boxes. Furthermore, there is room for improvement in maturity classification. The analysis suggests that an average improvement of over five percent is achievable with more accurate classification. Also, after performing a qualitative analysis of the detection results on the test set, we discovered that most classification errors occur in two scenarios: first, when a blueberry transitions between stages, such as partially ripe and unripe, and second, when the bounding box inadvertently includes background elements like leaves, affecting the color analysis. Figure 5 illustrates some of these common errors made by the object detectors.

(**a**) YOLOv7-tiny

(**b**) YOLOv7-w6

(**c**) RT-DETR-L

(**d**) Mask-RCNN

Figure 4. Qualitative comparison of detection results using YOLOv7-tiny, YOLOv7-w6, RT-DETR-L, and Mask-RCNN: this image presents a side-by-side visualization of the detection results from each model in identifying the maturity stages of blueberries. The color coding for the maturity stages is as follows: blue indicates pint blueberries, red represents ripe blueberries, and pink denotes unripe blueberries.

(**a**) Cls error

(**b**) Loc error

(**c**) Miss error

Figure 5. Examples of typical errors encountered by object detectors. Figure (**a**) shows two pint blueberries incorrectly identified as unripe. In Figure (**b**), the bounding box has a low intersection-over-union ratio, encompassing only a portion of a blueberry. Figure (**c**) illustrates a missed detection where the object detector fails to recognize a blueberry hidden among the plant's leaves.

Table 6. Comparative analysis of types of object detection errors for three classes as identified by TIDE. Each value indicates the contribution of a specific error type to the overall mAP.

Type	Cls	Loc	Both	Dup	Bkg	Miss	FP	FN
YOLOv7-tiny	5.71	32.65	0.30	0	0.27	10.66	10.35	43.30
YOLOv7-default	5.84	29.46	0.31	0	0.35	10.18	10.64	38.33
YOLOv7-w6	6.32	26.78	0.26	0	0.26	11.39	9.45	39.15
RT-DETR-L	6.44	38.24	0.33	0	0.54	6.63	11.34	41.08
Mask-RCNN	2.64	27.88	0.14	0	0.77	10.22	9.31	35.69

3.3. *Edge Computing*

As previously mentioned, Deep Learning models optimized for real-time processing on high-end GPUs often struggle to perform under similar conditions on edge devices, necessitating post-processing optimization. In this study, TensorRT was applied to the models evaluated in earlier sections, assessing their performance using 16-bit precision. The results of this optimization are presented in Figure 6. The figure shows that most methods could not achieve real-time performance on the Jetson AGX Xavier, an affordable embedded vision system. However, on the more expensive and high-end Jetson AGX, not only YOLOv7-tiny but also YOLOv7-default could be run in real-time. Networks like Mask-RCNN, despite their accuracy, proved unsuitable for real-time tasks. In this context, YOLOv7-default emerges as a balanced choice, effectively bridging the gap between accuracy and runtime performance.

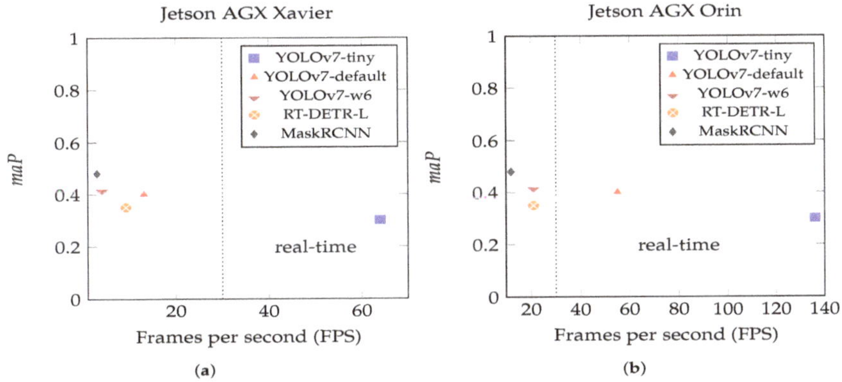

Figure 6. Comparative runtime results of the object detection models evaluated on Jetson AGX Xavier (**a**) and Jetson AGX Orin (**b**), following optimizations discussed in Section 2.3.

Regarding mAP, it is noteworthy that the optimization process did not significantly alter the mAP for any of the models. Only minor and relatively insignificant improvements were observed across most models, possibly due to the noise reduction during the optimization. The Mask-RCNN model exhibited the most noticeable change post-optimization, particularly when tailored for Jetson cards, though this change was also minimal. Table 7 details the variations in mAP for a selected model across different devices.

In conclusion, the results indicate that runtime optimization does not significantly alter the performance of the models in the task of detecting and estimating the maturity of blueberries, thus facilitating their deployment on embedded devices. However, it is noteworthy that only a select few models are capable of real-time operation, suggesting that further optimization may be necessary. This is particularly relevant as more modern, transformer-based models begin to gain prominence in this field.

Table 7. Comparison of model performance optimized with TensorRT on various devices: this table displays the mean average precision (mAP) at a threshold of 75 for different models, illustrating the impact of TensorRT optimization across multiple devices.

Model	RTX 3080TI 16bit	Jetson Orin 16bit	Jetson Xavier 16bit
YOLOv7-tiny	0.331	0.329	0.329
YOLOv7	0.433	0.434	0.435
YOLOv7-w6	0.436	0.419	0.421
RT-DETR-L	0.321	0.321	0.319
Mask-RCNN	0.462	0.462	0.460

3.4. Discussion

Detecting and estimating the maturity of blueberries remains a complex task, primarily due to background elements like leaves and natural occlusions inherent to the plant. Most of the research in this area, including our study, depends on static images from a single viewpoint, which may limit the accuracy potential. From our findings, we hypothesize that in field applications, capturing multiple images of the same blueberry cluster from various viewpoints could significantly enhance the detection process. Selecting images from multiple angles could reduce the number of occluded blueberries, and analyzing clusters from these different perspectives might provide critical supplementary information. This approach could improve detection algorithm performance by offering a more detailed view of each cluster. However, this hypothesis requires further exploration, as it involves challenges such as rapid processing speeds and sophisticated tracking capabilities, which have not been extensively investigated in blueberry research.

In this context, creating datasets that more closely mirror real-world conditions is essential for advancing research in this field. Shifting our focus from static images to video data is particularly important, as it aligns more directly with the practical needs of the industry. This change will allow future research to address the challenges in agricultural settings more effectively. The limitations highlighted in existing datasets, including the one used in our study, emphasize the urgency of this transition.

Finally, enhancing the performance of detection models on embedded devices presents distinct challenges. Our findings reveal that not all embedded systems can run advanced detection techniques in real-time. Furthermore, object detection and maturity estimation are often just part of a more extensive system, especially in robotic applications. This means the runtime must accommodate additional computations for functionalities such as tracking and navigation. Consequently, there is a critical need for ongoing advancements in these technologies aimed at boosting performance across a range of devices, including more cost-effective options. Future research should focus on optimizing processing efficiency to encourage broader adoption and practical implementation of these technologies. Such advancements could make these solutions more widely available and cost-effective, potentially transforming agricultural practices on farms of every size.

4. Conclusions

This study has conducted an extensive analysis of various advanced Deep Learning techniques for detecting and estimating the maturity of blueberries. Our investigations reveal that while current models are good at localizing individual blueberries, they face challenges from the inherent constraints of the detection techniques and from the natural characteristics of blueberry plants, where berries often remain partially occluded.

A significant observation from our research is the difficulty in achieving accurate localization due to the clustered nature of blueberries and the viewpoint from where the image was captured. These complexities often result in detection inaccuracies, such as misclassification or imprecise bounding boxes, as highlighted by our TIDE analysis, which indicates a significant prevalence of false negatives and localization errors.

Regarding edge computing, our experiments show that some models can perform in real-time on edge devices without significantly losing precision. However, the efficiency of these models varies, with some, like Mask-RCNN, exhibiting higher accuracy but longer runtime, which restricts their real-time application. In contrast, models such as YOLOv7-default strike a more effective balance between accuracy and processing speed, making them more suitable for real-time tasks.

Finally, our findings provide valuable insights into the capabilities and limitations of current techniques in blueberry detection and maturity estimation. They emphasize the complexity of this task, influenced by both the nature of the blueberry plants and the limitations of existing detection models.

Author Contributions: Conceptualization, C.A.A., C.F.-F. and C.A.; formal analysis, C.A.A. and C.F.-F.; funding acquisition, C.A.; investigation, C.A.A., C.F.-F., C.A. and C.N.; methodology, C.A.A. and C.F.-F.; software, C.A.A. and C.N.; supervision, C.A.; writing—original draft, C.A.A., C.F.-F. and C.A.; writing—review and editing, C.A.A., C.F.-F. and C.A. All authors have read and agreed to the published version of this manuscript.

Funding: This research was funded by the National Research and Development Agency through the FONDEF project ID21I10256 and Project INES I+D 22-14 Bio-Bio University.

Data Availability Statement: Data are contained within the article.

Conflicts of Interest: The authors declare no conflict of interest.

References

1. Ni, X.; Li, C.; Jiang, H.; Takeda, F. Deep learning image segmentation and extraction of blueberry fruit traits associated with harvestability and yield. *Hortic. Res.* **2020**, *7*, 110. [CrossRef] [PubMed]
2. Kamilaris, A.; Prenafeta-Boldú, F.X. Deep learning in agriculture: A survey. *Comput. Electron. Agric.* **2018**, *147*, 70–90. [CrossRef]
3. Gunawan, G.; Zarlis, M.; Sihombing, P.; Wage, S. Optimization of the CNN model for smart agriculture. *IOP Conf. Ser. Mater. Sci. Eng.* **2021**, *1088*, 012029. [CrossRef]
4. Jin, Y. Detection of Crop Leaf Diseases and Insect Pests Based on Improved Faster R-CNN. *Fresenius Environ. Bull.* **2021**, *30*, 7278–7290.
5. Mondal, S.; Banerjee, S.; Mukherjee, S.; Sengupta, D. Plant Disease Detection Using Ensembled CNN Framework. *Comput. Sci.-AGH* **2022**, *23*, 323–335. [CrossRef]
6. Manuel Lopez-Correa, J.; Moreno, H.; Ribeiro, A.; Andujar, D. Intelligent Weed Management Based on Object Detection Neural Networks in Tomato Crops. *Agronomy* **2022**, *12*, 2953. [CrossRef]
7. Zhang, X.; Cui, J.; Liu, H.; Han, Y.; Ai, H.; Dong, C.; Zhang, J.; Chu, Y. Weed Identification in Soybean Seedling Stage Based on Optimized Faster R-CNN Algorithm. *Agriculture* **2023**, *13*, 175. [CrossRef]
8. Qiao, M.; He, X.; Cheng, X.; Li, P.; Luo, H.; Tian, Z.; Guo, H. Exploiting Hierarchical Features for Crop Yield Prediction Based on 3-D Convolutional Neural Networks and Multikernel Gaussian Process. *IEEE J. Sel. Top. Appl. Earth Obs. Remote Sens.* **2021**, *14*, 4476–4489. [CrossRef]
9. Ju, S.; Lim, H.; Ma, J.W.; Kim, S.; Lee, K.; Zhao, S.; Heo, J. Optimal county-level crop yield prediction using MODIS-based variables and weather data: A comparative study on machine learning models. *Agric. For. Meteorol.* **2021**, *307*, 108530. [CrossRef]
10. Tellaeche, A.; BurgosArtizzu, X.P.; Pajares, G.; Ribeiro, A.; Fernández-Quintanilla, C. A new vision-based approach to differential spraying in precision agriculture. *Comput. Electron. Agric.* **2008**, *60*, 144–155. [CrossRef]
11. Behera, S.K.; Rath, A.K.; Sethy, P.K. Maturity status classification of papaya fruits based on machine learning and transfer learning approach. *Inf. Process. Agric.* **2021**, *8*, 244–250. [CrossRef]
12. Yang, C.; Lee, W.S.; Gader, P. Hyperspectral band selection for detecting different blueberry fruit maturity stages. *Comput. Electron. Agric.* **2014**, *109*, 23–31. [CrossRef]
13. Tan, K.; Lee, W.S.; Gan, H.; Wang, S. Recognising blueberry fruit of different maturity using histogram oriented gradients and colour features in outdoor scenes. *Biosyst. Eng.* **2018**, *176*, 59–72. [CrossRef]
14. Gonzalez, S.; Arellano, C.; Tapia Farias, J. DeepBlueBerry: Quantification of Blueberries in the Wild Using Instance Segmentation. *IEEE Access* **2019**, *7*, 105776–105788. [CrossRef]
15. Mu, C.; Yuan, Z.; Ouyang, X.; Sun, P.; Wang, B. Non-destructive detection of blueberry skin pigments and intrinsic fruit qualities based on deep learning. *J. Sci. Food Agric.* **2021**, *101*, 3165–3175. [CrossRef] [PubMed]
16. Obsie, E.Y.; Qu, H.; Drummond, F. Wild blueberry yield prediction using a combination of computer simulation and machine learning algorithms. *Comput. Electron. Agric.* **2020**, *178*, 105778. [CrossRef]
17. MacEachern, C.B.; Esau, T.J.; Schumann, A.W.; Hennessy, P.J.; Zaman, Q.U. Detection of fruit maturity stage and yield estimation in wild blueberry using deep learning convolutional neural networks. *Smart Agric. Technol.* **2023**, *3*, 100099. [CrossRef]
18. Wang, C.Y.; Bochkovskiy, A.; Liao, H.Y.M. YOLOv7: Trainable bag-of-freebies sets new state-of-the-art for real-time object detectors. *arXiv* **2022**, arXiv:2207.02696.

19. Qiao, Y.; Truman, M.; Sukkarieh, S. Cattle segmentation and contour extraction based on Mask R-CNN for precision livestock farming. *Comput. Electron. Agric.* **2019**, *165*, 104958. [CrossRef]
20. Lv, W.; Xu, S.; Zhao, Y.; Wang, G.; Wei, J.; Cui, C.; Du, Y.; Dang, Q.; Liu, Y. DETRs Beat YOLOs on Real-time Object Detection. *arXiv* **2023**, arXiv:2304.08069.
21. Hinton, G.; Vinyals, O.; Dean, J. Distilling the Knowledge in a Neural Network. *arXiv* **2015**, arXiv:1503.02531.
22. Aguilera, C.A. SBIN: A stereo disparity estimation network using binary convolutions. *IEEE Lat. Am. Trans.* **2022**, *20*, 693–699. [CrossRef]
23. Jeong, E.; Kim, J.; Ha, S. TensorRT-Based Framework and Optimization Methodology for Deep Learning Inference on Jetson Boards. *ACM Trans. Embed. Comput. Syst.* **2022**, *21*, 1–26. [CrossRef]
24. Bolya, D.; Foley, S.; Hays, J.; Hoffman, J. TIDE: A General Toolbox for Identifying Object Detection Errors. In Proceedings of the Computer Vision—ECCV 2020: 16th European Conference, Glasgow, UK, 23–28 August 2020.

MDPI AG
Grosspeteranlage 5
4052 Basel
Switzerland
Tel.: +41 61 683 77 34

Agriculture Editorial Office
E-mail: agriculture@mdpi.com
www.mdpi.com/journal/agriculture